Physics at a Research University
Case Western Reserve 1830 – 1990

William Fickinger

Professor of Physics Emeritus
Case Western Reserve University
Cleveland Ohio 44106 USA

The author, the Case Physics Department, and the College of Arts and Sciences thank Dr. Sherwood L. Fawcett for his support in making the publication of this history possible. Dr. Fawcett, in 1950, was the third student to earn a physics doctorate at Case. Under the direction of Professor Eugene Crittenden, Sherwood Fawcett designed the electron beam extraction system for the Case Betatron accelerator. Upon graduation, Dr. Fawcett spent his entire career at Battelle Memorial Institute, the last seventeen years as President and Chief Executive Officer.

Case Western Reserve University
Cleveland, Ohio 44106
2006

Library of Congress Cataloging in Publication Data
LCCN 2005933988

Fickinger, William J.
Physics at a research university: Case Western Reserve 1830-1990

Copyright © 2005 by William Fickinger
ALL RIGHTS RESERVED.
NO PART OF THIS BOOK MAY BE REPRODUCED
IN ANY FORM WITHOUT PERMISSION
IN WRITING FROM THE PUBLISHER.

The sources of all illustrations are listed in Appendix H.

ISBN 0-9773386-0-6

Table of Contents

Introduction

Chapter 1	**The Young Ivy Leaguers**	1
	1829 to 1866	
Chapter 2	**Edward W. Morley: Physical Chemist**	7
	1868 to 1906	
Chapter 3	**Albert A. Michelson: Light and Ether**	13
	Michelson Reid	
	1882 1889	
Chapter 4	**Dayton C. Miller: X-rays, Flutes and More Ether**	24
	Miller Hodgman Albright Nusbaum Wallace	
	1890 1906 1921 1922 1926	
Chapter 5	**University Physics: the Middle Years at WRU**	46
	Barrows Smith Freeman Whitman Mountcastle	
	1866 1870 1882 1886 1907	
Chapter 6	**Robert S. Shankland: Modern Physics Arrives**	53
	1930 to 1976	
Chapter 7	**Case Experiment Takes Off**	68
	Olsen Crittenden Shrader Smith Gregg Scharenberg Silverstein	
	1937 1938 1940 1942 1945 1961 1964	
Chapter 8	**Neutrinos and Cosmic Rays**	88
	Crouch Reines Frye Jenkins Woods Wang Albats Koga	
	1952 1959 1960 1960 1964 1966 1973 1974	
Chapter 9	**Case Theory**	111
	Foldy Klein Milford Winterberg Tobocman Thaler	
	1948 1949 1952 1959 1960 1960	
Chapter 10	**WRU Experiment Takes Off**	130
	McCarthy Beth Meeks Winter Major	
	1937 1946 1948 1951 1955	
Chapter 11	**WRU Theory**	137
	Tauber Kisslinger Machlup Weinberg Zilsel Goswami	
	1954 1956 1957 1959 1960 1963	
Chapter 12	**Case Experiment Round 2**	146
	Hoffman Benade Gordon Eck Schuele Chottiner	
	1947 1955 1955 1957 1963 1980	
Chapter 13	**Case Nuclear and Particle Theory**	175
	Kowalski Nagarajan Pearle Rix Kantor Brown Shakin	
	1963 1964 1966 1967 1967 1970 1970	
Chapter 14	**WRU Experiment Round 2**	194
	Casper McGervey B. Robinson Jha Reichert Huang	
	1960 1960 1960 1966 1966 1966	
Chapter 15	**Low Temperature Experiment**	209
	Green Chandrasekhar Sparlin Adler Farrell Dahm	
	1961 1963 1964 1965 1966 1968	

Chapter 16 Particle Experiment — 233
  K. Robinson Fickinger Kikuchi Willard Strelzoff
     1966   1967   1967   1967   1969
  Sullivan Bevington Baer Eisner
   1969   1970   1974   1974

Chapter 17 Materials Theory — 257
  Reitz Leff P.Taylor Coopersmith Silvert Segall
  1954 1964   1964   1964   1966   1968
  Petschek Mathur Lambrecht
    1983   1995   1996

Chapter 18 New People, New Physics — 279
  Modern Materials   Rosenblatt Singer Kash Shan
  Particle/Cosmology/Astro Theory CTaylor Krauss Starkman Vachaspati
  Astrophysics and Cosmology Experiment Akerib Covault Ruhl Shutt

Appendix A Bachelor Degree Recipients — 291

Appendix B Master and Doctoral Degree Recipients — 302
          Including Thesis Titles and Advisors

Appendix C Faculty List — 335

Appendix D Programs and Lecture Series — 337
          1949 APS Ohio Section Meeting
          1987 Michelson-Morley Centennial
          Michelson Lecturers
          Michelson Postdoctoral Fellows

Appendix E Foldy List of AIP Centennial Papers — 341
          with CWRU Connections

Appendix F Department Chairs and Endowed Chairs — 345

Appendix G L. L. Foldy Essay on FW Transformation — 347

Appendix H List of Illustrations — 352

Index — 355

Introduction

This is about physics research at Case Western Reserve University. Teachers of "natural science", later called "physics", have long understood that science is an ongoing process and that the pursuit of research in the university helps to keep the teachers up to date and their teaching interesting.

There are currently about three dozen private, research-oriented, American universities which have physics faculties of twenty or more members. With the exception of the Ivy League schools, most of these began in the 19th century as either church-affiliated or technical schools. Typically, each began with one or two professors of natural science. In the 20th century, all these departments expanded rapidly, partly to teach growing numbers of engineers and scientists, and partly to participate in the exciting developments in experimental and theoretical physics.

Case Western Reserve University was formed in 1967 by the federation of Case Institute of Technology and Western Reserve University. It is probably unique in having its origins in both types of progenitor institution. Its physics department is similar in history, size, and research activities to the majority of the departments in the group of three dozen. It grew from one professor in 1830 to about 25 today. Its story describes the passage from research in meteorology, astronomy, optics and acoustics in the 19th century to experimental and theoretical atomic, nuclear, condensed matter, particle and astrophysics in the 21st.

I retired at the end of 1999, after thirty-two years on the CWRU physics faculty. Since then, I have become interested in tracing the lines of experimental and theoretical research done by my predecessors and colleagues. Much of this interest comes from working in the building which has been a home to the department for 100 years and the fact that I have recently taken responsibility for many of the artifacts and archival materials which have collected in it.

This project was begun with the intention of making a list of people and research areas for the history page of our departmental website. It soon went beyond that as I read through materials in the University Archives and other documents which have survived within the department. This book is more an informal story than a scholarly project. I concentrate on the physics, rather than on the physicists; on research, rather than on teaching. I describe work our faculty did while they were members of the department, with only the few necessary comments on what they did elsewhere, before and after. I include experimental details and actual data from the principal publications to help the reader understand the techniques used and appreciate the significance of the results. I include in the running text the titles and references of selected publications. Each title gives the author's best description of the work and the journal name gives a measure of the importance of the research. I frequently insert italicized *"asides"* which offer definitions of technical terms or explanations of the theory or experiment to help the non-physicist reader make sense of the story.

The first job is to find out who the players were and when they were part of the department. The following *"departmental tree"* shows roughly when each physicist was on the scene. I include only those who were assistant professor or higher, and who remained for three years or more. I apologize to the hundred-or-so hard-working instructors and to the score of folks who picked up stakes after a shorter stay. The list is still as long and as hard to remember as the cast of characters in a Russian novel. The reader might find it useful to come back to these pages from time to time, better to place a name or a date. The lines under the names mark the years spent on the faculty. The corresponding chapter numbers are to the right of each name. The three separate lists include, respectively, the people who started at Western Reserve, at Case, or at Case Western Reserve.

WESTERN RESERVE UNIVERSITY PHYSICS FACULTY

1830	35	40	45	50	55	60	1865

Wright   Loomis   Nooney   Emerson Young   Ch 1

1865	70	75	80	85	90	95	1900

Barrows___Smith___Freeman_____Whitman____>   Ch 5
                        Morley_____>   Ch 2

1900	05	10	15	20	25	30	35	40	45	50

<Whitman_____   Ch 5
<_Morley_   Ch 2
                    Springsteen/Mountcastle_____   Ch 5
                                            McCarthy   Ch 10
                                                Beth   Ch 10
                                                Meek   Ch 10
                                              Winter   Ch 10

1955	60	65	70	75	80	85	90	95	00	05

     __Tauber_____   Ch 11
      _Major__   Ch 10
     _Kisslinger____   Ch 11
            _____Machlup_____   Ch 11
       _Weinberg___   Ch 11
        Casper   Ch 14
                  McGervey_____   Ch 14
        _B Robinson__   Ch 14
           _Zilsel__   Ch 11
           Green   Ch 15
              Chandrasekhar_____   Ch 15
            Goswami__   Ch 11
             Chew_   Ch 11
             Silverstein   Ch 7
              Sparlin_   Ch 15
               Adler_   Ch 15
                        Farrell_____   > Ch 15
                       DKRobinson_____   Ch 16

1955	60	65	70	75	80	85	90	95	00	05

CASE INSTITUTE OF TECHNOLOGY PHYSICS FACULTY

| 1865 | 70 | 75 | 80 | 85 | 90 | 95 | 1900 |

               Michelson   Reid   Miller   >   Ch 3, 4

| 1900 | 05 | 10 | 15 | 20 | 25 | 30 | 35 | 40 | 45 | 50 |

<_____Miller_____ Ch 4
                    _____Hodgman_____ Ch 4
                     __Albright_____ Ch 4
                               Wallace_____ Ch 4
                     Nusbaum_____ Ch 4
                              _____Shankland_____>
                            ____Olsen_____ Ch 7
                                Crittenden____ Ch 7
                                  Shrader_____>
                                    Smith_____>

| 50 | 55 | 60 | 65 | 70 | 75 | 80 | 85 | 90 | 95 | 00 |

<____Shankland_____ Ch 6
<_____Shrader_____ Ch 7
<____Smith_____ Ch 7
       Gregg_____ Ch 7
<____Hoffman_____ Ch 12
<_____Foldy_____ Ch 9
<___Klein_____ Ch 9
<_____Crouch_____ Ch 8
___Milford_____ Ch 9
        Reitz_____ Ch 9
                 Benade_____ Ch 12
                     Gordon_____ Ch 12
                      Eck_____ Ch 12
         Reines_____ Ch 8
              Frye_____ Ch 8
              Jenkins_____ Ch 8
          Thaler_____ Ch 9
               Tobocman_____ Ch 9
         Scharenberg_____ Ch 7
               PTaylor_____ > Ch 17
            Kowalski_____ > Ch 13
              Schuele_____ > Ch 12
       Nagarajan_____ Ch 13
         Coopersmith_____ Ch 17
           Leff_____ Ch 17

| 50 | 55 | 60 | 65 | 70 | 75 | 80 | 85 | 90 | 95 | 00 |

CASE WESTERN RESERVE UNIVERSITY PHYSICS FACULTY

60	65	70	75	80	85	90	95	00	05
	Woods		Ch 8						
	Blanpied		Ch 16						
	Frisken		Ch 16						
	Huang		Ch 14						
	Jha		Ch 14						
	Pearle			Ch 13					
	Wang			Ch 8					
	Reichert		Ch 14						
	Silvert		Ch 17						
			Fickinger					Ch 16	
		Willard		Ch 16					
		Kantor		Ch 12					
		Kikuchi		Ch 16					
		Rix		Ch 13					
			Dahm					Ch 15	
			Segall					Ch 17	
		Sullivan		Ch 16					
		Bevington		Ch 16					
			Brown					> Ch 13	
		Shakin		Ch 13					
		Albats		Ch 8					
			Baer	Ch 16					
			Eisner	Ch 16					
			Koga	Ch 8					
				Chottiner				> Ch 12	
					Petschek			> Ch 17	
					Rosenblatt			> Ch 18	
						CTaylor		> Ch 18	
						Singer		> Ch 18	
						Krauss		> Ch 18	
							Kash	> Ch 18	
							Mathur	> Ch 18	
							Starkman	> Ch 18	
							Akerib	> Ch 18	
							Lambrecht	> Ch 17	
							Vachaspati	> Ch 18	
								Covault > Ch 18	
								Ruhl > Ch 18	
								Shan > Ch 18	
								Shutt > Ch 18	

60	65	70	75	80	85	90	95	00	05

Acknowledgements

Members of the staff of the Case Western Reserve University archives have been especially helpful in finding information on the pre-1950 departments. I much appreciate the cooperation of the former archivist Dennis Harrison and of current staff members, Thomas Steman and Helen Conger.

Portions of the text have been read by many of my colleagues, some emeritus and some still teaching. Let me list here the faculty members who, by email or in person, were kind enough to read selections of my text and to make valuable suggestions. I especially appreciated the "first-person" stories and anecdotes which have added to the narrative. Among the original Western Reserve faculty, I have been able to contact Leonard Kisslinger in Pittsburgh, Berol Robinson in Paris, and Chandrasekhar in Munich. Stefan Machlup and David Farrell suggested many improvements. Emeriti readers from the Case side included particle experimentalists Marshall Crouch, Glenn Frye, and Tom Jenkins. From the "middle generation" Case condensed matter folks were Bill Gordon, Tom Eck, and later Arnie Dahm, Gary Chottiner and Jie Shan. I appreciate the guidance of theorists Bill Tobocman, Ken Kowalski, Phil Taylor, Bob Brown, Rolfe Petschek, Ben Segall, and Walter Lambrecht.

I thank the family of Les Foldy for permission to include among the appendices two documents which he made available to the department: his essay on the Foldy-Wouthuysen transformation (Appendix G) and his list of CWRU connections to the 1993 APS centennial list of "most cited papers" (Appendix E). I thank Lawrence Krauss, Cyrus Taylor and the other members of the physics faculty for their support.

Finally, I thank Dr. Sherwood L. Fawcett for making it possible for the physics department to send a gift copy of the book to each of our physics alumni.

William Fickinger, Cleveland 2006

Chapter 1. The Young Ivy Leaguers

In the mid-nineteenth century, there were no "physics departments", no "graduate programs", and whatever research was done was pretty much up to the initiatives of the individual academicians, occasionally with funding from wealthy sponsors interested in the science. The roots of most university physics departments in America go to the relatively small number of professors of natural philosophy who taught alongside their colleagues in classical studies. Most of the action in physics in the early eighteen-hundreds was in Europe, where the experiments of Coulomb, Ampere and Faraday would provide the foundation for Maxwell's theory of electricity and magnetism. While the Americans Benjamin Franklin and Joseph Henry had made significant contributions in this area, most American physicists were teaching Newton's mechanics and studying its applications.

Some of the following information on Western Reserve's early teachers of physics comes from the Centennial History of CWRU by C. H. Cramer (Little Brown, 1976) and from a long article on the sciences at Western Reserve by Frederick Clayton Waite. It appeared in the Reserve Record (the newspaper of Western Reserve Academy) of 13 May 1938. Waite, a WRU professor emeritus of biology, later wrote a detailed history of the early years of the college: "Western Reserve University: the Hudson Years" *W. R. University Press 1943.*

Western Reserve College was founded in 1826 by a group of farmers and land developers in the area around Hudson, Ohio. The northeast corner of Ohio had been designated as the "Western Reserve of the State of Connecticut". According to Cramer: "Connecticut was given permission by Congress to reserve 3,500,000 acres in northeastern Ohio for the purposes of reimbursing those of its citizens who had suffered losses from British depredations during the Revolution." Moses Cleaveland, surveyor, had come out to the area in 1796 and left the settlement which bears his name (approximately). Ohio became a state in 1803 as large numbers of families from Connecticut traveled west to set up farms and towns. Some of their sons were sent back to Yale and other eastern universities. They soon saw the need to create a college closer to hand. "Reserve's" first graduating class, in 1830, consisted of four students. A section of a contemporary map of the Western Reserve is shown in **Fig. 1-1**.

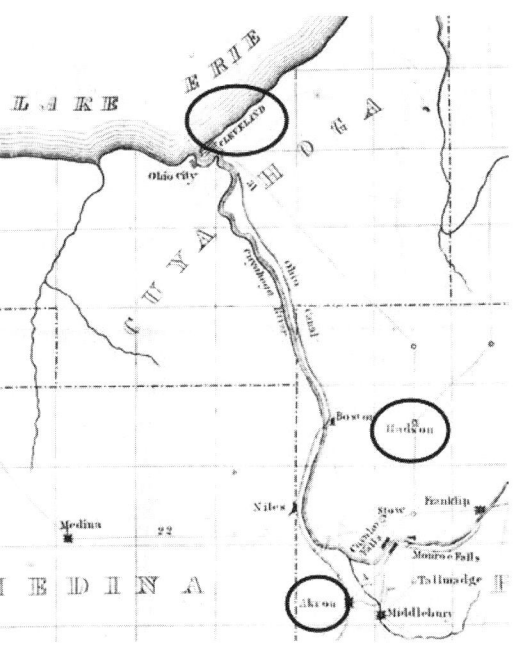

Fig. 1-1. An 1830 map showing the location of Hudson relative to Cleveland and Akron.

As was the case in the older Ivy League schools, most of the colleges established in this period were staffed by faculty trained in theology, philosophy, and the classics, and who were affiliated with a particular denominational group. Western Reserve College was founded by Congregationalist clergymen educated at Yale.

The first professor of physics at Western Reserve College was **Elizur Wright, Jr**, a twenty-five year old graduate of Yale who was hired in 1829. As the third member of the college faculty, he taught mathematics and natural philosophy. By 1832 the young professor had become preoccupied by the controversy between the abolitionists who favored immediate emancipation of the slaves and the colonizationists who argued for the return of slaves to Africa. This controversy was hotly argued by students, faculty, and trustees and the resulting publicity so negative that the survival of the college was at risk. Wright's frequent absences from campus as he lectured around the country in favor of abolition and the general commotion he caused on campus led to the trustees' requesting his resignation in 1833, before he had a chance to accomplish very much in the way of natural philosophy. Making his way back east to New York and then to Massachusetts, he continued his fight against slavery as editor of several emancipation journals. Wright used his rare understanding of mathematics to bring about the reform of the insurance industry in Massachusetts, eventually being known as the "friend of the widow". He was, until his death at age 81, continuously associated with the advancement of liberal causes. A good number among later professors of physics at CWRU would, in their own eras, emulate Wright in promoting social reform.

Filling in after his son's departure was **Elizur Wright**, **Sr**. (AB Yale 1781) who was one of the three founders of the college. The elder Wright clearly had better credentials for teaching than his son, having been ΦBK at Yale. He was an astute astronomical observer and expert in the fluxions of calculus. He was determined that the new college should succeed and was responsible for all of the science and mathematics courses for the next three years, without pay.

Loomis: astronomy on the frontier

In 1836, the second twenty-five year old Yale graduate to be made professor of natural philosophy and mathematics was **Elias Loomis**. **(Fig. 1-2)** He was appointed by President George Pierce who had come from Yale two years earlier. Loomis had established himself as a promising astronomer at Yale, where he was the first to determine the orbit of Halley's comet and to observe its return. Another experiment was done in collaboration with colleague Alexander Twining who was about 75 miles away at West Point, NY. Each measured the angular position in the sky of meteor trails, recording the time each appeared. They were thus able to estimate, by triangulation, the altitude of the luminous trajectories.

Loomis was sent by Western Reserve College to Paris and London to study the teaching of astronomy and to arrange for the purchase of a telescope and chronometer.

The $4000 appropriated for this purpose represented a very generous and ambitious move by the trustees. Loomis spent a year abroad and purchased top-of-the-line instruments: a transit-circle with a 2.7-inch lens and an equitorial telescope with a 3.3-inch lens, both bought from Troughton and Simms, and a clock purchased from Robert Molyneaux, all of London. The clock would provide a time standard at Western Reserve for over 50 years; its accuracy was better than 3 seconds per day.

Fig. 1-2.
Astronomer Loomis.

In a speech delivered at the college in August, 1838, on the importance of higher mathematics, the 27 year old professor described the impact of such astronomical measurements on world commerce. He explains with great admiration the work of Isaac Newton on the orbital motion of the moon, an analysis which took into consideration the gravitational effect of the sun: the first attack on the "three body problem". In his speech, Loomis went on to describe the mathematical analysis of comet appearances, explaining that science triumphs over ignorance (referring to the belief that comets were omens of earthly disasters), saying that we can through science "emancipate our race from the thralldom of superstition". He then outlined advances in the construction of ever better telescopes, and how the newly acquired telescope and chronometer which he had brought from Europe would bring important scientific work to America. Finally, having suitably prepared his audience, he arrives at his main point, a plea for monetary support for the construction and operating expenses for his observatory.

In this he was successful, for the trustees were convinced and provided the necessary funds. The excellent instruments were installed in the newly built observatory - the third such in the nation. Loomis was especially interested in the use of telescopic observations of the moon and stars in the determination of latitude and longitude.

This facility would play an important role in the practice of modern science at WRC.. Loomis, over the course of the next six years, measured great numbers of occultations, lunar culminations and cometary orbits. In a series of articles in the Transactions of the American Philosophical Society, he reports the latitude and longitude of his observatory at Hudson as 41° 14' 38.1" north and 5h 25m 39.5s west of Greenwich. (A tenth of a second is about ten feet.) This measurement provided the basis for subsequent wide-ranging surveys in the newly opened western territories.

With the most precise clock west of the Appalachians, Loomis arranged for a telegraphic signal to be sent daily at noon which rang a bell, thirty miles away, on Cleveland's Public Square. During this same period, Loomis compiled measurements made in 13 states east of the Mississippi of magnetic declination. This involved measuring the component of the earth's magnetic field perpendicular to the ground. ("Observations of the magnetic dip in the United States", American Philosophical Society, Philadelphia 1843] This work, and a later compilation of data on auroral displays, followed from his interest in the shape of the earth's magnetic field. His aurora map appears on the NASA website at http://www-istp.gsfc.nasa.gov/ Education/wloomis.html

Loomis invents the weather map

Loomis is best known for his pioneering work in the science of meteorology. He was especially interested in how storms were structured and how they move. At the time there were conflicting theories for the direction of the winds associated with a storm: either rotational about the storm center or centripetal, toward the center. Loomis sent a paper to be read at the 1843 meeting of the American Philosophical Society in which he described his analysis of data collected over the course of a few days in 1836 at a large number of stations in the US and Canada. In an 1890 *American Journal of Science* memorial to Loomis, H. A. Newton describes Loomis' technique of presenting the data: "Professor Loomis drew on the map a series of lines of equal barometric pressure…. A series of maps representing the storm at successive intervals of twelve hours were thus constructed….A series of colors represented respectively the places where the sky was clear, overcast, rain or snow…A series of lines represented the places at which the temperature was at the normal, or 10 or 20 or 30 degrees above or below the normal. Arrows of proper direction and length represented the direction and intensity of the winds. These successive maps for the three or four days of the storm furnished to the eye all its phenomena in a simple and most effective manner." Loomis' map, as shown here in **Fig. 1-3**, appears on the Yale website:

http://love.geology.yale.edu/kgl/Dept_Information/History/loomis-map.gif

It was not until 1871 that the United States Signal Service began to provide weather maps following Loomis' design.

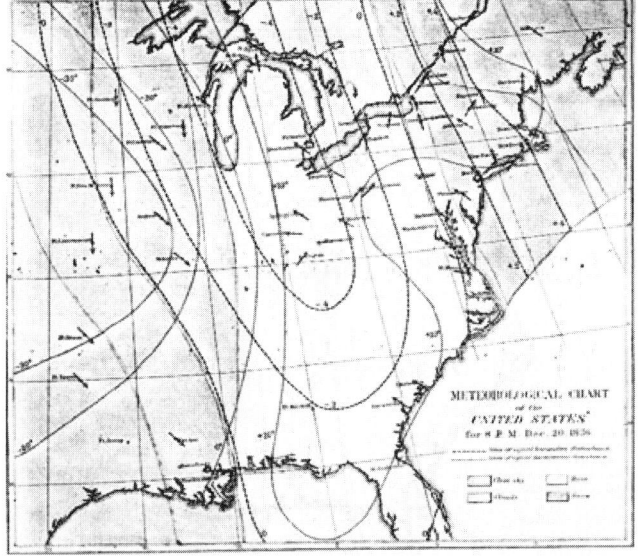

Fig. 1-3. Loomis' 1843 weathermap.

Loomis was in contact with other scientists interested in meteorology and the nature of storms. Some of his correspondence with William C. Redfield in New York (first president of the AAAS) and with James P. Espy in Philadelphia has been reproduced in a book by Nathan Reingold (*Science in Nineteenth-century America* Univ. of Chicago Press 1964). Here are two excerpts. One describes doing science on the frontier, the other makes one wonder if the writer were joking: Loomis to Redfield (Redfield papers, Yale University): "Being so far from New York, I find it difficult to obtain regular information from Europe. This is one of the greatest inconveniences of my position and I feel it sensibly. It takes a great while to get a box from New York, as it is very liable to be mis-sent or to be cast aside in some ware-house on the route." Espy to Loomis (Loomis papers, Yale University): "The information which I have received concerning the chickens and turkies *(sic)* remaining alive after being stripped of their feathers in a tornado is verbal and I have no doubt of the fact though I cannot refer you to the authority by name."

It was about this time that the trustees approved the construction of a new and ample building. Funds for this new facility came from a gift of General Simon Perkins, one of the founders of the Western Reserve. The handsome three-story building, eventually called the "Athenaeum", would house lecture rooms and spaces for student organizations. It is here that chemist Morley would set up his teaching laboratories, as described in the next Chapter.

Loomis remained at Western Reserve until 1844, when he accepted a position at the University of the City of New York. He subsequently moved on to Yale in 1860. He is perhaps best known as the author of *A Treatise on Meteorology*, published by Harper and Bros., 1868.

Following Loomis in the position of Professor of Natural Philosophy was **James Nooney** (AB Yale 1838) who had served as professor of mathematics in the U.S. Navy – at a time when the teaching was done aboard ship. At Western Reserve, Nooney taught all the mathematics and physics courses from 1844 until 1848, when he left the college. For the next five years, rather grim ones on the economic side, there was no one on the faculty trained to teach physics. The college did eventually find another young Yale graduate, **Alfred Emerson**, AB 1834, who served from 1853 until 1856. I have not yet found evidence that these two young men had the opportunity to do any astronomical or other physics research. However, the Loomis astronomical observatory is featured each year in the Western Reserve catalogs until 1856.

Young and geomagnetism

During the 1850's there was a movement among American universities away from degrees based solely on the classics, philosophy, religion and literature and toward the study and application of the sciences. Henry Hitchcock (president of the college 1855-1871), himself an practicing clergyman, inaugurated an enhanced scientific program at Western Reserve. The first designation of "physics" in the college catalogue was in 1855. In 1856 Hitchcock hired **Charles Young**, a 22 year old graduate of Dartmouth, who took over mathematics, natural philosophy, and astronomy. **(Fig. 1-4)** Young continued the work of Loomis in meteorology and geomagnetism at Western Reserve. In fact, Loomis wrote to Young in 1859, asking him to repeat the magnetic declination measurements at Hudson in order to compare them with those which Loomis had made seventeen years earlier. Loomis published the

Fig. 1-4.
Charles Young.

results, claiming an increase in the dip angle by 14 minutes, evidence for the dynamic nature of the earth's magnetic field.

In 1862, the youthful Prof. Young was placed in charge of a company of student "soldiers" who served in the Union Army for a few months by guarding confederate prisoners. In this way, the college was spared the loss of most of its students to the draft. In 1865, Young was appointed the first Perkins Professor of Natural Philosophy and Astronomy. (This professorship, which ensured the continued teaching of physics at Western Reserve, was established with a gift of $5000 by Joseph Perkins in memory of his father, Simon. The Perkins Professorship continues to the present day; a list of its holders is included in Appendix F.) In spite of this honor, Young left Hudson the following year to return to the East after ten years at Western Reserve.

In later years at Dartmouth and Princeton, Young became a world authority on solar physics. Among his more important accomplishments was the measurement of the Doppler shift of spectral lines in light coming from opposite equatorial regions of the sun and consequently of the rate of solar rotation. His 1881 book, *The Sun*, and his 1898 introductory text on astronomy, *Lessons in Astronomy*, were used worldwide for more than fifty years. More on Young as a researcher and teacher can be found at the Princeton website:
(http://mondrian.princeton.edu/CampusWWW/Companion/young_charles.html).

Young died in Hanover, New Hampshire on January 3, 1908, a day of a total solar eclipse.

Physics in the Reserve: a beginning

Important aspects of the history of physics research at Western Reserve are the amounts and sources of funding that made it possible. In the early years, it was the enlightened trustees and the persevering presidents who sought out the dollars to be spent on salaries, buildings, and equipment. It was they who got the college through the difficult years of the 1830's and 40's. As commerce in northeast Ohio expanded exponentially, especially with the coming of the canals, the railroads and shipping on the Great Lakes, significant funding for professorships and memorial buildings became more available. The annual salary for a professor rose from $400 in 1829 to one thousand dollars in 1864.

There is a common element within this sequence of five young Ivy League teachers of natural philosophy: Wright, Loomis, Nooney, Emerson, Young. Each of them returned to the East after his stay in the Western Reserve. Almost as missionaries, they came to the Midwest to bring science to sons of farmers and land developers, and then left to pursue their various careers, having spent their allotted time on the American frontier.

Chapter 2   Edward Williams Morley Physical Chemist
1869 - 1906

Morley's background

Fig. 2-1. E. W. Morley

In 1869, the 30 year old Edward Williams Morley, from Williams College in Massachusetts, was appointed professor of chemistry and "natural history". The young Morley is pictured in **Fig. 2-1**. He had graduated from Williams in 1860. He then completed three years of theological studies at the Andover Theological Seminary, followed by a few months as a medical assistant serving the army in the civil war and a three year stint teaching in a private school. In 1868, he took a position as pastor of a church in Twinsburg, Ohio. His talents as a scientist were soon recognized by the administrators of the nearby Western Reserve College, and he was quickly offered a faculty position. He did not return to the ministry, though he did his share of preaching in the local church.

Morley would spend his entire scientific career on the Western Reserve faculty, until his retirement in 1906. While he ultimately held the position as chairman of the Chemistry Department, he must also be counted among the *physicists* of the institution. As pointed out by C.H. Cramer in his history of CWRU, of the 52 major papers published by Morley, 23 were in pure physics. His chief research concerned precision measurements of the atomic masses of oxygen and hydrogen (published in 1895). Of course, the ether-drift experiment which Morley performed in 1887 with A. A. Michelson was 100% physics. Michelson's successor, Dayton C. Miller, said, "The physicists have always felt that they had a strong claim on Prof. Morley as one of their group."

Between 1868 and 1882, Morley was Mr. Science in Hudson, and indeed in all northeast Ohio. His interests were wide-ranging. He set up the first teaching laboratory at the college; he was a pioneer in offering hands-on chemistry experience to the students. For fifteen years he commuted by train to Cleveland to teach chemistry at the Cleveland Medical School; he acted as scientific consultant for industrial firms and the civil courts.

Ratio of oxygen and hydrogen atomic masses

Morley's experiments to measure atomic masses were spread over many years, from the 1870's in Hudson into the 1890's at the new Cleveland campus. He worked very much alone, designing ever more refined techniques. Like his colleagues, Michelson and Miller, Morley was more concerned with making precise measurements than with the underlying "theory".

The ratio of the various atomic masses to the mass of the hydrogen atom was a controversial subject in the latter half of the 19th century. It was widely believed, as proposed by William Prout in 1815, that each element should have an atomic mass exactly equal to an integral number times the hydrogen mass. As measurements became more precise, that simple picture became less viable, and Morley was determined to settle the issue beyond any doubt. Ultimately, with the publication of his results in 1895, Morley was able to quote an oxygen to hydrogen mass ratio of 15.879 with an uncertainty in the last decimal place, a ratio clearly different from 16.000. ("On the Densities of Oxygen and Hydrogen and the Ratio of their Atomic Weights", *Smithsonian Institution Contribution to Knowledge* 980, 1895. Much of this material is expertly described in the doctoral dissertation of Ralph R. Hamerla (PhD, History of Science, CWRU May 2000. I had the pleasure of serving on Dr. Hamerla's committee.) Morley's result was not understood until the discovery of the neutron and Einstein's theory on the equivalence of mass and energy quantified the role of binding energy in nuclear masses.

Method 1: T P V & M

How, then, does one go about determining M_O/M_H? In one set of experiments, Morley measured the densities of each gas. He expressed his results in terms of the ratio of the two densities measured at 0° C and 760 mm Hg pressure. (The assumption is that each gas is ideal, so that the ratio of densities equals the ratio of the atomic masses.) The trick, then, is to determine the absolute temperature and pressure, the volume of the gas and its mass, and its purity, all to a precision of one part in 10,000. Morley used rather large volumes of gas, typically 10 to 20 liters in a glass globe. Much of his work was done in the basement of Adelbert Hall which today houses the top CWRU administrators.

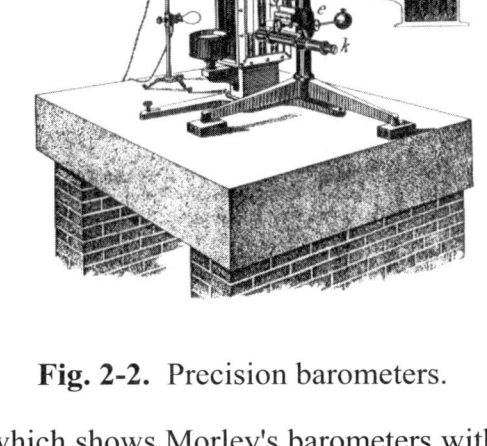

Fig. 2-2. Precision barometers.

Fig. 2-2, which shows Morley's barometers with their telescope readouts, comes from the 1895 Smithsonian Institution report.

It is interesting to step through part of such a measurement, noting all the minute corrections he made along the way. For example, here are the steps he followed to determine the internal volume of the glass globe. Evacuate the globe and weigh it in air by suspending it below one of the pans of a precision balance. **Fig. 2-3** shows Morley's drawing of the weighing spheres and balance. The balance (one microgram sensitivity up to a loading of 1.2 kg) was constructed by

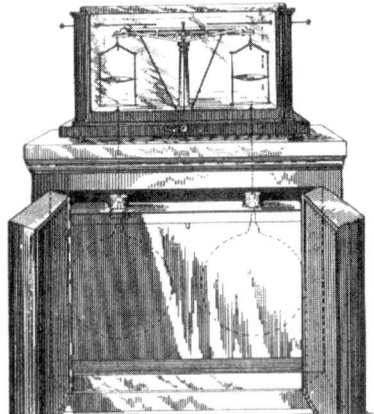

Fig. 2-3. Gas globes suspended from balance.

Rüprecht of Vienna and loaned to Morley by the Smithsonian. Next, following the example of Archimedes, submerge the evacuated globe in water at known temperature, adding weights to a pan which hangs in the water below the globe. Determine how much weight is required just to submerge it. This gives the mass of the water displaced, and thus the external volume of the globe. Then fill the submerged globe with water and weigh it in water again to determine the buoyant force on the glass itself. This, along with the known density of the glass, gives the volume of glass. The difference between the exterior volume of the globe and the volume associated with the glass walls gives the interior volume of the globe.

Note that one never has to weigh the large mass of water which fits in the globe; this would be far beyond the capacity of the balance. The brass weights used to sink the globe were typically about one kilogram each, so they could be weighed one at a time. Morley weighed them (and the pan) in water to remove the effect of the buoyant force on the weights themselves. Morley even determined the effect of the hydrostatic pressure on the volume of the globe, which contracted slightly when the globe was evacuated. To illustrate the quantities he was measuring, the precision of his measurements, and the careful corrections he applied, I give a typical set of numbers taken from his 1895 publication: This is a measurement of the interior volume of a 10 liter glass globe.

	grams
total weight of brass weights added to pan to submerge globe	8425.64
reduction to vacuum *	-1.17
reduction to temperature **	-0.11
corrected weight of brass	8424.36
weight of cage and pan	178.32
weight of evacuated globe	1015.22
sum of weights suspended from left balance pan	9617.90
balancing weights on right pan	351.09
weight of water displaced	9266.81

	cubic cm
divide by density of water at 17.24 C $\rho = 0.998765$ g/cc to find volume of water displaced	9278.27
contraction of globe to 0 C	-4.47
exterior volume of glass globe at 0 C	9273.80
repeat above measurements with globe filled with water resulting volume occupied by glass after similar small corrections	443.07
subtract to get interior volume of globe	8830.73

* correct for buoyancy in air
** correct for thermal contraction when weighing in water

The globe is then filled with purified hydrogen or oxygen gas and similar measurements are made to determine the mass of the known volume of gas, thus determining their densities. Morley gives detailed descriptions of his calibration of thermometers and barometers. In the latter case, since the barometric pressure is determined by the height of the mercury column and by the value of g, the acceleration due to gravity, he introduces a factor of 0.999627 to correct g to an altitude of 216.1 m above sea level at New York and a latitude of 41° 30' 15". (Do not conclude that this represents a correction of four parts in 10,000 in the gas densities. The absolute pressure is used only indirectly to do some small corrections. The weights of the precisely measured brass masses *do* depend on the local value of g, but when they are used with a balance, the correction is the same on both sides and thus cancels out.) Morley's expertise in the techniques of chemistry was essential to the production of pure gasses and the removal of contaminants.

Method 2: V & M

The set of measurements of the type outlined above was only a beginning for Morley, who continued to seek ways to reduce the uncertainties. For example, the precision of the measured densities depends linearly on his ability to measure temperature and pressure and to extrapolate his results to zero Celsius and one atmosphere. Morley completely redesigned his apparatus so that the hydrogen and oxygen densities could both be measured at the *same* temperature (13.5 C) and the *same* pressure (736.49 mm Hg). Furthermore he devised a way to manipulate the globes and weights without ever touching them. This included an arrangement which allowed him to reverse the loads from one balance pan to the other to correct for imperfections in the balance itself. The possibility of loss of gas through leaky stopcocks was avoided by use of fusible metal plugs.

Method 3: M_H M_O & M_{water}

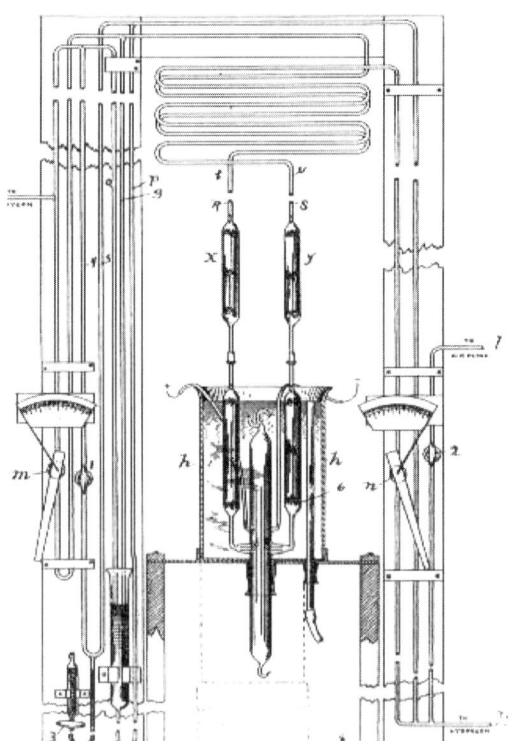

Fig. 2-4. Weighing H, O and H_2O.

The third type and ultimate set of measurements of the atomic mass ratio was done with an apparatus in which measured masses of hydrogen were burned with measured masses of oxygen to produce measured masses of water. The apparatus is illustrated in **Fig. 2-4**. The gasses were combined in a container and ignited by high voltage electrodes. To quote Hamerla: "Morley conducted 12 experiments using 42 liters of hydrogen and 21 liters of oxygen which produced 34 grams of water in each experiment." Here are the numbers, in grams, from one of the runs: Hydrogen taken 3.8392; hydrogen residue 0.0010; hydrogen used 3.8382; oxygen taken 30.4741; oxygen residue 0.0041; oxygen used 30.4700; O/H atomic mass ratio 15.877; water formed 34.3151. (This latter value differs by only 0.007 grams from the sum of O used and H used.)

The density ratio from this water-synthesis method as published by Morley is 15.879 ± 0.00032 and may be compared to the currently accepted value (from a table in Krane's 1988 book on nuclear physics). This is 15.8731 (when one includes the contributions of naturally occurring hydrogen and oxygen isotopes). The Morley value differs from the current value by about 4 parts in 10,000, so that the published uncertainty was a bit optimistic.

The most remarkable fact about this long series of experiments is that Morley found the **identical** result in all the different methods employed: the early density measurements, the isothermal, isobaric measurements of isolated systems, and the water-synthesis measurements. Morley was nominated for the Nobel Prize in 1902, and according to C. H. Cramer, "stood second in the voting" (though I thought these votes are usually secret).

It has been suggested that if Morley had been able to continue and further refine these measurements, he might very well have discovered deuterium, *the isotope of hydrogen which makes up about one part in 10,000 in terrestrial hydrogen, or the isotope* ^{18}O *which is two parts per thousand in oxygen gas. It is not clear to me how such discoveries could follow from Morley's measurements since there were no theoretical predictions for the masses with which to compare. These discoveries had to await magnetic or mechanical separation of the isotopes.*

Beyond atomic masses: with Michelson and Miller

In the world of physics, Morley's name is usually associated with that of Michelson. Morley had been at Western Reserve since 1868. In 1882 Reserve moved from Hudson up to Cleveland to be a neighbor of the new Case School. Morley and Michelson got together when the latter arrived in Cleveland in 1883. Michelson was 31 years old, Morley 45. Their famous 1887 ether-drift experiment, and Morley's role in it, will be described in the next chapter. In a second, less well-known collaboration, Morley worked with Michelson's successor, Dayton Miller, in a continuation of the ether-drift experiment. More on this work in Chapter 4.

Well respected for his uniquely skillful experimental work, Morley was elected president of the AAAS (American Association for the Advancement of Science) in 1895. A biography of Morley was published in 1957. *"Edward Williams Morley: his influence on science in America"* by Howard R. Williams, Chemical Education Publishing Co., Easton, Pa.

Morley's relationship with the college administration, in the person of President Charles Thwing, deteriorated when, in 1896, after a year's leave of absence spent in Europe, Morley returned to Cleveland to find his laboratory and equipment cruelly disassembled. It would seem that Dr. Thwing was not appreciative of Morley's world-renowned work or, perhaps, of academic research in general. By 1906, when Morley was 68, he found it impossible to remain at Western Reserve, and he resigned with a one-sentence letter, returning to the east, sadly following the lead of his five young physicist predecessors. He lived in West Hartford until his death in 1923. His ample estate was bequeathed to Williams College. President Thwing is memorialized today in the name of Charlie's Place pizza and sandwich shop in the student union.

Chapter 3   Measuring Light
Albert A. Michelson
1882 – 1889

When the Case School of Applied Science (CSAS) was formed, John Stockwell, a self-educated astronomer, was its *de-facto* president. He essentially single-handedly "formulated a curriculum, bought equipment and established laboratories and hired a faculty that was small but brilliant" (according to historian C. H. Cramer). The first faculty, in 1881, consisted of Stockwell, Charles Maberry (chemist-Harvard), John Eisenmann (civil engineer-University of Michigan), and Albert Michelson (physicist-U.S. Naval Academy). Michelson was an extraordinary catch for the brand-new school. In the eight years following his graduation from Annapolis, Michelson had caught the attention of the international physics community, beginning with his measurements of the velocity of light at Annapolis and at the Naval Observatory in Washington.

Interferometry

Michelson's youth was very much different from those of the New Englanders described in the first two chapters. He was born in Prussian Strelno (now Strzelno in Poland), brought by his parents to America at age two, raised in Nevada, finished high school in San Francisco, and appointed to the Naval Academy by President Grant, graduating in 1873. **Fig. 3-1.** After a stint at sea and another teaching at the Academy, he was transferred to the Nautical Almanac Office where he and astronomer Simon Newcomb used the rotating-mirror method to measure the velocity of light. (His subsequent work on the velocity of light which was done at Case will be detailed later.) He traveled to Europe in 1880 where he was to remain for two years, first in Berlin, then in Paris. It was during this period that Michelson invented his interferometer.

Fig. 3-1. Midshipman Michelson portrays himself.

An aside on what an interferometer does. This device consists of a light source, a telescope, and some mirrors. The incident light beam is split at a partially silvered mirror, the two beams traveling by different paths to be subsequently rejoined into a single beam which is observed in the telescope. To put it rather simply, if the two paths differ in total length by one-half wavelength or three-halves or any odd number of half wavelengths, the two beams will cancel one another upon rejoining, and no light is observed in the telescope. As one slowly changes the length of one of the paths, the observed light will vary between bright and nil. The device could thus be used to measure the length of an object in terms of the wavelength of a selected spectral line, or conversely to measure a wavelength in terms of a standard length.

The money for the construction of this first interferometer was provided by Alexander Graham Bell who had heard of the young naval officer's talents. Michelson demonstrated his invention in Paris to Marie Alfred Cornu and other distinguished French physicists. Michelson made the acquaintance of Lord Rayleigh when they met in Germany. Rayleigh encouraged him to continue his velocity of light measurements. He was especially interested in having Michelson check on recent claims by other experimenters of significant dispersion of light traveling through air. *(Dispersion refers to different light-speeds for different wavelengths – resulting, for example, in the separation, or dispersion, of colors by a prism.)* Michelson and Rayleigh continued their friendly correspondence for many years.

Michelson's first attempts to use the interferometer to detect the motion of the earth through the ether were made in Berlin and then in Potsdam in 1881. It was generally believed that light must have a medium of some kind through which it moves, just as sound propagates by moving the molecules of the air. This medium was referred to as the ether (or aether).

*An aside on detecting motion through the ether. If the speed of light relative to an all-pervading ether is a constant, and if the earth moves through the ether in its travels around the sun, then the speed of a beam of light relative to the earth should have one value when the light travels parallel to the earth's orbital motion, and another when the light travels at right angles to that motion. This effect can be observed if one sets up an interferometer so that one light beam travels parallel to the earth's orbital motion and the other travels at right angles. One then watches the two interfering light beams in the telescope while at the same time rotating the whole device so that the beams exchange directions. The amounts of time it takes to travel along the two paths will also be switched, and there will be a small change in the pattern of the recombined light as seen in the telescope. This is usually referred to as a shift in the interference pattern, or simply a "fringe-shift". A photograph of what one sees in the telescope is shown in **Fig. 4-7** in the next chapter..*

When Michelson tried this experiment in Germany, he did not observe the expected shift in the interference pattern. The famous repetition of this experiment by Michelson and Morley will be described later in this chapter.

Velocity of light measurements at Case

Michelson received word of his official appointment to the CSAS faculty while still in Europe and he was given a year's leave of absence to conclude his work there. He arrived in Cleveland in September 1882. **Fig. 3-2**. He quickly began setting up for a new measurement of the velocity of light. What is remarkable is the speed with which Michelson got his work underway: he arrived in September, made his measurements in October and November, and wrote his first report in January.

Fig. 3-2. A.A. Michelson

This was possible because most of the principal components of the setup were the same as those he had used with Newcomb two years earlier in Washington. In the report which he sent to Newcomb, Michelson described exactly what he had done, including details of several significant improvements of his own design. *Proc. of the AAAS* **28** 124-160 1879.

With the help of his new colleague, the engineer John Eisenmann, Michelson laid out an optical path along the railway tracks which still border the south edge of the Case campus. Eisenmann's survey indicated that the baseline was 2049.532 feet in length. This was the distance between X-marks inscribed on the heads of copper tacks set in two brick piers, one to hold a revolving mirror, the other to hold a fixed mirror. The separation was determined by using a 300-foot steel tape to measure the distance between two marks inscribed directly on the top of the steel rail. Corrections were made for temperature for both the tape and the rail and for tension applied to the tape. This measurement was then transferred by theodolite to the X-marks on the stone caps on the two brick piers which were placed some 50 feet north of the railroad track. *(No uncertainty is quoted in this report from Eisenmann, but I have my doubts about the significance of the last two decimal places.)*

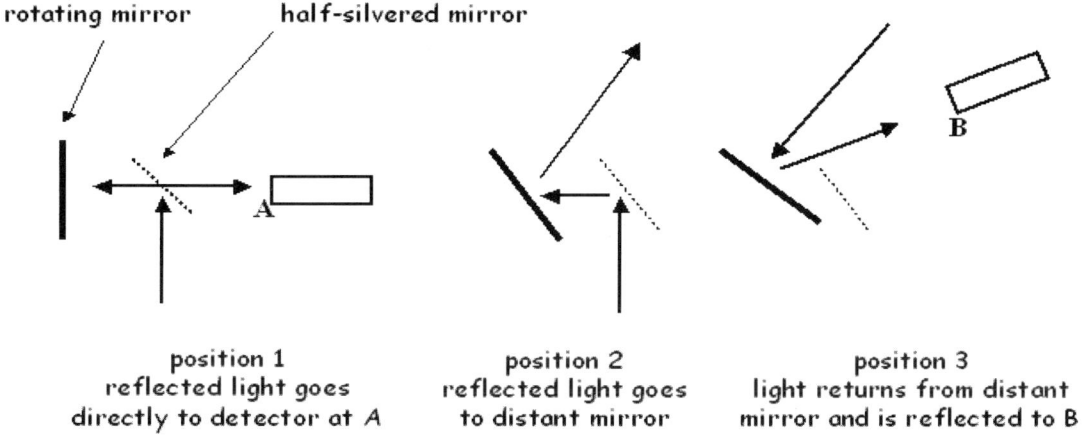

Fig. 3-3. Rotating mirror system for measuring the speed of light.

The following is a much-simplified description of the technique Michelson used. Light, from an electric arc, or from the sun, is incident on a slit. It is then directed via a half-silvered mirror to a mirror which is rotating around a vertical axis. This mirror is actually a polished nickel box with four reflecting faces, driven by compressed air.

In the three plan-view drawings of **Fig. 3-3**, the heavy line represents the rotating mirror which is spinning counterclockwise; the dotted slanted line is a fixed half-silvered mirror. The little rectangle is the viewing telescope. When the rotating mirror is in position #1, the light is reflected back through the half-silvered mirror toward the viewing telescope A placed about 33 feet away. When the mirror is in position #2, the light is reflected toward a distant fixed 15-inch diameter mirror (L=2050 feet away). When the rotating mirror is in position #3, the light returning from the distant mirror (at a time $2L/c$ later) is reflected back toward the viewing telescope, but at an angle about half a degree different from the prompt beam. This deflection angle is twice the angle through which the mirror rotated between positions #2 and #3. Thus there are *only two* positions of the viewing telescope (pos 1 and pos 3 in the sketch) at which light can arrive from the rotating mirror: one beam traveling a short distance, the other traveling over 4000 feet. Michelson would move the telescope until he found the two places where he could see the blinking light. Knowing the angle between these two beams and the rate of rotation of the mirror, one can then determine the time it took the light to make the round trip to the distant fixed mirror, and thus calculate the speed of light. The angular separation between the two beams of light which reach the viewing telescope is measured by a micrometer which moves the telescope across the beams. The displacement between the two beams, combined with a precise measurement of the radial distance (~33 ft) from the rotating mirror face to the axis of the screw of a micrometer, defines the desired angle.

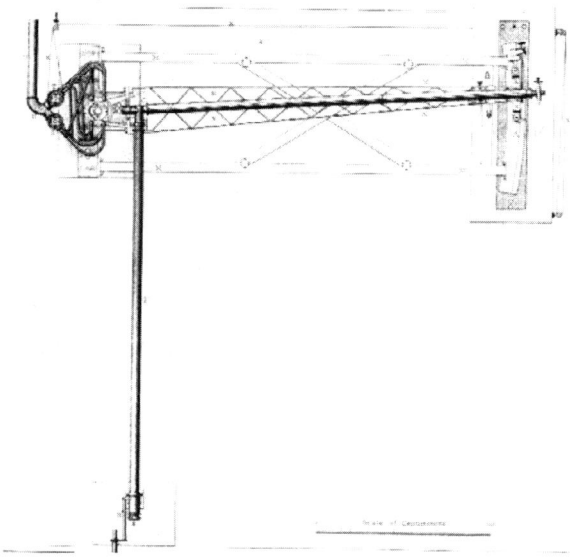

Fig. 3-4. Source, rotating mirror and viewing telescope.

The beautifully drawn plan of the experiment in **Fig. 3-4** is from Newcomb's report on the earlier Washington measurements. It shows the light source at the bottom, the four-sided rotating mirror at the left, and the viewing telescope with micrometer at the right. The distant fixed mirror is far away off to the upper right. **Fig. 3-5** shows two views of the viewing telescope and the micrometer with its two reading microscopes. Details of the rotating mirror and its position relative to the light tubes are shown in **Fig. 3-6.** The half-silvered mirror is enclosed in the right-angle junction of the tubes at the right.

The rate of rotation of the mirror was controlled by adjusting the flow of compressed air directed at vanes attached to the mirror's shaft; one valve controlled the main flow, and a second, "fine-tuning" valve controlled a small flow opposing the main flow.

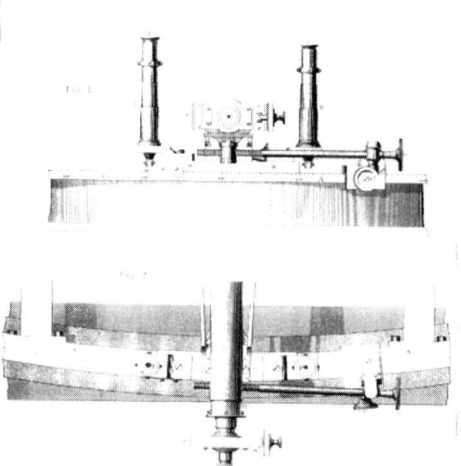

Fig. 3-5. Two views of viewing and reading telescopes.

The angular speed of the mirror was measured by comparison with a calibrated tuning fork. *(The 128 Hz fork, made in Paris by the famous instrument maker Rudolph Koenig, is preserved in the CWRU physics department archives.)* To compare the mirror rotation rate with the tuning fork, Michelson attached a tiny mirror to the fork and directed a narrow light beam first onto the spinning mirror and then onto the "fork mirror". The fork is positioned so that the two motions are mutually perpendicular. The twice reflected light beam moves with two components: a linear (sawtooth) sweep from the rotating mirror and a sinusoidal sweep from the fork.. When the frequencies of the two motions are equal, the light beam is locked in a stationary pattern.

Three rotation rates were used: 1.0, 1.5 and 2.0 times 128 rev/s. The tuning fork was calibrated by a very elaborate scheme involving beating it against an electrically driven fork which in turn was compared to an astronomical "seconds" clock accurate to one-half second per day. (The temperature dependence of the fork's frequency was determined to be 0.0079 Hz/degree F.)

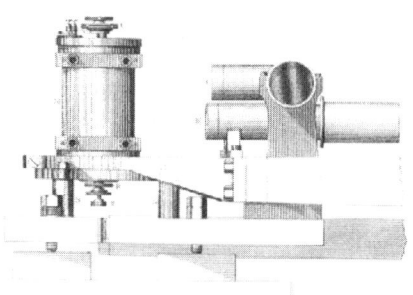

Here are some typical numbers from one trial. The frequency of the standard fork 129.127 Hz; the rate of rotation set at twice the fork frequency or 258.254 Hz. The linear deflection of reflected beam is 137.920 mm at a radius of 33.350 feet, giving an angular deflection of 2788.7 arcsec; resulting in a velocity of light for this measurement of 299 883 km/s. The weighted mean of 23 such measurements was 299 853 ± 60 km/s. Thus, the quoted uncertainty is 2 parts in 10,000, a remarkable accomplishment for a purely mechanical measurement. (This corresponds to about one inch in 2000 feet, so the Eisenmann survey was more than sufficient.) This value of c was the accepted standard until Michelson's later measurement in 1926 (299 796 ± 4 km/s; the uncertainty on the earlier measurement uncannily just reaches the later value). ("Albert A. Michelson at Case", R. S. Shankland, *Amer. Jour. of Phys.* **17** 487 (1949))

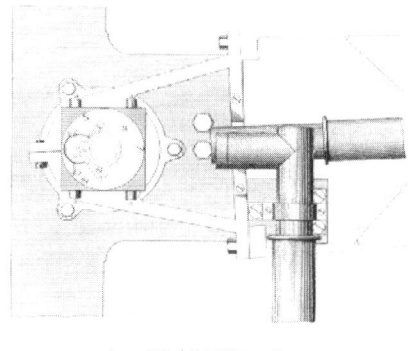

Fig. 3-6. Side and top view of rotating mirror.

At the end of his report to Dr. Newcomb, his mentor at the Navy Department, Michelson writes, "I would take this opportunity of expressing my obligation to ....several of the students of the Case Institute for their cheerful aid in carrying out the work ...." Thus begins the long history of research experiences for Case undergraduate physics students.

The velocity of light in water and in a dispersive medium

In a further report to Newcomb sent the following August, Michelson describes an experiment in which the light path to the fixed mirror included ten feet of distilled water. The same arrangement of rotating mirror, fixed mirror, and telescope were used as before in this quick "one part per thousand" measurement. The resulting ratio of c/v_{water}

= 1.330 ± 0.003 was in good agreement with the measured index of refraction of water (1.334). This was the first quantitative demonstration of that equality.

Next, Michelson replaced the water by 3.07 m of carbon disulphide. (All the data for this experiment are given in metric units, in contrast to the inches and feet of his earlier reports.) Different mirror rotation rates were used (from 128 to 320 per sec), as well as different distances from rotating mirror to telescope (from 3 to 6 m), and the results for the velocity of light v in CS_2 were consistent, giving a mean value of c/v of 1.77 ± 0.02. The measured value of the refractive index of CS_2, however, is 1.64, six standard deviations away. Nonetheless, Michelson stood by his measurement. The disagreement had in fact been resolved by Rayleigh's theory (1881) for the propagation of light in a dispersive medium. Rayleigh had shown that the index of refraction is related to the *phase* velocity (ω/k) while the propagation speed is the *group* velocity ($d\omega/dk$). (Here $\omega = 2\pi$ times the frequency and $k = 2\pi$ over the wavelength.) His calculations agreed with the Michelson result. In their famous textbook on physical optics, Jenkins and White mention that, since water is also slightly dispersive, Michelson should have measured something like 1.35, rather than 1.33; Michelson must have wondered about this, given the better agreement for the CS_2 measurement.

Completing this series of measurements, Michelson looked for a dependence of the speed of light in CS_2 on the wavelength of the light. He placed a small prism in front of the slit, with which, by small rotations of the prism, he illuminated the slit with blue and then with red light. His result was $v_{red}/v_{blue} = 1.014$, the first quantitative measurement of dispersion.

Michelson and Morley team up: Light in a moving medium

Michelson had made the acquaintance of Edward W. Morley, the eminent physical chemist at the neighboring institution, soon after his arrival at Case in 1882. The two men shared a passion for the development and application of precise measuring instruments. They traveled together to scientific meetings in Baltimore and Montreal. Their first joint effort was to repeat an earlier (1859) experiment by Fizeau to check on the theory developed by Fresnel (1818) for the propagation of light through a moving medium. Michelson devised an arrangement in which a beam of light was split into two beams, one of which traveled through water in the direction of the water flow, the other against the water flow. The beams were recombined to give an interference pattern. The pattern shifted as the water velocity was changed. (*Amer. J. Sci.* **31** 377 (1886))

The result was extremely gratifying: the predicted change in the speed of light was $v_{water} (n^2-1)/n^2 = 0.438\ v_{water}$ where n is the index of refraction of water; the measured coefficient was 0.434 ± 0.02.

In pursuit of the ether: the Michelson-Morley experiment

It was time for Michelson to give the ether-drift experiment another try. He realized that he had overestimated the expected effect when he was in Germany. He had believed that the shift in the light speed should be in the order $v_{earth}/c = 10^{-4}$, but, as was pointed out by H.A. Lorentz, the size of the effect should actually be only 10^{-8}. Lorentz showed that the light travel-time in the interferometer arm *perpendicular* to the earth's motion should *also* be affected by the motion of the apparatus through the ether. The displacement of the fringes in the 1881 attempt at Potsdam should have been only four hundredths of a fringe, well below the sensitivity of the experiment.

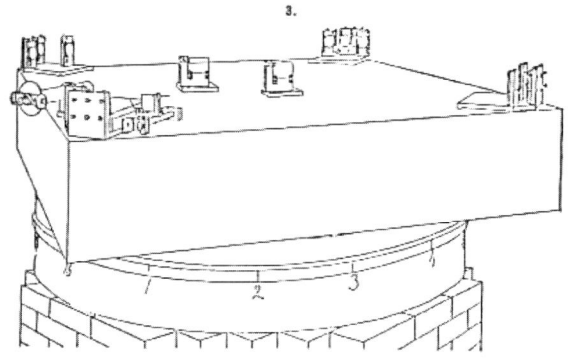

Fig. 3-7. MM experiment setup.

With Morley, Michelson designed a much improved arrangement. Here is the Michelson-Morley description of the situation: "In the first experiment one of the principal difficulties encountered was that of revolving the apparatus without producing distortion; and another was its extreme sensistiveness to vibration .... Finally .... the quantity to be observed, namely, a displacement of something less than a twentieth of the distance between the interference fringes may have been too small to be detected when masked by experimental errors. The first named difficulties were entirely overcome by mounting the apparatus on a massive stone floating on mercury; and the second by increasing, by repeated reflection, the path of the light to about ten times its former value." The expected fringe shift was thus increased by a factor of ten, to four tenths of a fringe, well within the experimental sensitivity.

The line-drawings of the setup (**Fig. 3-7, 8** and **9**) are from their famous paper. (*Amer. Jour. Sci. 3rd Series* **34** 273 (1887)) The paper is available on the internet at:
http://www.aip.org/history/gap/Michelson/Michelson.html

The authors' description of the procedure is straightforward: "The observations were conducted as follows: Around the cast-iron trough were sixteen equidistant marks. The apparatus was revolved very slowly (one turn in six minutes) and after a few minutes the cross wire of the micrometer was set on the clearest of the interference fringes at the instant of passing one of the marks. The motion was so slow that this could be done readily and accurately. The reading of the screw-head on the micrometer was noted, and a very slight and gradual impulse was given to keep up the motion of the stone; on passing the

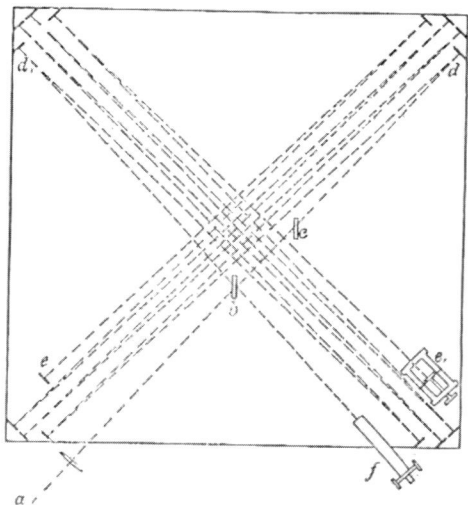

Fig. 3-8. The light paths in the MM experiment; source lower left, telescope lower right.

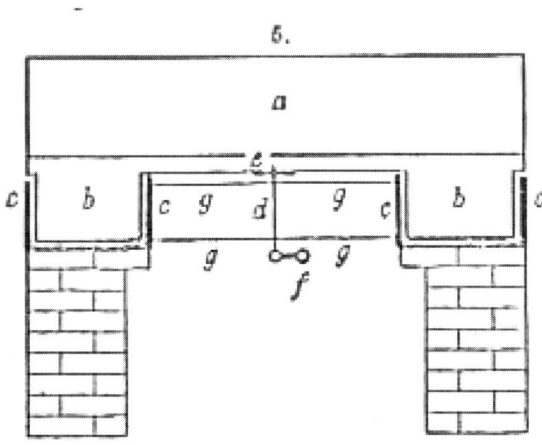

Fig. 3-9. Cut-away side-view of stone block floating on mercury.

second mark, the same process was repeated, and this was continued till the apparatus had completed six revolutions. It was found that by keeping the apparatus in slow uniform motion, the results were much more uniform and consistent than when the stone was brought to rest for every observation; for the effects of strains could be noted for at least half a minute after the stone came to rest, and during this time effects of change of temperature came into action."

The resulting data consisted simply of several series of micrometer readings. The entire set is shown in the table (**Fig. 3-10**), as printed in the 1887 publication. Readings were taken at noon and later at six in the evening, allowing the earth to rotate through 90 degrees. The readings are in divisions on the micrometer. The approximate width of a single fringe is about 50 divisions. The mean readings are then converted to wavelengths (for sodium light). The data are first shown for the sixteen positions, and then, in the bottom line, are folded around the 180 degree position. Thus, the noon and evening runs are reduced to 8 numbers each. These are constant to within 0.02 wavelengths, twenty times smaller than the expected variation of 0.4 wavelengths. The two sets of eight numbers are shown graphically in the bottom half of **Fig. 3-10**, along with the expected signal **reduced by a factor of eight**. *(If I were teaching a class now, I would repeat the last phrase two or three times!)*

The conclusion is straightforward, without any complicated calculations or error analysis. The clear difference between the observed and the predicted effect allows the authors to state, "the relative velocity of the earth and the ether is probably less than one

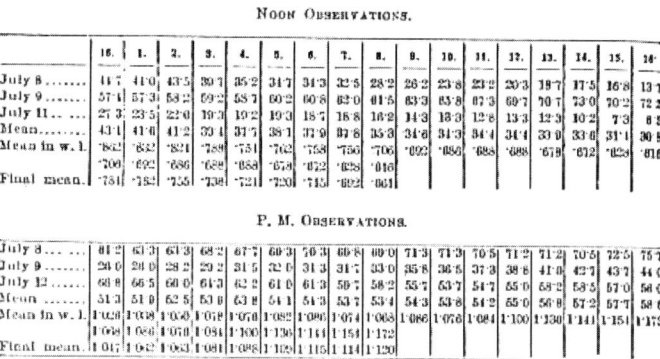

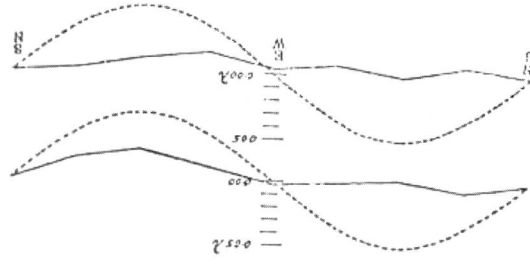

Fig. 3-10. Original 1887 MM data. Expected signal (dashed curve) **reduced** by a factor of 8.

sixth the earth's orbital velocity, and certainly less than one-fourth." It is remarkable that this very straightforward experiment, involving rather simple instrumentation and essentially no data analysis, opened a new era in physics.

Michelson and Morley pointed out at the end of their paper that they planned to repeat their observations at three month intervals to sample different relative velocities of the earth through space. They also mention the possibility that the ether may be carried along with the earth (ether drag). In a supplementary paragraph, they add that one might want to repeat the experiment at higher altitudes to reduce the effect of ether drag. In addition, they discuss the possibility of designing experiments to measure the speed of light with a precision sufficient to observe a direction-dependent variation. Interestingly, they further suggest a precision repetition of the Römer experiment (1676) which determined the speed of light (to within about 40%) by observation of the moons of Jupiter.

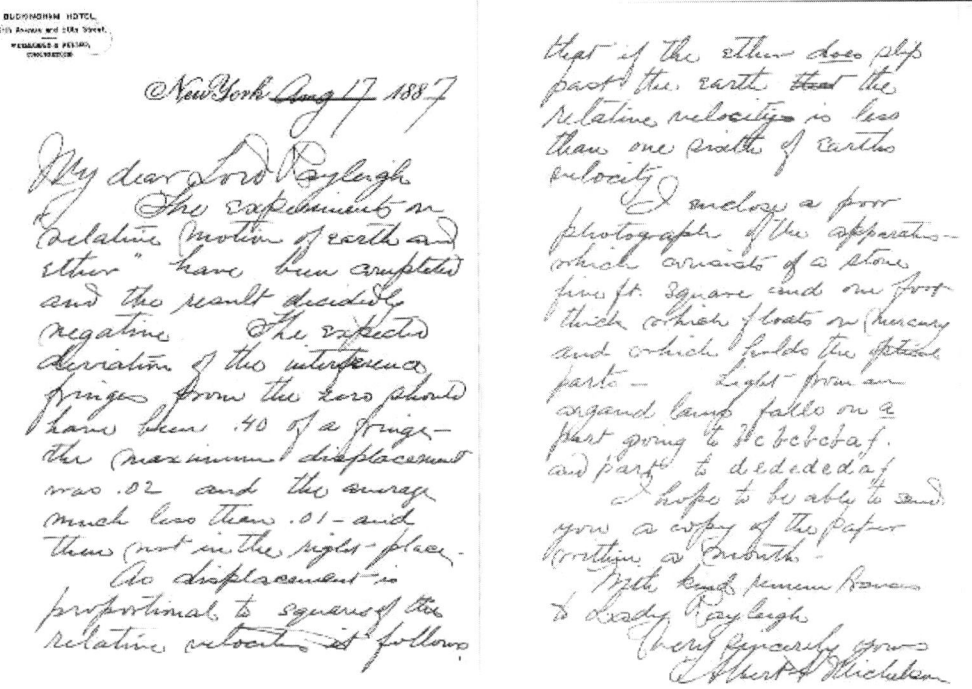

Fig. 3-11. Michelson to Rayleigh – August 1887.

Michelson soon wrote to Lord Rayleigh from New York. **(Fig. 3-11)** Here is the text of his letter:

Aug 17 1887

My dear Lord Rayleigh
The experiments on "relative motion of earth and ether" have been completed and the result decidedly negative. The expected deviation of the interference fringes from the zero should have been 0.40 of a fringe – the maximum displacement was .02 and the average much less than .01 – and then not in the right place.

As displacement is proportional to squares of the relative velocities it follows that if the ether does slip past the earth ~~that~~ the relative velocity is less than one sixth of earths velocity.

*I enclose a poor photograph of the apparatus – which consists of a stone five ft. square and one foot thick which floats on mercury and which holds the optical parts - Light from an argand lamp falls on **a** part going to **bcbcbcbaf** and part to **dedededaf**.*

I hope to be able to send you a copy of the paper within a month.

With kind remembrances to Lady Rayleigh.

    Very sincerely yours

       Albert A Michelson

While the results of the 1887 experiment appear to be convincing, this is far from the end of the story. Repetitions of the experiment, with various instrumental and analytical improvements, performed in Cleveland, on mountain-tops, and in laboratories in many countries, would continue for the next fifty years. We shall follow some of these efforts in Chapter 4 on Michelson's successor, Dayton Miller.

Michelson left the Case School two years later, in 1889, moving on to Clark University and thence to the new University of Chicago. He continued his work in applied optics until his death in 1931, including refinements of the velocity of light and ether-drift experiments, as well as standardizing the meter in terms of light waves and the first measurement of the diameter of a star. He received the Nobel Prize in 1907. The prize was awarded for his precision optical measurements of the standard meter and of spectroscopic wavelengths. There was no specific mention of his velocity of light work or of the ether-drift experiment. This was only two years after Einstein had explained the null result.

Fig. 3-12 is a photo taken on the occasion of the 1960 dedication at Case of a portrait of Michelson. Celebrating the event are Frederick Reines (chairman and Nobelist to be), Robert Shankland (chairman emeritus), and Michelson's daughter and biographer, Dorothy Michelson Livingston. (Her book, *Master of Light*, Scribner 1973, contains many interesting insights into the life of her famous father.)

Fig. 3-12. Reines, Shankland and Michelson's daughter, Dorothy M. Livingston.

Harry Fielding Reid - Geophysicist

After Michelson had gone, professor of mathematics, **Harry Fielding Reid**, was appointed professor of physics and took over as chairman of the CSAS physics department in 1889. **Fig. 3-13**. Reid had earned his doctorate in physics at Johns Hopkins in 1885 and had spent time at the Universities of Berlin and Cambridge. During his tenure

at Case, Professor Reid's principal research interest was glaciology. In 1890 he took four Case students with him to study the Muir Glacier in Alaska, incidentally taking the opportunity to name a mountain after Case (either Leonard or the school). Reid's work on glaciers is nicely summarized in the following excerpt from the website of the Milton S. Eisenhower Library at the Johns Hopkins University.
http://archives.mse.jhu.edu/mss/ms367.txt

"Because of his proficiency in both physics and geology, Reid has been recognized as the first American geophysicist. His work in science fell into two main categories: the study of earthquakes and the study of glaciers. Reid's published writings between 1892 and 1907 deal exclusively with the subject of glaciers. His first glacial study was an extended survey of the Muir Glacier (Alaska) in 1890. With a team of five others, he mapped the glacier area of over 900 square miles. His research continued for several years, and his findings were published in a paper on the Glacier Bay area of Alaska in 1896. (16th Annual Report., U. S. Geological Survey). He was interested in the problems of glacial accumulation, motion, and wastage."

Fig. 3-13. Henry F. Reid physics chair 1889-1893.

Reid left Case in 1893, eventually to become a world-renowned expert on earthquakes as Professor of Dynamic Geology at Johns Hopkins. Today, most of the internet references to Dr. Reid concern his analysis of the 1907 San Francisco earthquake in terms of "elastic rebound" and his model for slippage along tectonic plates. He was appointed in 1915 by President Wilson to study the possibility of landslides into the Panama Canal, and ways to prevent them. Reid died in 1944.

Chapter 4   Xrays, Flutes and More Ether: Dayton Clarence Miller

Miller,   Hodgman,   Albright,   Nusbaum,   Wallace
1890-1941   1906-52   1921-1943   1922-53   1926-60

Dayton Miller was the first member of the department who was born west of the Appalachians. A search of the internet shows that Miller is best known today for two widely different activities: first, as a collector and investigator of musical flutes and second, as an experimentalist who found evidence for ether drift. Some of the material in this chapter comes from William J. Maynard's master's thesis (Long Island University, 1971) which is available in its entirety on the internet.
http://memory.loc.gov/ammem/dcmhtml/may12.html

Fortunately, most, perhaps all, of Miller's publications were preserved by his successor, Robert S. Shankland, who had them bound into a single large volume which is now part of the departmental archives.

Miller was born on a farm in Strongsville, Ohio in 1866 and raised in nearby Berea. His dual interests in science and in music were already in place by the time he was 14. He built several telescopes and became an accomplished flute player. At age 16, he entered Baldwin College (now Baldwin Wallace), graduating with a bachelor of philosophy degree in 1886. At his graduation ceremonies, Miller gave a lecture, "The Sun", and played a flute solo with the university orchestra. His decision to become an astronomer rather than a musician was made after he read in *Scientific American* about plans for a new 23 inch refractor telescope at the Halsted Observatory at Princeton. He applied to study there, and became a student of the by-then-famous solar astronomer, Charles A. Young, the fellow who had begun his teaching career at Western Reserve (see Chapter 1).

Fig. 4-1. The Faculty of the Case School. DCM at far right.

Miller completed his doctoral degree in 1890: "Observations of Comet 1889 V and an Investigation of its Orbit with an Ephemeris". Transit times of the comet and of nearby reference stars were made with the big 23 inch telescope between November '89 and March '90. "The times of the transits were recorded with a chronograph in electrical connection with the standard sidereal clock at the School of Science Observatory." The "ephemeris" consisted of a table of predicted positions for certain dates in 1890 and a comparison of these with subsequent observations; his quoted deviations are in the order of a few seconds of arc. (The original manuscript of this dissertation is in the CWRU Physics Archives.)

After a summer back home in Berea, Miller planned to return to Princeton to work with Professor Young on a new spectrograph for the 23-inch. However, because there was to be a considerable delay in the delivery of the device, Miller decided he'd better find some temporary, local employment. He paid a visit to Dr. Cady Staley, president of the 9-year-old Case School, and Professor Charles S. Howe, a newly appointed professor of mathematics and astronomy. Staley and Howe were delighted by Miller's wide interests and enthusiasm and quickly hired the 24-year-old to begin teaching the following month, September 1890. Miller was to spend the next fifty years at Case. Chairman Reid left Case in 1893 and Miller was appointed chairman two years later. He is the slight fellow on the right end of the table in **Fig. 4-1**. Note that the CSAS building boasted both gas and electrical lighting.

Applying the newly discovered X-rays

Within two months after the discovery of x-rays by Röntgen in November 1895, Miller began experimenting with the Crookes tubes he had bought a few years earlier at the Columbian Exposition in Chicago. *(These are glass tubes, filled with various low pressure gases. When a high voltage is applied across the electrodes at each end of the tube, the gas glows as electrons travel through it. This was all rather mysterious at the time because the electron had not yet been discovered.)* Within a week, Miller and his wife Edith had produced a series of x-ray photographs and his services were soon made available to the medical profession. There is some question as to whose arm bones were published first, Mrs. Miller's or Frau Professor Röntgen's. From Miller's article in *Science* in March of 1896: "The arm was photographed with an exposure of twenty minutes…The Crookes tube ... was excited by an induction coil giving about a six-inch spark in air, when using a current of three amperes and twenty volts, obtained from eleven cells of storage battery." In **Fig. 4-2**, Miller is

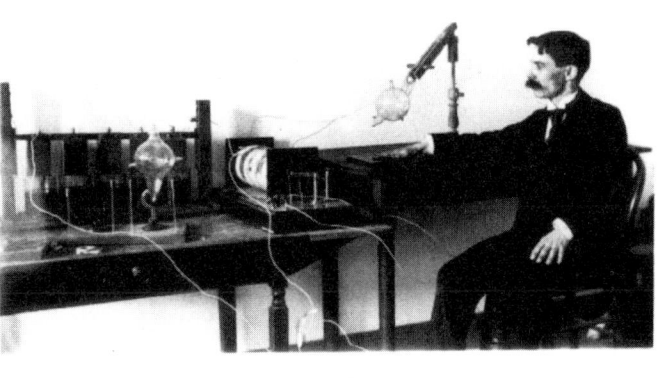

Fig. 4-2. "D. Miller experimenting with X-Rays Jan 1896".

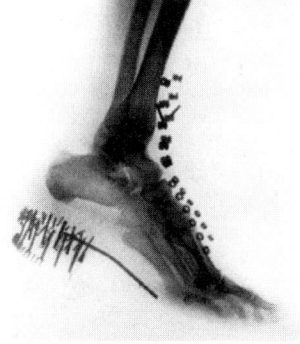

Fig. 4-3. X-ray of foot in shoe.

shown making a "roentgenogram" of his own hand, in **Fig. 4-3** an early x-ray of a foot in a boot, and in **Fig. 4-4**, an extraordinary composite of Miller's whole body, including the buttons on his boots and the change purse in his pocket.

Miller took his x-ray demonstrations on the road, signing with a Chicago booking agent. He and his talented assistant, Dudley B. Wick, Jr., hit the lecture circuit with a show which included the making of an X-ray. From the handbill: "Prof. Miller takes an X-Ray Negative before the audience, the subject being chosen from those present." **Fig. 4-5.** Both Miller and Wick subjected themselves to repeated exposure, and while Miller survived for almost fifty years more, Wick was less fortunate. He died at age 28 in 1905, possibly from radiation sickness.

In April of 1896, Miller lectured to the Cleveland Medical Society. His talk was published in *The Cleveland Medical Gazette*. One section is of particular interest to this reader. In reference to the "rays" within the tube (not the x-rays) Miller writes: "The cathode rays of themselves are not luminous, but seem to be intimately connected with light. Physicists are compelled to believe that there is an all-pervading elastic medium which fills all space, and penetrates all bodies, however solid they may be, and that light is a wave motion of this ether." He goes on to mention the theory of Maxwell and the experiments of Hertz and Lenard, but not a word about the experiments of Michelson and Morley performed at Case nine years previous. Later, in speculating about the nature of the x-rays themselves: "another theory would make them longitudinal ether vibrations, thus connecting them with light, but distinguishing them from light". Miller's interest in the ether would last all his life, as we shall describe later.

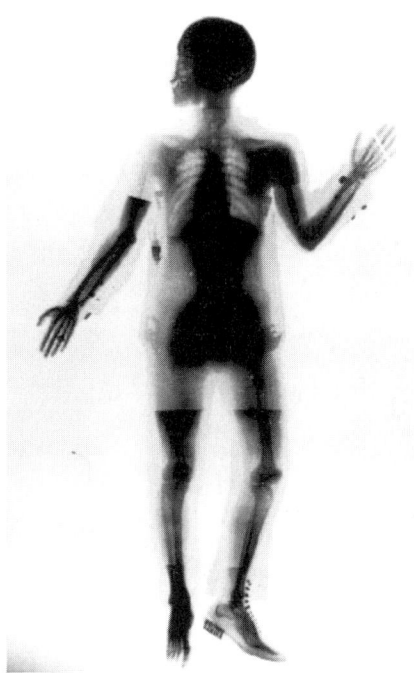

Fig. 4-4. Whole body composite x-ray.

Fig. 4-5. "A Lecture on the Marvelous X-rays".

Teaming up with Morley

We backtrack momentarily to Edward Morley and a contact he made in 1889 and some research he did which made use of the Michelson interferometer. He had attended a AAAS (in this case, the American Academy of Arts and Sciences) meeting in Toronto where he heard a talk by Henry T. Eddy, a mathematician at the University of Minnesota. Eddy discussed the proposal that the speed of light in polarizing materials may change with the application of an external magnetic field. Morley must have proposed to Eddy that they should try to observe this effect by using the Michelson interferometer technique.

The following year, Eddy came to Cleveland to work with Morley on an experiment to search for such an effect. The experimental arrangement consisted of the usual interferometer with a 38 cm-long tube inserted in each light path. These tubes were filled with liquid carbon bisulfide and were surrounded by water-cooled copper coils to produce axial magnetic fields. The current directions were chosen so that the sought-after effect would slow the beam in one arm and speed it up in the other. While one man observed the interference fringes, the other threw a switch which reversed the electric currents. This was certainly a lot easier than following the rotating stone of the ether-drift experiment. The light source was a Bunsen lamp with sodium salts in the flame. In the AAAS meeting in Indianapolis that same year, Morley and Eddy reported that they saw no effect greater than "one part in one hundred millions".

Now we return to Miller, who, after his work with x-rays, teamed up with Morley and Eddy in 1898 to repeat the magnetic field experiment. It was nine years after their first effort. Morley had been very busy during the intervening period with his atomic mass work. Several improvements were made to the apparatus: the tubes lengthened, the temperature of the liquid better controlled. Their conclusion, as reported in the *Physical Review*, **7** 283 (1898): "we are confident that when light corresponding to the solar D line is passed through one hundred and twenty centimeters of carbon bisulphide in a magnetic field which produces rotation by half a circumference in the plane of its polarization, there is no such change of velocity as one part in sixty million." Once again, the Michelson interferometer was shown to enable measurements of extraordinary precision.

The search for the ether continues

No one had been particularly pleased with the null results of the 1887 Michelson-Morley experiment, including especially the two gentlemen themselves. When Miller and Morley and their wives traveled to Paris in 1900 to attend the International Science Congress, they met with Lord Kelvin who encouraged them to repeat the ether-drift experiment. Morley and Miller began in 1902 a series of experiments to try to improve the measurement. Morley was 67 and Miller 39; in Zurich the 26 year-old Mr. Einstein was contemplating a paper on a related subject.

For Miller, this was the beginning of an effort which would challenge him for more than thirty years. Experiment after experiment would give small indications of ether-drift. Miller was driven to improve both the statistics and the experimental technique so that the result would be unambiguous and conclusive, either for or against the ether hypothesis.

Morley and Miller published a paper in the *Philosophical Magazine* in May 1905 in defense of the validity of the technique used by Michelson and Morley 18 years earlier. In particular, they addressed claims by W. M. Hicks of University College, Sheffield, that the earlier results were in doubt because of neglected geometrical optical effects associated with moving mirrors. (*Phil. Mag.* **6** iii. 9 1902) Morley and Miller include in their paper eight rather complicated line drawings illustrating Dr. Hicks' arguments, but at the end conclude that "the theory of 1887 is correct to terms of the order retained, which were sufficient; that Dr. Hicks's theory agrees with it precisely as to numerical amount...." Having disposed of Dr. Hicks' arguments, Morley and Miller submitted a second paper which described their own new measurements.

Rather than presenting here the details of each separate set of ether-drift measurements, it might be more efficient to insert a table with dates, places, results, etc. In the table, the three entries between 1902 and 1905 are for the Miller-Morley Cleveland experiments. The column marked "Result V_{rel}" lists the reported values of the relative velocity of the earth and the ether, which one may compare with the earth's orbital velocity, 30 km/s. The values quoted are the authors' best estimates, based on the observed fringe shifts. In no case is a lower limit stipulated, or a chi-square goodness-of-fit parameter given, or even a calculated probability that the result is zero. More on this later. Most of this information comes from Miller's "final word" on the subject, his forty-page paper in the *Reviews of Modern Physics* (*Rev. Mod. Phys.* **5** 203 July 1933).

Date	Personae	Location	Features	Light path	Data set	Result V_{rel}
1887	Michelson Morley	Cleveland	stone base, mercury float	11 m	6 turns	8 km/s
1902-1903	Miller Morley	Cleveland	White pine frame	33 m	505 turns	10 km/s
1904	Miller Morley	Cleveland	Steel frame	64 m	260 turns	7.5 km/s
1905	Miller Morley	Euclid Heights	Higher location (300 ft), less material	64 m	230 turns	8.7 km/s
1921 April	Miller	Mt.Wilson	1750 m altitude, minimal enclosure	64 m	350 turns	10 km/s
1921	Miller	Mt.Wilson	Cork insulation on interferometer	64 m	273 turns	10 km/s
1921 Dec	Miller	Mt.Wilson	Nonmagnetic interferometer	64 m	422 turns 24 hour readings	10 km/s
1924 Sep	Miller	Mt.Wilson	Protected site, lightweight materials, new optics	64 m	136 turns	10 km/s
1925-1926	Miller	Mt.Wilson	Four "epochs"	64 m	6402 turns	from 5 to 10 km/s

The various improvements made over the course of these 24 years addressed many issues. The substitution of wood for stone in 1902 was to test the 1894-5 Lorentz-Fitzgerald hypotheses of the contraction of moving objects and whether it depends on the nature of the material. The moving of the experiment from the campus up hill to nearby Euclid Heights, and subsequently to a mountain top in California, was to determine whether the ether was dragged along by the earth any less at higher altitudes. The opening of the experiment to the "outdoors" was to expose the interferometer to the unobstructed ether. Special efforts were made to address vibrations, winds, thermal, and magnetic effects, and even gravitational inhomogeneities. Readings were taken hourly around-the-clock to sample all directions in the equitorial plane. The "four epochs" entry refers to extending the measurements to four locations on the earth's orbit (i.e. four seasons) to search for any contribution from the motion of the solar system through the ether.

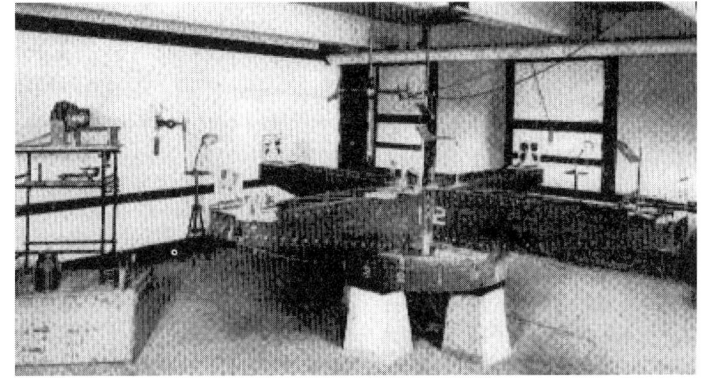

Fig. 4-6. Miller's ether-drift interferometer set up in the Rockefeller Building.

(Among the letters in Robert Shankland's files is one from Samuel G. Hibbin, Case class of 1910. Attached was a four-page essay by Hibbin in which he described "a single-room, low, wood-frame, unpainted shed, off to the side of the Chem. Lab., that contained little-known and seldom seen apparatus." He explains that in 1907 it was his duty as a 20¢/hr work-study student to tidy up the equipment which Miller and Morley had used there.)

What does one see in this Table? For one thing, one sees a very determined man who tried any number of things to improve the experiment, and at the same time, one sees a serious lack of quantitative analysis. Miller seemed to believe that vast quantities of data would make his results more convincing. He notes that he walked 160 miles around his interferometer to take over 200,000 observations of the fringe patterns.

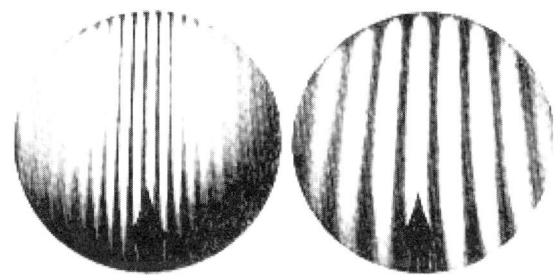

Fig. 4-7. Miller's photographs of fringes as seen in telescope.

The four figures in this section come from the 1933 paper. The photo **(Fig. 4-6)** was taken in the Rockefeller Physics building at Case. The sandstone, floating on mercury, turned at the slightest touch and would continue to coast for an hour or more. The photo of the fringe-pattern allows one to judge the claim of a precision of one-tenth of a fringe. **(Fig. 4-7)**

Each rotation of the interferometer yields 16 fringe positions. With each 22½ degree increment, an electrical contact would sound a bell and the observer would call out the position (in tenths of a fringe-width) of the pointer. In the later experiments, each set of data was analyzed by a mechanical device, the Henrici Harmonic Analyzer, which extracted the amplitudes for the first five harmonics. (Miller had published a paper on the use of this device in the *Journal of the Franklin Institute*, Sept. 1916. This clever machine allows one to move a scribe along a plotted curve, the x- and y-motions of which drive a series of pulleys, spheres and gears which in turn produce a readout of the amplitude for each harmonic component. A photo of the analyzer and a description of how it works will appear later in this chapter, where its use in the analysis of sound is described.)

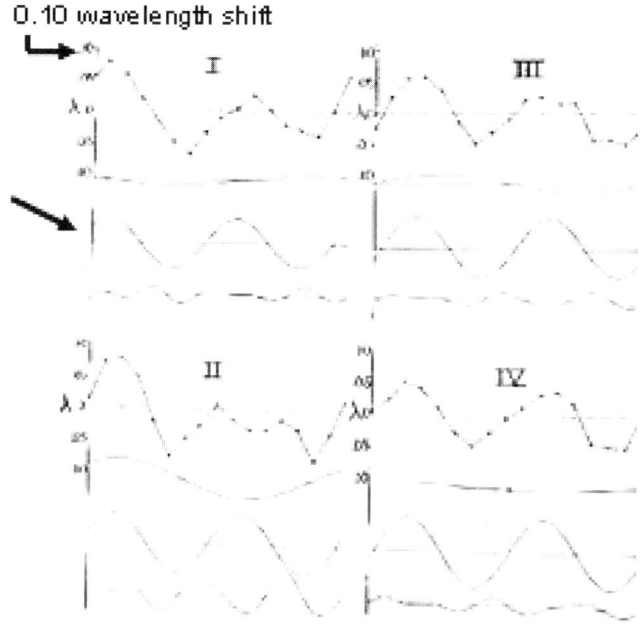

Fig. 4-8. Four sets of data: shifts in wavelengths vs. time over 24 hours. Expected signal would be about 0.3 λ in the double hump

Fig. 4-8 shows four such harmonic analyses. In each case, the top curve is the data plotted vs. the angular position of the interferometer, the next is the extracted contribution of sin θ, the next sin 2θ, and the bottom the sum of the next three terms. The expected ether-drift signal would be in the double-peaked sin 2θ curve (dark arrow). The ordinates are in wavelengths (rather than fringe widths or relative velocities). The relative ether-drift velocity is determined from the amplitude of the second harmonic, and its azimuthal direction from its phase. A shift of 0.1 wavelength corresponds to a velocity of about 9 km/s, about a third of what was expected.

The combined data taken over several days six months apart are shown in **Fig. 4-9**. The ordinate is the sidereal time covering the 24 hours of the earth's rotation. The upper curves of each set are the magnitudes of the velocity through the ether; the lower curve the azimuthal angle of that motion. (The velocities and azimuthal angles have been averaged over the individual runs to give the twenty dark data points on each plot.) What one would like to see is all curves sweeping through some maximum once each 24 hours. This is hardly the case.

Through a sequence of averaging, Miller arrives at ether-drift speeds of 9 to 11 km/s (with a probable error of 0.33 km/s). He discusses at some length the fact that any motion of the solar system through the ether must be added vectorially to the orbital motion of the earth, but he does not find a solution which fits the data. He summarizes: "This procedure", the inclusion of all observations, with no selections or corrections, "has been adopted as the only safe one in the first search for a hitherto unidentified effect. The present results strikingly illustrate the correctness of this method, as it now appears that the forty-six years of delay in finding the effect of the orbital motion of the earth in the ether-drift observations has been due to efforts to verify certain predictions of the so-called classical theories and to the influence of traditional points of view." This statement is rather strange, since Miller was one of many physicists who, through the 1920's and even 1930's, were clinging to the ether idea. What he probably meant was that it was incorrect to throw out the whole idea of the ether just because a full 30 km/s effect was ruled out by experiment. He insisted that the size of the effect was "not zero".

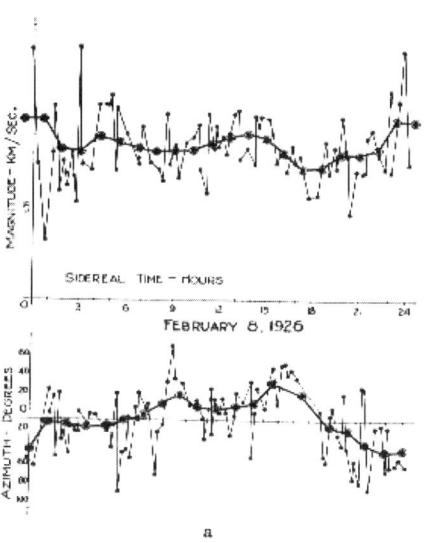

Einstein visited Cleveland in 1921, in the company of the Zionist leader, Chaim Weizmann; they were on a fund-raising tour of America. Einstein took the opportunity to visit Miller and to discuss his experiments. According to Robert Shankland, Einstein encouraged Miller to continue to improve upon his measurements, hopefully to arrive at an unambiguous result. Another famous visitor from Europe, the Dutch theorist, H. A. Lorentz, came to Cleveland in the spring of 1922 to meet with Miller. Lorentz played an important role in the development of relativity theory. He, too, felt that Miller should continue his efforts at Mt. Wilson. The photo of Miller and Lorentz **(Fig. 4-10)** was taken at the front door of the Rockefeller Building. Einstein and Miller stayed in touch over the subsequent years, so that in 1925 Miller wrote to Einstein in Berlin. His letter was apparently in response to a letter he had received from the theorist. Miller agreed with Einstein that variations in temperature would certainly introduce a false signal in the interferometer experiments, but he assured him that utmost precautions were taken to control the temperature.

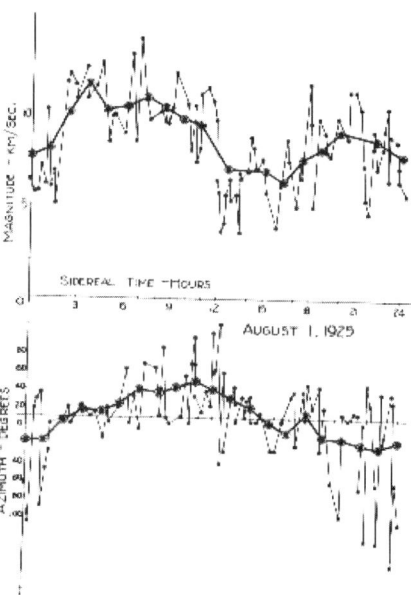

Fig. 4-9. Compilation of hundreds of runs six months apart.

Fig. 4-10. Lorentz and Miller meet at Case.

A conference on the ether-drift experiments was held at the Mt. Wilson Observatory in Pasadena in 1927. Among the participants were Michelson, Lorentz, and Miller. Michelson describes the original 1887 experiment and mentions that ether-drag might explain the negative result. He goes on to say, "This assumption, however, is a very dubious one because it contradicts some other important theoretical considerations." He continues, "Lorentz then suggested another explanation which in its final form yielded as a result the famous Lorentz transformation equations. These contain the gist of the whole relativity theory. The Michelson-Morley experiment was continued by Morley and Miller, who again obtained a negative result. Miller then continued alone, and seems now to get some positive effect. This effect, however, has nothing to do with the orbital motion of the earth. It seems to be due to a velocity of the solar system relative to stellar space, which may be much greater than the orbital velocity."

The CWRU Kelvin Smith Library recently (2002) acquired a letter written by Miller to F. B. Vesper which indicates how Miller felt about the ether vs. relativity issue. Here it is in its entirety.

March 16, 1926

Dear Sir,

The final report of my ether-drift experiments has not yet been published. I think it will appear within the next three or four weeks. I am sending you a reprint of the article which appeared in *Science* for June 19, 1925. This gives the history and the general conclusions.

I am hardly qualified to enter into an extended discussion on the effect of the Theory of Relativity. I prefer to leave this for the mathematical physicists. It seems to me, in a word, that my experiments will invalidate one of the fundamental postulates of the Theory of Relativity which will require a considerable modification of the latter.

Very truly yours,

Dayton C. Miller

Miller was certainly not alone in the belief that the luminiferous ether would someday be detected. The physics community did not quickly accept or even understand what Einstein was proposing. Since Miller had the equipment and the experience, and especially since he had been explicitly asked by Kelvin, Einstein, and Lorentz to continue the experiment, how could he not do so?

In 1925 Miller was awarded a thousand dollar prize by the American Association for the Advancement of Science at their annual meeting in Kansas City. The prize was given for the most interesting and important results reported at the meeting. On the committee was the physicist Karl Compton. According to Robert Shankland, the committee really thought that Miller had found something.

If one goes today to the internet and does a search for Dayton Miller, one finds most of the hits refer to his flute collection or his work with the analysis of sound, but one also finds many references to his "observation of the ether". Of these sites, many are sponsored by persons or groups who maintain today that Miller's work proved Einstein wrong. One often finds a quotation from the *Cleveland Plain Dealer* in which Miller responds to a remark by Einstein that Miller might have had a problem with thermal effects: "The trouble with Professor Einstein is that he knows nothing about my results…. He ought to give me credit for knowing that temperature differences would affect the results. He wrote to me in November suggesting this. I am not so simple as to make no allowance for temperature."

A related quotation is the one which appears in Abraham Pais' biography of Einstein. It seems that when Einstein heard in 1921 of Miller's ether drift claim, he made the now famous remark: "Subtle is the Lord, but malicious he is not."

In Chapter 6 we shall take a look at the 1955 reanalysis of the Miller data and what one hopes would be the resolution of the problem (notwithstanding ongoing claims to the contrary by the few who continue to reject relativity theory).

Miller builds a building

A gift of $50,000 from J.D. Rockefeller made possible the construction in 1905 of a splendid, state-of-the-art building. The architect was Charles Schneider. Miller had visited laboratories in Europe and incorporated many modern features such as clean-room, dark-room, balance room, mercury room, spectroscope room, extensive distribution of AC and DC electrical power, gas, compressed air, and even masonry columns reaching from bedrock up to the third-floor lecture hall for vibration-free platforms. On the frieze which encircles the building, he placed the names of 34 physicists, from Archimedes to his friend, instrument maker Rudolph Koenig. *[for the record: Maxwell, Helmholtz, Carnot, Huyghens, Newton, Archimedes, Galileo, Chladni, Fresnel, Rowland, Joule, Fraunhofer, Arago, Kelvin, Kepler, Koenig, Henry, Wheatstone, Ampere, Pascal, Gilbert, Regnault, Stokes, Young, Oersted, Fizeau, Ohm, Hertz, Mersenne, Volta, Franklin, Foucault, Faraday, and Boyle].*

Fig. 4-11. Case Main, Rockefeller Physics, and Case Chemistry from the Doan Brook valley.

The photo **(Fig. 4-11)** shows the location of the Rockefeller Building, in center background with a flat roof-line, relative to the Case Main Building, the great Victorian hulk on the left. The building on the right with the two chimneys was the home of the Case Chemistry Department. It is the opinion of many visitors today that Rockefeller is among the most handsome on the campus, 100 years after its construction.

Fig. 4-12. Rockefeller in the 1940's.

In 1912, Miller established an official substation of the U.S. Weather Bureau in the Rockefeller Building. Some of the instrumentation was eventually placed on a tower on the roof of the building. In the 1920's, cameras for a study of lightning would be installed by John Albright (see below). The tower platform did not add much to the attractiveness of the building, but it remained in place until at least the 1940's. **Fig. 4-12**.

The floor plans for each of the three floors are shown in **Fig. 4-13**. Note the double-walled "constant temperature" room and "clock room" on the first floor. The large lecture hall on the third floor had seats for 175 and looks today much as it did in 1905.

Fig. 4-13. Original 1905 Rockefeller floor-plans.
first floor   second floor
third floor

Miller gets some help: Hodgman, Nusbaum, Albright, and Wallace

Fig. 4-14.
Charles Hodgman

In 1906, Miller hired 25-year-old **Charles David Hodgman** as an instructor to help him with the growing teaching load. **(Fig. 4-14)** Hodgman was to teach physics at Case for the next 46 years, probably longer than anyone else, before or since. Having completed his BS at Dartmouth in 1905, Hodgman earned a Masters degree at Case in 1920. His work on the specific heat of iron-chromium alloys at high temperatures provides probably the earliest example of condensed matter research at Case. Later, Hodgman's main interests were threefold: the *Handbook of Chemistry and Physics*, the technology of color photography, and the cultivation of roses. The handbook, published annually by Cleveland-based Chemical Rubber Company, is today found in just about every physics and chemistry laboratory. Hodgman was its Editor-in-chief from 1913 to 1949, his associate editor being R. C. Weast of the Case department of chemistry. As early as 1908, he became interested in the technology of three-color transparencies. In 1933, Hodgman published a paper on the transmission of ultra-violet radiation through water. (*J. Opt. Soc. of Amer.* **23** 436 1933)

The department expanded once again when, in 1922, Miller hired **Christian Nusbaum. (Fig. 4-15)** Born in Ohio, Nusbaum had done his BS at Ohio State and a PhD at Harvard in 1915. He held positions at Harvard, MIT and the National Bureau of Standards before joining the CSAS faculty. During his first decade at Case, Nusbaum did research on magnetic susceptibility and on diamagnetism of single crystal elements. One of his students, John Richard Martin, completed an MS on the magnetic properties of iron in high frequency alternating fields. Martin subsequently spent a few years in the department as assistant professor.

Fig. 4-15. Christian Nusbaum.

Fig. 4-16. John Albright and his lightning camera.

John G. Albright joined Miller, Hodgman and Nusbaum in 1923 to teach the growing number of Case engineering students. His principal interest was meteorology and he took advantage of the weather "observatory" which Miller had built on the roof of Rockefeller. He was especially interested in the propagation of lightning bolts. A paper in the "laboratory and shop notes" section of the *Reviews of Scientific Instruments* described his "apparatus for photographing lightning flashes". It consisted of eight box cameras attached around the circumference of a horizontal bicycle wheel and three stationary cameras. The wheel was spun at 4 turns per second, in time with a ticking metronome. A given bolt could be picked up by several cameras. By comparing the traces of lightning on the film he could determine the direction and speed of propagation of the "leader stroke" from the ground up and a series of "component strokes" which follow the same path from the cloud down. *Rev. Sci. Instru.* **8** 36 1937 and *J. Appl. Phys.* **8** 36 1937. Albright published a textbook, *Physical Meteorology (*Prentice Hall 1939). He had another rather interesting avocation which was somewhat related to instrumentation. He raised spiders to harvest their webs to provide filaments for the eye pieces of microscopes. He left the department in 1943, right in the middle of the war, to become chair of the College of Rhode Island (now URI).

Fig. 4-17. Case physicists 1938. Nusbaum, Wallace, Boyer, Hodgman, Shankland, Albright, and Miller (seated).

Clarence William Wallace spent 43 years of his life in the Rockefeller building, from his time as a freshman in 1917 until his retirement in 1960. While most of his efforts were in the classroom, he was a participant in several areas of Miller's research, including the study of wave-forms in resonating pipes and the photography of the shock-waves of bullets. He is mentioned in Miller's little book *Sparks, Lightning, Cosmic Rays* (Macmillan 1939) as having successfully designed and built a "projection alpha ray track chamber" for use in Miller's popular "show and tell" lectures. *(This was a demonstration version of the Wilson cloud chamber in which one could see tracks of liquid drops caused by alpha particles emitted by a tiny radioactive source.)* Wallace was also associate editor of Hodgman's *Handbook of Physics and Chemistry*. During the second world war, he put together a 60-lecture "Defense Course in Physics". In his letter accepting Wallace's resignation, acting President Kent H. Smith wrote, "With respect to your generous offer to assist those students in physics who may be in academic difficulties, I suggest you confer with Dr. Reines." Nice fellow, Wallace. He and his comrades are shown in Miller's office in **Fig. 4-17.**

Back to Miller: equipment for the analysis of sound

Miller was very much interested, from childhood, in music and musical instruments. He was an accomplished flautist; he composed arrangements of his favorite melodies from opera, mostly from Wagner. In his travels to Europe, besides meeting with Röntgen and Kelvin, he made the acquaintance of Rudolph Koenig, an instrument-maker in Paris. Koenig specialized in laboratory equipment for the analysis of sound waves, a subject of great interest to Miller. They were to establish a long-lasting friendship, and Miller bought a significant selection of apparatus for use in his laboratories and lectures in his new physics building at Case. His carefully maintained departmental accounts ledgers (now in the CWRU Physics Archives) list regular purchases from European instrument makers, especially Koenig and Max Kohl in Berlin. Koenig had earlier shown off his wares at the 1876 Centennial Exposition in Philadelphia, and subsequently supplied all manner of acoustic equipment to North American institutions such as the

Smithsonian, Johns Hopkins, Case, and the University of Toronto. Thomas B. Greenslade, Professor Emeritus of Physics at Kenyon College in Ohio, has created an extraordinary website which features a vast collection of photographs of nineteenth century scientific instruments. Included are photos and descriptions of dozens of Koenig's devices.
http://physics.Kenyon.edu/EarlyApparatus

The departmental ledgers run from 1887 (before Miller's arrival) to 1939. Every allocation and expenditure is recorded. The major purchases were signed off by Eckstein Case, the school's treasurer. Each annual allowance for department expenses, which grew from about $500 in 1903 to $2000 in 1930, was signaled by a short note from President Charles S. Howe. (In 1917, the sum dropped to half that of the preceding year, the note from Howe saying that he expected normal funding would resume when the war was over and the students returned. It did, the following year.) Occasionally, rather large supplementary sums were made available for the purchase of equipment, including four lots of about $500 each, bought from Koenig between 1893 and 1898. The allotment for one expense item gives us some normalization to today's dollar: $75 per month for a research assistant in 1913, to compare with $3000 or more in 2005.

Much of the acoustic equipment, as well as later purchases of electrical and electromagnetic devices, has survived the six decades since Miller's retirement and are now part of the collection in the Physics Department Archives. Starting in 2002, selected pieces have been on display in showcases in the Rockefeller Building. I have taken an interest in preserving these historical instruments. I plan to collaborate (when this book is finished) with similarly interested people at other institutions by cataloging the collection and displaying it on a new website: the "Case Collection of Physics Instruments". I had the pleasure of reporting in 2004 on some of Miller's acoustic research at an international conference at Dartmouth, sponsored by the Scientific Instrument Commission of the International Union of the History and Philosophy of Science (www.sic.iuhps.org).

Miller invents the phonodeik

In parallel with the ether experiments, Miller took up the study of the production of sound in musical instruments and the analysis of sound waves. One of his major contributions was his invention, in 1908, of the phonodeik (Greek for *show sound*). This device (**Fig. 4-18**) consists simply of a horn (H) to collect the sound, a tiny mirror (M) coupled to a diaphragm (D) which is moved by the sound, and a beam of light which, after being reflected by the jiggling mirror, traces out a wave pattern on a moving photographic film. The mirror, only about one millimeter square, is attached to a rotatable shaft which is mounted in jewel bearings. A fine thread attached to the diaphragm makes a single turn around the shaft and terminates at a

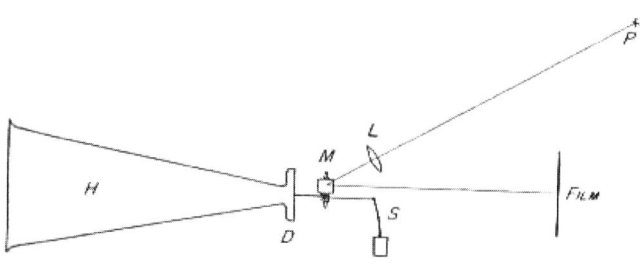

Fig. 4-18. Schematic of phonodiek.

spring which keeps it under tension. Vibrations of the diaphragm cause the shaft to rotate slightly, thus turning the mirror. The resulting photographic traces reproduce extremely fine details of the sound wave.

The photograph of the central element of the phonodeik (**Fig. 4-19**) shows the lenses at the left, and diaphragm frame to the right. A wider view of the arrangement of the apparatus in the "phonodeik laboratory" is shown in **Fig. 4-20**. The device was placed on one of the vibration-free masonry columns which reached to bedrock under the physics building.

Much of Miller's acoustics work was done with the help of a young fellow who was his faithful assistant from 1912 until 1924. Ralph F. Hovey was not only responsible for the phonodeik sound-analysis lab, but traveled with Miller on his many lecture tours and even to Mount Wilson for the ether-drift experiments.

Fig. 4-19. Miller's phonodeik.

What did Miller seek to learn from the resulting sound traces? Here is his list, from a talk given in 1915: "Among the subjects under investigation are the characteristics of tones from different musical instruments, the effects of changes in material or construction of musical instruments, the nature of vowel tones and other sounds of speech, the nature of noises and their prevention." He was very much aware, from his long-standing interest in flutes, that there are subtle differences among instruments of different design or material. It was his hope that the mathematical analysis of the phonodeik pictures would lead to a better understanding of sound and sound production.

Fig. 4-20. The phonodeik laboratory.

A series of traces of tuning fork sounds is shown in **Fig. 4-21**, a page taken from Miller's book *Sound Waves, their Shape and Speed* (Macmillan 1937) One must appreciate the detailed and reproducible patterns which this elegant mechanical device can produce. The traces are right up there in quality with the best oscilloscope! The phonodeik lab, **Fig. 4-22**, as set up for the comparative study of mechanical phonograph machines, illustrates one of Miller's applications of the technique. Miller seems to have been contracted to determine and compare the sound reproduction characteristics of commercial phonographs, as well, later, of radio receivers.

Analog computers for wave analyses

Miller had at his disposal three instruments which he used to study soundwaves. The first was his **phonodeik** which records the shape of the waves. The second was the **Henrici harmonic analyzer** which was used to determine the amplitudes for each harmonic overtone as recorded in the phonodeik pictures. The third was the **harmonic synthesizer**, whose origins lay in Lord Kelvin's 1872 device for predicting the tides. This machine was used to combine specified harmonics, amplitudes and phases into drawings of the composite waveform

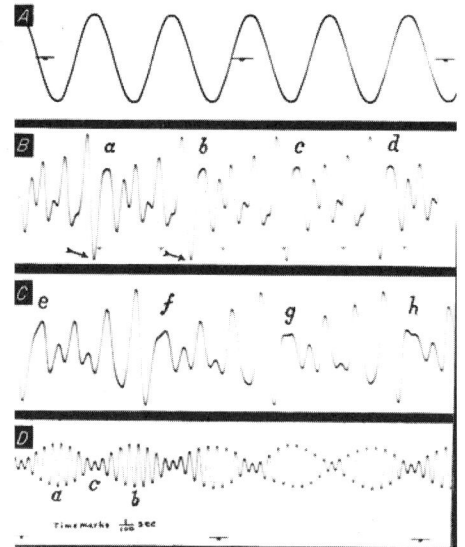

Fig. 4-21. Phonodeik traces of tuning fork sounds. top: simple tone; next two:. 4-fork chord; bottom: 2- fork beats.

The Henrici Harmonic Analyzer.

A harmonic analyzer allows one to determine how much of each partial wave is present in a complex wave form. For example, the wave-form may be a simple sine function, y = A sin θ. When this is plotted from θ = 0 to θ = 180 degrees you have a smooth curve with a single hump, like curve **a** in **Fig. 4-23**; if the wave form has two humps, like curve **b**, it can be written y = B sin 2θ. If the wave form is a combination of one hump and two humps, like curve **c**, then it can be written y = A sin θ + B sin 2θ. The relative amounts of the two contributions are given by the ratio of A to B.

The Henrici harmonic analyzer can determine the contributions to a given wave form from one hump, two humps, three humps, etc. all the way up to ten humps, that is, it can determine A and B and C, etc. for a function y = A sin θ + B sin 2θ + C sin 3θ ..... all the way to the tenth term J sin 10θ. In the case of musical sounds, these various contributions to the sound wave form are called overtones. Different musical instruments, playing the same basic note, will have different relative amounts of these overtones.

Fig. 4-22. Comparing sound quality of phonographs in the phonodeik lab.

*A rather long aside on how it works. Mathematician Olaus Henrici (Philosophical Magazine **38** 110 1894) designed a mechanical device which allows one to move a pointer along a plot of the curve to be analyzed from θ = 0 to θ =*

180 degrees, and which, through a clever array of pulleys, balls and wheels, allows one to read out the values of A and B and C, etc. The next few paragraphs explain how.

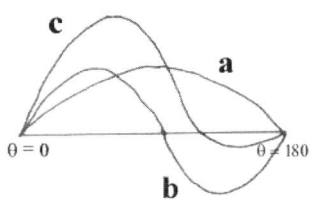

Fig. 4-23. Adding two sine waves.

On the plot of the waveform to be analyzed, let the x-axis run from $\theta = 0$ to $180°$ and let the y-axis measure the height of the curve. The operator moves an "indicator" (the "mouse-like" device) along the plotted curve. This can be seen in the foreground of **Fig. 4-24**. As the indicator is moved up the page (in the y direction), each in a set of glass balls rotates about a fixed axis parallel to the x-axis. In **Fig. 4-25** you see two of these balls toward the bottom of the photo. They roll away from the viewer. (The photo, supplied by Prof. Roger Hanson of the University of Northern Iowa is from an operating device in the instrument collection of the University of Iowa.)

Each glass ball is girdled by a light circular frame which holds two rubber wheels which press against the equator of the ball and which turn as the ball turns. The wheels are 90 degrees apart on the frame. Each wheel has a counter which records how much it turns as the ball rotates. In **Fig. 4-26** is a sketch of the view from above the ball, where one can see the two wheels which touch the ball and the counters attached to each. The little retainer wheel at "m" keeps the wheels snugly against the ball.

Fig. 4-24. A Henrici wave form analyzer.

Fig. 4-25. Closeup of Henrici mechanism.

As the indicator is moved in the x direction, a series of wires and pulleys causes the frame to rotate so that the locations at which the wheels contact the ball move around its equator. Some of these pulleys are visible at the top of **Fig. 4-25**. In **Fig. 4-27**, the frame has been turned clockwise through an angle α. The rate at which each wheel is turned by contact with the glass ball changes as it contacts different points on the equator of the ball. When α is zero, the wheel marked **C** turns rapidly as the ball turns, and the wheel marked **S** turns not at all.

As the indicator moves across the plot from $\theta = 0$ to $180°$ the frame turns through one full turn (α goes from 0 to $360°$). The C wheel counter *gives a measure of the amplitude of the curve multiplied by the cosine of θ and the S wheel counter gives the amplitude times the sine of θ. These two readings taken together give the amplitude and phase (since you have both sine and cosine contributions) of the first term of the expansion.*

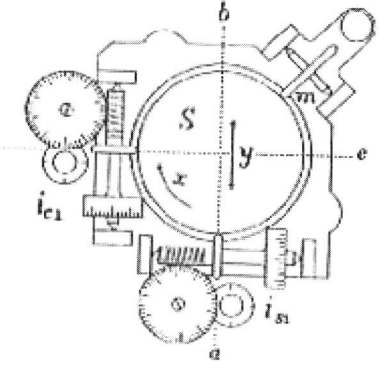

Fig. 4-26. Top view of rollers in contact with glass ball.

Now how do we get the higher terms? Easy. There are actually 10 glass balls, each with its own set of counter wheels. The diameters of the pulleys which rotate the rings are all different, so that for the ball measuring the sin 2θ contribution, the frame rotates two full turns as the indicator is moved from 0 to 180°. The next ring turns three full turns as the indicator is moved from 0 to 180°, etc. **Figs. 4-26 and 27** *are from Miller's article (*Journal of the Franklin Institute*, p285 September 1916).*

A good test of the Henrici is to have it analyze a curve whose components are known in advance. Miller first analyzed a "saw tooth" wave, i.e. a straight line rising from θ = 0 to θ = 180°. One can calculate the partial amplitudes and compare them with the Henrici readings. Here are the first five terms: (calculated, measured) (127.32, 127.30) (63.66, 63.55) (42.44, 42.47) (31.83, 31.85) (25.46, 25.50). The agreement is astonishing! Then Miller analyzed a simple sine curve, i.e. all terms should be zero except the first. He found for the first five terms: (250.03, 0.97, 0.69, 0.27, 0.19).

The Henrici Harmonic Analyzer will be resurrected at least two times in later chapters of this story: in Chapter 6 when Miller's young colleague, Shankland, uses it to study the shapes of electronic pulses produced by Geiger counters, and again in Chapter 9 when Shankland's protégé, Leslie Foldy, writes a classified paper in 1944 for the Navy Department on its potential use in the analysis of underwater signals.

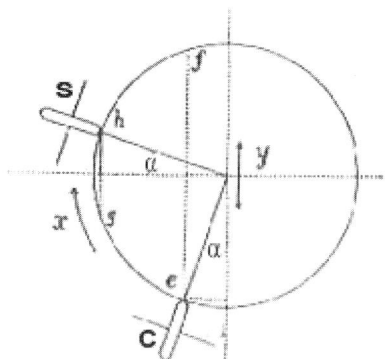

Fig. 4-27. Schematic of "sine" and "cosine" rollers.

The Kelvin harmonic synthesizer

Miller's version of this device was designed and built in the Case physics shop. **Fig. 4-28**. It could combine 32 components into a single curve. A large piece of paper is placed on a drafting table. The table moves smoothly under an inked stylus, along the x-axis. The y-position of the stylus is controlled by a light cable which pulls it up or down over the page. The cable loops around a complicated set of 32 wheels. The wheels are turned by a system of gears which track the motion of the paper under the stylus. The wheels turn from one time up to 32 times as the paper is moved from one end to the other (i.e. from θ = 0 to 180°.). The amount that each wheel moves the stylus up or down can be set beforehand, so that all 32 amplitudes and phases can be programmed by locking-in 64 set screws. The resulting motion of the stylus up and down over the paper produces the desired composite curve.

Fig. 4-28. Harmonic synthesizer built in the Case shop.

Miller used this mechanical curve-drawing machine to verify that his Henrici numbers were true representations of his phonodeik tracings. He published a detailed description of its workings in the *Journal of the Franklin Institute*, January 1916 p 51.

A fourth instrument

Miller no doubt knew about a fourth instrument, one which completes the sequence just described: 1. the phonodeik which plots the sound, 2. the Henrici which analyzes the plots, and 3. the harmonic drafting machine which draws new plots. I'll include a description of a fourth device which reproduces the sounds. In the spring of 2004 I visited the University of Toronto. I had been introduced to the physics department there by David Pantalony, the historian of science mentioned above. I was treated to a demonstration of a device built by Helmholtz which is more literally a harmonic synthesizer. This one can be programmed with the amplitudes for a selection of frequencies, so that the corresponding sound is produced. **Fig. 4-29.** A 128 Hz "master" tuning fork is driven by a pair of coils and an interrupter. The current from these coils can be selectively directed through the driving coils of up to nine other forks. The frequencies of the forks are multiples of 128 Hz. Each fork is placed at the mouth of a cylindrical resonator which enhances its sound. Between each fork and its resonator is a little swinging gate, controlled by a small keyboard. The amplitude of each component can be controlled by the gate or by moving the resonator relative to its fork. In his book, *On the Sensations of Tone*, (1895 translation by A. J. Ellis), Helmholtz describes his attempt to use this device to produce the vowel sounds. He includes a table specifying the recipe for each vowel, e.g. the vowel sound "ahh" is fork #1-piano, #2 pianissimo, #3 piano, #4 forte, #5 forte. I made a recording of a few "vowels" – but I concluded that artificial speech would have to wait for much more sophisticated technology.

Fig. 4-29. Helmholtz sound synthesizer at U. Toronto.

Miller was not alone in using mechanical devices to analyze periodic behavior. In his 1916 book *The Science of Musical Sounds* (Macmillan), he describes many other applications; for example he includes a photograph of a monstrous device built by the US

Coast and Geodetic Survey which predicted the tides! While he was successful in bypassing a great deal of tedious mathematics, it is not clear what could be done with the results, i.e. a table of amplitudes and phases for each example of sound. Builders of musical instruments and even students of the human voice would certainly be interested in attempts to characterize the subtleties of complex sounds, but ultimately the ear and brain would provide the desired guidance. Miller was aware of this, writing in his book: "a scientific investigation and analysis of the sound from a violin or a piano cannot determine whether it is the ideal."

Today, computerized voice recognition and artificial speech systems are direct descendents of the phonodeik, the Henrici analyzer, and the Kelvin and Helmholtz synthesizers.. Even though the electronic oscilloscope would quickly win out over the phonodeik, the useful analysis of complex sounds would have to wait for the electronic computer.

WWI, Students, and National Prominence

During World War I, Miller worked with the War Department on a study of the sounds made by large artillery pieces. The military were especially interested in the effect the loud blasts of the guns would have on the personnel firing them. He took his traveling version of the phonodeik to Sandy Hook, NJ for extensive measurements of artillery explosions. He also experimented with high-speed photography of bullets in flight; some of these tests were made in the attic of Rockefeller. (**Fig. 4-30**)

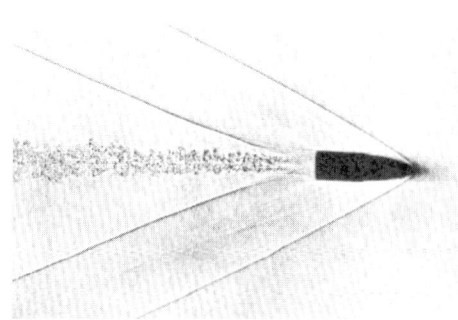

Fig. 4-30. Photo of bullet shockwaves taken by Miller.

Miller played a role in the original design of Severance Hall, the home of the Cleveland Orchestra. This magnificent concert hall was built in 1931 just across the street from the Case campus. Miller was engaged by architects Walker and Weeks as the acoustical consultant. (We shall have a bit more to say about Severance and Case physicists in Chapter 12.) Miller reported on the implementation of his recommendations at the 1932 national meeting of the Acoustical Society of America which took place, incidentally, in his lecture hall in Rockefeller. (*J. Acoust. Soc. Amer*. **3** 312 1932)

Work in acoustics in the 1920's and 1930's may not have been on the cutting-edge of physics, but it had many technical and engineering applications at the beginning of the electronic age. Miller's books and his public lectures, illustrated with delightful visual and audio demonstrations, were very popular. They included presentations on x-rays, electrostatics, acoustics combined with musical performances, and later of such modern devices as the cloud chamber. He played an important role in taking science to the public and in creating interest in Case. (In this area, Miller was to be emulated seven decades later by one of his successors as Andrew Swasey Professor of Physics.)

One of Miller's triumphs as a prominent leader in American science was his hosting, along with colleagues at Western Reserve, of the 1934 Fall meeting of the National Academy of Sciences. This honor served to put the two Cleveland institutions "on the map". The five preceding meetings had been at MIT, Michigan, Yale, California and Princeton. Featured was a public lecture in a packed Severance Hall by Harvard's Harlow Shapley on "Evolution Among the Stars".

Two of Miller's students were to become world-renowned physicists. The first of these was **Edwin C. Kemble** (1889-1984) who was the only physics major in the Case class of 1911. Kemble had transferred to Case from Ohio Wesleyan when he decided to go into engineering rather than the ministry. He worked with Miller in his phonodeik laboratory, learning a great deal about waves and vibrations. Miller contacted his fellow acoustician, Wallace Sabine, at Harvard, recommending Kemble for advanced study. Sabine not only arranged for Kemble's admission to Harvard, but paid his expenses as well. Kemble went on to be the principal spokesman for the new quantum mechanics in the United States. He dedicated his 1937 book, *Fundamental Principles of Quantum Mechanics*, to Miller. According to Gerald Holton in his 1988 book, *Thematic Origins of Scientific Thought* (Harvard Press 1988), Kemble had become intrigued by the new, mostly European, quantum mechanics when he heard a lecture given at Case in 1912 by Robert A. Millikan. The occasion was a joint meeting of the APS and AAAS. Quoting Holton: "Just back from a six-month excursion to Europe, Millikan appears to have given at Case the earliest analysis by an American physicist in a scientific society meeting of the new science that was taking shape abroad."

The second of Miller's extraordinary students, **Philip M. Morse** (1903-1985) was in the Case class of 1925. In chapter 2 of his autobiography (*In at the Beginnings: A Physicist's Life* MIT Press 1977), Morse recalls his undergraduate days at Case. He describes his professor as "tiny and neat and polished, with an imposing mustache, beautiful white, wavy hair, and a pleasant but very formal manner. His lectures were clear but not theatrical; his lecture demonstrations were carefully planned and always worked". Morse describes the workings of Miller's phonodeik and Henrici analyzer. In 1924, his junior year, he was enlisted by Miller to help in the analysis of the Mount Wilson ether drift data. His was the chore of plotting up all the data and of struggling to extract an ether-drift signal by using the Henrici. He remarks, "by the time all the circuits in the run were plotted, it looked as though a gaggle of beetles had trailed across the paper…" (as in **Fig. 4-9**). He describes a terrible weekend which began by his telling Miller that he had found a strong effect and which ended by his confession (his word) that he had erred. In 1927, Morse published a paper with Jason Nassau, professor of astronomy at Case, on their use of the Henrici in an attempt to extract the motion of the sun relative to 476 stars. ("A Study of Solar Motion by Harmonic Analysis" *Astrophys. J.* **65** 73 1927.) Morse went on to earn a Princeton doctorate in 1929; he served as a distinguished professor at MIT from 1931 to 1969. He is best known to physics grad students worldwide through the text he wrote in 1953 with Herman Feschbach: *Methods of Theoretical Physics*, McGraw Hill.

Miller's life-long acquisition of an important collection of flutes is the subject of another study. The entire collection of 1500 flutes, 10,000 pieces of music for the flute, and 1200 books relating to the flute is on display at the Library of Congress. http://memory.loc.gov/ammem/dcmhtml/dmhome.html

Miller wrote seven books, one of the more interesting of which is *An Anecdotal History of the Science of Sound* (Macmillan 1935): a non-technical description of the work of acousticians from the ancient Greeks to researchers in the 20th century. Miller was elected to the National Academy of Sciences in 1921 and to the presidency of the American Physical Society in 1925, a great honor for Case. *(The APS began in 1899 with 59 members, growing to 1760 members by 1925, and 43 thousand today.)* In 1929, as one of the founders of the Acoustical Society of America, he was the featured after-dinner speaker at their inaugural meeting, held in New York City. In his demonstration lecture, titled "The Science of Musical Sounds", Miller described the creation and analyses of his phonodiek traces. He was elected president of the ASA in 1931. Miller was chairman of the Case department from 1895 until his death in February 1941.

Fig. 4-31. Miller's library – 1910.

*I am sitting in my office, in the room which was once Miller's beautiful wood-paneled library (**Fig. 4-31**), and I can imagine an insistent voice whispering, "it's not zero!"*

Chapter 5   University Physics: The Middle Years at WRU

Barrows, **Smith,** **Freeman,** **Whitman,** **Mountcastle**
1866-1870 1870-1881 1881-1886 1886-1919 1907-1945

As described in the last chapter, Dayton Miller took the Case physics department almost halfway through the twentieth century, so we must now backtrack seventy years and pick up the story of the successors of the solar astronomer Charles Young at the college in Hudson. Young left Reserve in 1866 and the trustees quickly chose a replacement. **Allen Campbell Barrows** had graduated from Western Reserve in 1861. He spent three years in the army and two teaching at Phillips Andover. He would be the first Western Reserve alumnus on the WRU physics faculty. He was appointed the second Perkins Professor of Natural Philosophy and Astronomy. Today, the tenth holder of this endowed chair is my colleague Philip Taylor. While the more recent holders have retained the title until retirement, Professor Barrows did so for only five years, at which time he was appointed professor of Latin and English literature.

There would, in fact, be three more Perkins professors appointed over the following 16 years. **Charles Josiah Smith** was appointed in 1870, just a few days before he received his BA from WRC. Three years later he completed a master's degree. His principal interests did not lie in physics or astronomy and he switched to the department of mathematics in 1881, heading that department for two decades. (**Fig. 5-1**) Smith (according to his obituary) teamed up with Edward Morley in putting together the "first" classroom demonstration of the telephone. In his book, *"Western Reserve University – The Hudson Era"* (WRU Press 1943), Frederick Waite noted that following the departure of Young in 1866, all research in and teaching of astronomy at WRC essentially came to an end, and "has never been extensively revived".

Fig. 5-1. Charles J. Smith.

The Perkins went in 1881 to **Spencer H. Freeman** (MA Johns Hopkins 1878). There were great expectations for this talented young professor, but he died at age 31 after only 5 years on the faculty. Young Freeman must have been considerably occupied with the move of the college from Hudson to Cleveland.

In 1886 the Perkins chair was given to **Frank Perkins Whitman**, who would four years later be awarded a D.Sc. from Johns Hopkins. Whitman would lead the Western Reserve University physics department for more than thirty years His picture is shown in **Fig. 5-2**.

Teaching and research at WRU

The general picture of the WRU physics department, from its inception in 1830 until 1907, was that of a single professor assisted by one or two instructors who would remain with him for only two or three years. (The same was true for the neighboring Case department under Michelson and Reid, and under Miller until 1905.) The bulk of the physics teaching at WRU was in the two-semester introductory physics course with laboratory (about 50 students), and advanced classes for perhaps two or three students who were enrolled in the advanced physics courses.

Fig. 5-2. Frank Perkins Whitman.

The CWRU Archives have a complete collection of both WRU and CSAS annual catalogues and bulletins which expanded from a brochure of a dozen pages in the 1830's to thick soft-cover books of several hundred pages by 1920. They include names of trustees, officers, faculty, various functionaries, and even the names and addresses of all the students and alumni, descriptions of all the departments and courses, facilities, clubs, and prizes. Information on research is a bit more difficult to come by, although some clues can be found in the annual departmental reports.

In many instances, text books are mentioned in the course descriptions, and as an indication of what physics was being taught, I list here those titles which appeared in the catalogues most often: 1860 Lardner's *Natural Philosophy*, Jackson's *Mechanics*, Snell's Olmsted's *Natural Philosophy (*a translation by Mr. Snell of Mr. Olmsted's book*)*; 1870 Tyndall's *Heat and Sound*, Loomis' texts on astronomy and meteorology; 1880 Atkinson's Ganot's *Frictional and Dynamic Electricity*; 1890 Deschanel's *Physics*, Glassbrooks' *Physical Optics*; 1900 Hastings and Beach *General Physics*. Chairman Whitman in fact wrote a detailed and thoughtful review of the Hastings and Beach text for the Physical Review. (*Phys. Rev. Series I* **9** 313 1899*)*.

Buildings and facilities

From the very beginning, *natural philosophy*, later *physics*, was provided with state-of-the-art accommodations. Loomis' Observatory in Hudson was built in 1838 and the Athenaeum completed in 1843. (From the 1938 article by Dr. Waite: "In the Athenaeum on the first floor was provided a room for physics and one for chemistry, the first definite laboratory rooms to be provided. The greater part of the third floor of this building was devoted to a museum for scientific specimens of various types and some curios, chiefly contributions of alumni who had become foreign missionaries.") After the move in 1881 from Hudson to Cleveland, the department occupied space in Adelbert Main, the new building which housed administration, classrooms, labs and even dormitories on the top floor. Today it accommodates the CWRU administrative offices. The new Western Reserve undergraduate college for men was given the name "Adelbert" to honor the memory of Adelbert Stone, the son of the college's major benefactor, Amasa Stone. The young fellow, a student at Yale, had drowned while trying to swim across the Connecticut River.

Meanwhile, on the other side of the fence, the Case physicists occupied rooms in the new Case Main Building until it was gutted by fire in 1886. For the next few years, Case borrowed space in the basement of Western Reserve's Adelbert Main and in its large dormitory building (later called Pierce Hall). No doubt the friendship between Michelson and Morley facilitated this arrangement. It was in the basement of this dormitory building that the 1887 MM experiment took place. By 1890, the Case physics department had moved back to the restored Case Main building.

In 1894, Western Reserve celebrated the dedication of a splendid three-story Physical Laboratory on Adelbert Road just south of Adelbert Main. This fully equipped teaching and research facility, funded by a gift from Mr. Samuel Mather, was built under the direction of Chairman Whitman. The architect was Charles F. Schweinfurth.

Reserve's physics building was completed a full decade before the construction, only a hundred yards away, of Case's Rockefeller Building which was described in the previous chapter. Miller was no doubt determined to match and surpass the first-class facility at the neighboring institution. It is interesting that each of these large and well equipped buildings was built for a department with one professor! Of course, there were various instructors and assistants. No doubt the growing numbers of engineering students at Case and pre-med students at Reserve required extensive classroom and lab space.

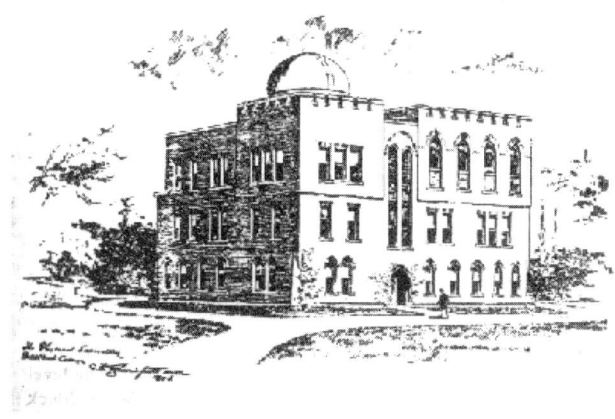

Fig. 5-3. The Western Reserve physics building on Adelbert Road.

A drawing of Whitman's building is shown in **Fig. 5-3.** It featured a large lecture hall and generous teaching and laboratory space. Floorplans of the building were included in the 1894-5 university catalogue along with a full description of the modern facility. A domed astronomical observatory was constructed on the roof. Later an equatorial telescope (10.5 inch aperture, 15 foot focal length), a gift from the Warner and Swasey Company, was installed under the rotating dome.

A stone-slab platform outside the southwest corner of the third floor was included as a mounting place for a heliostat. This device is a spring-driven mirror which is designed to follow the sun through the day so as to provide a steady beam of sunlight for the "optics and photography" laboratory. Because sensitive magnetic measurements were to be made in this building, no iron, not even nails, was used in its construction.

The building was demolished in 1969. In 2005, the last two weather-worn red sandstone remnants of this building sit hidden in the shrubbery along the south wall of Adelbert Main: the first is inscribed PHYSICS MDCCCXCIV, the other appears to have a curious mix of Aleph and Omega. א Ω

Whitman and Springsteen-Mountcastle 1886 to 1935

Whitman was interested in photometry and the measurement of the reflectivity of colored surfaces. In a paper titled "On the Photometry of Differently Colored Lights and the "Flicker" Photometer" (*Phys. Rev. Series I* **3** 241 1896), he describes a clever technique for comparing two surfaces. A photometer consists of a track (like an optical bench) with light sources placed at each end. Placed on a movable carriage on the photometer track is a spinning disk with its plane at a 45-degree angle with the track. The disk is shaped like two half-circles as in the right-hand half of **Fig. 5-4**. The left-hand side of that figure shows a view looking down upon the track DE and the spinning disk AB and a cardboard "standard" surface C. When an observer looks through tube F, he sees the surface C alternating with the surface of the spinning disk. The visible surface of the disk is illuminated by the light at D and that of the standard is illuminated by the light at E. With one light and the spinning disk held at fixed positions, the second light is moved along the track until the observer no longer sees a flicker through the tube. In this way, the relative reflectivities of various colored surfaces may be determined. This type of "applied research" would be of interest for example to printers and manufacturers of paints and dyes.

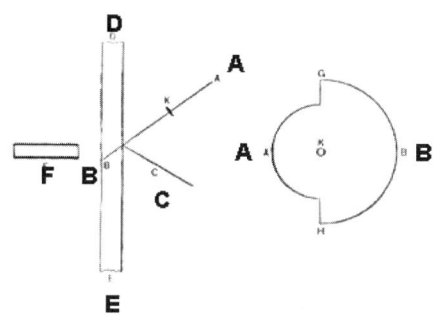

Fig. 5-4. Sketch of Flicker Photometer.

In his 1902 report to the Western Reserve trustees, chairman Whitman mentions research being done by **Harry William Springsteen** on the thermal conductivity of glass. Mr. Springsteen held a BS from Case and an MS from Western Reserve (1901), another example of WRU-CSAS cross-fertilization. The young Springsteen appears again in the 1907 catalog as Assistant Professor of Physics, with a 1904 PhD from Johns Hopkins. He would be a member of the Western Reserve faculty until his retirement in 1945. He became Perkins Professor in 1914 and took over the chairmanship from Whitman in 1918, the same year he changed his name to **Harry W. Mountcastle**. *(The CWRU Archivist found the official announcement: Harry William Springsteen having used the name of his step-father for twenty-two years announces that after July fifth nineteen hundred and eighteen he will resume the name of his father and will be known as Harry William Mountcastle.)* Mountcastle's photo is shown in **Fig. 5-5**.

Fig. 5-5. Harry Mountcastle.

We mentioned above Springsteen's interest in the thermal conductivity of glass. A year after I wrote that paragraph, I came across a copy of a doctoral thesis among Dayton Miller's documents. The 1898 thesis was from the University of Göttingen and written in German. The topic was the thermal conductivity of glass, which had little to do with Miller's interests. However, on the last page, the new young doctor, Theodore Moses Focke, writes that he was born in Massilon, Ohio, and that he received his BS degree at the Case School of Applied Science in 1892. It would seem that there was some connection, presumably in the person of a Case professor, between Focke and Springsteen. Focke was possibly the first CSAS physics BS to earn a doctorate.

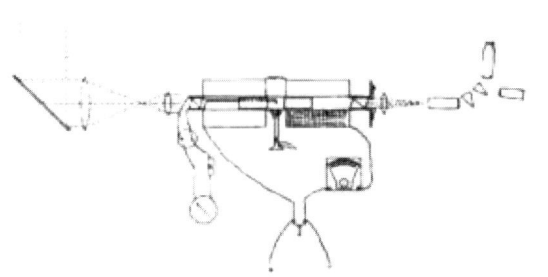

Fig. 5-6. Schematic of polarization spectrometer.

Springsteen published a paper titled "The Magnetic Rotation of Sodium Vapor" (*Phys. Rev. Series I* **21** 41 1905), coauthored with R. W. Wood of Johns Hopkins. Three years earlier, Pieter Zeeman in Amsterdam had reported on the behavior of light passing through a sodium vapor in a magnetic field. He had found that the plane of polarized light would be rotated as it passed through the vapor, in one direction at wavelengths just below the sodium D lines and in the opposite direction just above these lines. (The sodium D lines at 5889 and 5895 Ångstroms are produced when the sodium atom returns from its first two excited states to the ground state.)

Springsteen and Wood improved upon the Zeeman experiment by using a much denser sodium vapor. They used white sunlight from the heliostat mentioned above, and looked at the absorption lines. They used nicol prisms to polarize and analyze the light and selected the desired wavelengths by using diffraction gratings and a prism spectrometer. The applied magnetic field was about 2800 Gauss. Their sketch of the apparatus is shown, along with their results. **Figs. 5-6 and 5-7**. The "rotation in degrees" is plotted against the wavelength, and the plot is symmetric about the D lines – i.e. with no change in sign. (The vertical scale is essentially arbitrary; being based on the high density vapor measurements. The very large rotations, e.g. above 100 degrees, were determined by extrapolating upward from measurements made with low density vapor.) Their result was clearly in contradiction to that of Prof. Zeeman.

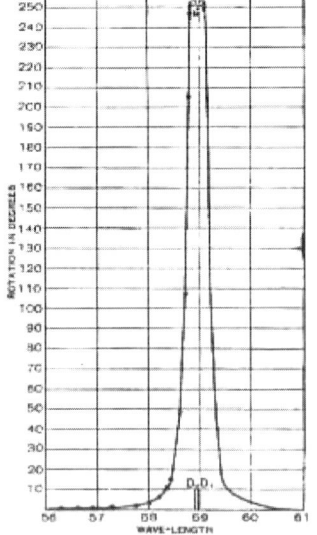

Fig. 5-7. Rotation (in degrees) of plane of polarization vs. wavelength in sodium vapor.

This 1905 experiment is a benchmark of sorts in that Springsteen's work was less "applied" than the work which Miller (x-rays and acoustics) and Whitman (photometry) had been doing It was closer to the hot-topics of the day: electromagnetism and the beginnings of atomic physics.

Graduate research

From 1920 to 1935, Mountcastle was Reserve's only professor of physics. He directed the research of eight students working on their MA degrees. The earliest, Lawrence Henry Ott, studied the production of single crystals of iron. The author thanks the scientists at the General Electric lab at Nela Park for their help. (A decade later, physicists at Case would develop strong ties to Nela Park. See Chapter 7.)

Another MA student was Abe Offner whose 1931 thesis was titled "A Critical Survey of Ether Drift Experiments". This was two years before Miller, over at Case, published his "definitive" paper on his own ether drift measurements (See Chapter 4.) The Offner thesis includes detailed descriptions of experiments ranging from the 19th century work of Arago, Fizeau, Maxwell, Michelson-Morley, and Rayleigh, as well as a dozen or more optical and electrical attempts up through the 1920's. Every one of these experiments resulted in ether drifts compatible with zero. Young Mr. Offner then turns to the data published by the chairman of the neighboring department. He thanks Professor Miller "for the prints which appear in this paper." He concludes diplomatically: "Miller's individual results are not impressive and his curves show that the errors of observation were almost as large as the result sought for. However the consistency of the results found at different times of the year and at different places shows that an effect has been observed which cannot be ascribed to errors of measurement." In his summary, Offner says "…only one experimenter, Miller, has attempted to detect the motion of the earth through the ether by means of a sufficiently extended series of observations." He no doubt had heard Miller's argument that the great extent of his data-set somehow justified his claim of a small effect. Finally, and this is 1931, Offner concludes, "…it is not necessary to abandon the hypothesis of an ether in order to explain the results of ether drag experiments, as was done by Einstein…"

The challenges of the new physics (relativity and the quantum) were further explored in an MA thesis by Herbert E. White in 1934. "The Laws of Radiation: a Critical Survey". This paper, too, included a detailed history of the subject: Kirchhoff, Maxwell, Boltzmann, Planck. The final sentence of the summary might very well have been written 70 years later by the typical perplexed student in the sophomore "modern physics" course. White concludes: "Thus the experimental evidence seems to indicate that both theories are true simultaneously. In spite of the vast amount of data bearing on the subject we are apparently still unable to answer the question: Is radiation undulatory or corpuscular? Any hope of compromise between the two theories appears to involve concessions fatal to either. Thus we are left confronted with the riddle of modern physics."

Later thesis topics give some clues to Mountcastle's very wide interests. These included the measurement of dielectric constants, the properties of dental amalgams, crystallography, and infra-red spectrometry. A 1932 thesis on fluoride gases has *two* authors: a curious departure from the usual arrangement. I wonder how their "defense" was handled. Appendix B lists all master's and doctoral thesis titles.

Between 1936 and around 1941, Mountcastle was joined in directing the master's level research program by a young colleague, **Cassius W. Curtis**. Curtis had just received his doctorate from Princeton where he did atomic spectroscopy. His dissertation was on the spectrum of manganese. His single-author paper on this subject, published after his arrival at WRU, included the classification of over 700 spectral lines. "The First Spark Spectrum of Manganese" *Phys. Rev.* **53** 474 1938 His first MA student, Paul Spremulli, wrote his thesis on an extended analysis of these same data.

Fig. 5-8. Cassius Curtis.

Curtis directed the master's level research of four other students, on a variety of topics: dielectric constants of alcohols, striations in Kundt's tubes, ultrasonic emulsifiers, and supersonic oscillators. The photo in **Fig. 5-8**, from the CWRU Archives, identifies Curtis as the WRU tennis coach. He moved on to Hamilton College, and subsequently to Lehigh University where he was on the faculty for over thirty years. Curtis died in 2004 at age 98.

Chairman Harry Mountcastle was elected fellow of the AAAS. He retired in 1946 after almost 40 years on the faculty. Soon after, Richard Beth took over the department as described in Chapter 10. Mountcastle died in 1955.

Chapter 6   Shankland
1930 – 1976

In the fall of 1925, a young fellow from Willoughby, an eastern suburb of Cleveland, signed up for the mechanical engineering program at Case. Robert Sherwood Shankland (1908-1982) soon switched his major to physics, becoming a student of Professor Miller. Shankland remained connected with Case until his death. His lifelong interest in various aspects of acoustics, stemming from his work with Miller, manifested itself in his wartime underwater-sound research and later in his extensive studies of the acoustic properties of large concert halls and churches.

After an eighteen month stint in Washington at the National Bureau of Standards (as part of the radio-wave transmission research group), Shankland returned to Cleveland to work as an instructor in Miller's department and at the same time to complete a Master's degree. His 1933 thesis title was "The Dispersion of X-rays". It describes measurements made of the index of refraction of calcite for x-rays. The equipment, including an evacuated spectrometer, x-ray tube, and photographic detectors were of professional quality and the analysis of the data (including, of course, some curve-fitting with the Henrici analyzer) clearly presented. Shankland determined the dependence of the index of refraction on the wavelength of the x-rays, i.e. the "dispersion". The original of the thesis, with skillfully hand-drawn figures is part of the departmental archives. Perhaps not coincidentally, Shankland was to be looking at a very different aspect of x-rays three years later in his doctoral research.

Shankland entered the graduate program at the University of Chicago where he became a student of Arthur H. Compton. Compton had lectured at Case in 1927 and Shankland had decided at that time that this was the physicist with whom he would like to work. He spent four summers at Chicago, while still an instructor at Case, and then two full years in residence. One of his closest friends at Chicago was fellow grad-student and future Nobelist, Luis Alvarez. Starting with his arrival in Chicago, Shankland began to keep detailed personal journals, a custom he would continue for several decades. (The twenty volumes of these journals are, in 2005, in the possession of his family.) In the first of twenty volumes he writes how he was nervously waiting for the posting of the results of his doctoral oral examination (with Compton and Gale) when Alvarez came in with a long and sad face, feigning bad news – and then how they celebrated.

Shankland completed his doctorate in 1935, writing his dissertation on "The Photon Theory of Scattering." His results were published in the *Physical Review* in a paper entitled "An Apparent Failure of the Photon Theory of Scattering". (*Phys. Rev.* **49** 8 1936) Shankland returned to Case, grateful to Miller for holding a position for him. This was not easy at the beginning of the Depression, when it was very difficult to find any kind of job.

An aside on photons and the quantum theory of light. Physicists knew that light is a wave in which a combination of electric and magnetic fields carries energy through space, all beautifully described by Maxwell's equations of the 1870's. Then Einstein came along in 1905 and explained the photoelectric effect, in which light shining upon a metal causes electrons to be knocked out of the metal. Einstein showed that light must be treated as bundles of energy called photons. Consequently, light may be described in two ways: as waves or as photons. Later it was found that ordinary things like electrons and protons also must be described both as waves and particles. Many physicists resisted these new ideas, but experiments, like the one done by Compton which we shall now describe, went a long way to clinching the argument. Even today, there are some who are, to say the least, uncomfortable with the wave-particle duality of quantum mechanics, relativity and other not-particularly-intuitive ideas in modern physics; we'll meet some of these "holdouts" later.

Compton and Compton Scattering

Twelve years before Shankland's experiment, Compton had reported his pioneering work on the scattering of x-rays by matter. He presented evidence that the process involved a quantum of electromagnetic radiation, the photon, interacting with a single loosely bound small electron in the atom. (*Phys Rev* **21** 483 1923) He argued that the photon has a well defined energy $h\nu$ and momentum $h\nu/c$ (ν is the frequency of the radiation, h Planck's constant, c the speed of light), and that both energy and momentum are conserved in the collision. This classic experiment appears today near the beginning of any "modern physics" textbook and is offered as evidence for both the quantum theory of light and of special relativity. Compton used conservation of momentum and energy, along with Einstein's relativistic expressions for the momentum and kinetic energy of the outgoing electron, to derive his famous formula: $\Delta\lambda = (2h/mc) \sin^2 \frac{1}{2}\theta$. (m is the mass of the electron.) This equation states that $\Delta\lambda$, the *change* in the wavelength of the scattered photon, depends only on θ, the angle through which it is deflected. (It. is proportional to the "square of the sine of half the scattering angle".) Compton was able to test this prediction experimentally by measuring the wavelength of the incoming and outgoing radiations. (He used the well-established technique of "Bragg" scattering.)

This view of what we now call "Compton scattering" was quite innovative at the time, and alternative "classical" models suggested that the radiation would be absorbed by the atom, and an electron and longer wavelength radiation would later escape from the excited atom. Another model proposed that the target electron is an extended object and that the radiation observed might result from (in Compton's words) "the interference between the rays scattered by different parts of the electron".

Compton calculated not only the expected shift in wavelength, but also the probability that the photon would scatter into a given angular range. As both predictions were unambiguously supported by his measurements, the competing theories were rejected, and the quantum theory of the photon was given a strong boost.

Shankland and Compton Scattering

Compton's experiments did not, however, include any detection of the recoiling electron. Therefore, the young graduate student set out to further test the Compton theory by detecting the outgoing electron. Shankland knew that the Compton theory required that the photon and electron must appear at the same time and that there must be a well defined correlation between their directions. He used the new technique of "coincidence counters". A radioactive source (radon) provided the beam of energetic photons. Geiger-Müller counters were placed so as to catch both the outgoing photon and the recoil electron. The electron detector included two counters which had extremely thin walls and were placed one behind the other. (**Fig. 6-1**).

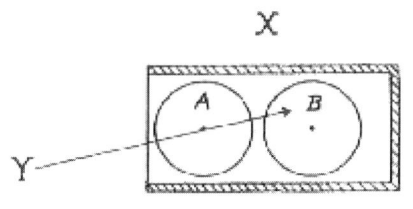

Fig. 6-1. Electron detector box.

An electronic circuit was devised which produced a "coincidence" signal only when both electron counters were hit. This greatly reduced the very large number of accidental electron counts. This signal was subsequently combined with one from the photon detector, so that all three counters must have been hit within an interval of about 0.3 milliseconds. He used several targets (air, Al, paraffin, Be) and fixed the photon counter at $\theta=35°$. In **Fig. 6-2**, the photon counters are labeled P and the recoil electron counters R. The radon source is at γ. He placed the electron counter at first where the electrons were expected, and then at two positions where they are not expected: *viz.* in the correct plane but on the photon side, and then 90° out of the plane. He presents a table of results giving the expected number and the observed number of "triple coincidences" for the various configurations and targets. Typical numbers for observed/expected (in counts per hour) in three runs are: 2.4/23, 4.7/48, 14/69, with uncertainties on the observed counts of about 30%. Shankland concludes: "the photon theory in its present form does not agree with the experiments reported here." (*Phys. Rev.* **49** 8 1936) He acknowledges Compton: "This experiment was suggested to the writer by Professor Arthur H. Compton and its completion has been possible because of his generosity and stimulating advice."

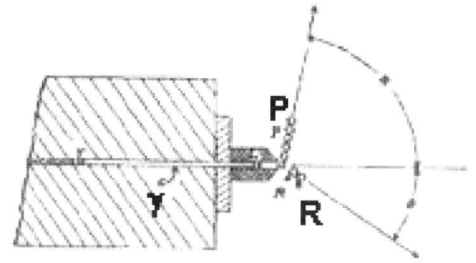

Fig. 6-2. Shankland's Compton scattering experiment.

It would be interesting to learn how Compton reacted to this result, but no doubt he urged Shankland to pursue the matter further. One might also speculate about discussions between Shankland and his mentor, Miller, since each had now published experimental data which called into question an essential aspect of the new theoretical physics. Within 8 months, however, Shankland sent a short "Letter to the Editor" describing a new attempt, featuring a new configuration of the counters. (*Phys. Rev.* **50** 571 1936) (**Fig. 6-**

3) First, the electron detector was reduced from two counters to one. Second, the photon counter was fixed at 90° to the beam direction, and the single electron counter was placed either at the expected angle (+Θ) or at the same angle, but on the wrong side of the beam (-Θ). Then coincidences with the photon counter were recorded for these two positions and for target in place and target removed. The results: in counts per hour, [target in: 69.1 ± 5.4 on correct side and 10.1 ± 5.0 on wrong side] [target out: 10.7 ± 4.4 on correct side and 12.5 ± 5.6 on wrong side]. Conclusion: "there is a coincidence in time between the appearance of a recoil electron and the scattered gamma ray which liberated it." There is no mention of the results of the earlier experiment or what may have been wrong, although it is probable that his "triple coincidence" signal was faulty. The 28-year-old researcher, who was perhaps the first to do experiments involving the use of Planck's constant in Cleveland, thus finessed a somewhat shaky start.

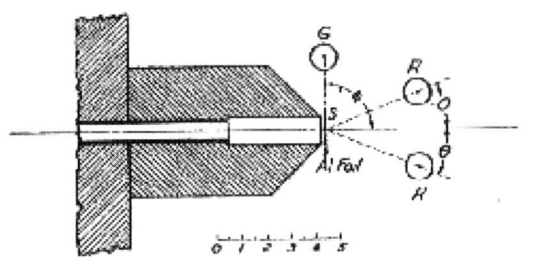

Fig. 6-3. The configuration that worked.

A longer paper ten months later presents a more detailed account of the ongoing experiment, including a quantitative discussion of uncertainties such as the effects of multiple scattering and accidental coincidences. (*Phys. Rev.* **52** 414 1937) In the introductory paragraph of this third paper, Shankland mentions that his earlier results did not support the Compton theory, but then he writes: "The publication of these findings aroused an active interest in the subject which resulted in several new experiments and theoretical discussions that have added greatly to the knowledge of these phenomena." Our young researcher skillfully finds the bright side and moves on to his new results.

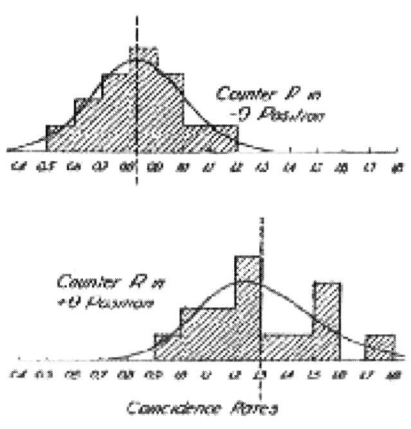

Fig. 6-4. Coincidence rates for wrong side (upper) and correct side (lower).

The revised experimental setup has the photon counter at 90° and the single electron counter at 22.5° or at the same angle on the wrong side. The 22.5° electron angle corresponds to a 90° scatter of a 350 keV Ra C gamma ray (i.e. photon). The average rates from 15 runs (about 200 coincidences) were, in counts per minute, (1.30 ± 0.10 for correct side and 0.83 ± 0.09 for wrong side), about a 5 standard deviation effect. For some reason, the "wrong side", i.e. accidental, count is much higher here than that mentioned in the Letter to the Editor. Histograms of the number of coincidence counts for each run, for the "wrong" and "right" positions are shown in **Fig. 6-4**. The paper goes on for four more pages discussing the origins of the high accidental rate and making comparisons with other researchers' results. The final paragraph states that "no time lag as great as 10^{-4} s can exist in the Compton scattering process, and ... the angular relationship given by the theory... is verified to within ± 20°". The young experimenter is pictured (standing) in **Fig. 6-5**.

In a recorded interview with Loyd S. Swenson, Jr. in 1974, Shankland commented on this work. Swenson was acting for the American Institute of Physics and its Niels Bohr Library Archives and the interview is part of the AIP history of physics collection. From the interview: "At first, we obtained results that seemed inconsistent with the Compton theory, and we were inclined to say that they supported the Bohr-Kramers-Slater theory, which ascribed a statistical view to the Compton interaction. But we found very soon that our counting rates were too high, and that we were experiencing what we now know as the dead-time of the counters, and we were missing a number of true coincidences. So when we used weaker sources of gamma rays, better circuits, then we got the coincidences that we reported." Having made an important contribution to the experimental basis of the photon quantum theory, Shankland seemed ready to move on to other pursuits.

Fig. 6-5. Shankland (standing).

In an interesting combination of his work with Geiger counters and his work with Miller, Shankland used the Henrici analyzer in the Fourier analyses of oscilloscope traces of pulse shapes for different counter designs and circuits.

Shankland in World War II

For the almost fifteen years that Shankland had studied under and worked with Dayton Miller, he had participated to some extent in his mentor's research in musical and architectural acoustics. He describes, for example, his tedious efforts to produce phonodeik traces which were up to Miller's high standards. It was only natural that, when most academic physicists found ways to serve their nation in World War II, Shankland should find himself working with the US Navy on underwater sound detection. He worked for various government agencies from 1942 through to the mid-1950's. Highlights of this work were the year he spent (1943) as the representative in England of the US Office of Scientific Research and Development and the time he spent as director of the Columbia University headquarters of the Underwater Sound Reference Laboratories.

Shankland and Architectural Acoustics

Shankland developed an interest in the acoustic properties of such large enclosures as churches and concert halls. As a well known author on acoustics, Miller collaborated with architects, preachers, and musicians on how best to design and outfit a hall for ideal acoustics. Shankland would eventually be equally recognized as a knowledgeable consultant. The main goal of this work is to have just the right amount of reverberation to provide richness to the sound without the garbling that too many echoes or too persistent a sound will cause. Another desired property is the ability of musicians on a stage, for example, to hear themselves and one another. Shankland's principal contribution to this, as it is sometimes described, "black art" was to visit scores of halls and to determine quantitatively the acoustic response to standard sound intensities. He describes this work (once again from the AIP interview tape transcript) as the "accumulation of as much information as we can about buildings that are successful". And this he did, traveling with his wife, Hilda, to make measurements all over the United States and Europe.

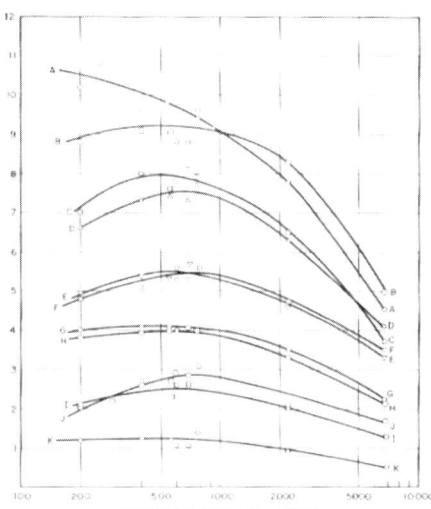

Fig. 6-6. Reverberation times vs frequency for 11 churches.

In Chapter 12, we shall describe how Shankland and his young colleague, Arthur Benade, would contribute in the 1950's to plans for the redesigning of Severance Hall, the home of the Cleveland Orchestra. Recall that Miller had participated in the original acoustical design in 1931.

At the 1967 meeting of the Acoustical Society of America in New York, Shankland presented a paper entitled "Quality of Reverberation". He discussed the results of measurements of reverberation times as a function of frequency for a dozen famous churches and theaters in Italy. Reverberation times in seconds are plotted against frequency from 200 to 7000 Hz in **Fig. 6-6**. They range from ten seconds in the basilica of San Paulo fuori le Mura – the top curve - down to only one second in the San Carlo Opera Theater in Naples. The floor plan of the long, highly reflective interior of "St. Pauls outside the Walls" is shown in **Fig. 6-7**. Shankland even went so far as to have sound absorbing curtains and carpeting installed in one of the smaller halls, successfully reducing long high-frequency reverberation times, which he said caused confusion. He remarks that long reverberation times are desirable in large halls, but not necessarily in smaller ones. Furthermore, he points out that reflected sounds from directions in the horizontal plane including the source and the listener are preferable to those bouncing off floors and ceilings. (*J. Acous. Soc. of Amer.* **43** 426 1968 and **50** 389 1971). An extensive article summarizing the measurements made during his European trips appeared in

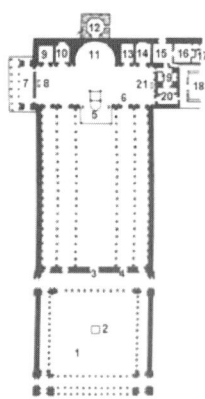

Fig. 6-7. Floorplan of San Paulo in Rome.

the Sigma Chi journal: *Amer. Scientist* **60** 201 1972.

In 1973 the widely circulated magazine of the AIP, *Physics Today,* featured a cover-article by Shankland on the acoustical properties of open-air classical Greek theaters. (*Physics Today* October 1973). Illustrated in this article are "articulation scores" plotted against distance from center of stage to listener. The articulation score is the percent of words spoken at the stage which are understood by the listener (English in this case, not Greek!)

Another Shankland quotation from the Swenson interview: "I have a great body of data and information about the acoustics of buildings that I hope to use to write a book on architectural acoustics one of these days. I think it will be a book that has a lot in it that the conventional books do not have, because these are largely based on buildings in the United States, and fairly recent buildings." There are twelve Shankland papers in the Journal of the Acoustical Society on these studies. I don't know if Shankland began organizing this book; his health seemed to be quite good during the years after the AIP interview. His files on the subject, now in the departmental archives, are extensive, but there is no sign of a draft.

Accelerator Physics at Berkeley

Shankland spent the summer of 1946 at the Radiation Laboratory of the University of California at Berkeley, doing a proton-proton scattering experiment at the 37-inch synchrocyclotron. This accelerator-based program, mounted so quickly after the end of the war, would lead the race in the 1950's toward higher and higher energy machines. The development of accelerators and their associated particle detectors would make possible today's picture of particles and forces: a.k.a. the "Standard Model."

The Berkeley paper appeared in the Physical Review. Shankland's collaborators included Robert R. Wilson (future Nobelist from Cornell). They presented the differential scattering cross section for 14 MeV protons incident on protons. *(This means they measured the probability that a proton will scatter into a given angular range when it bounces off another proton. The resulting angular distribution leads to an understanding of the forces between the protons. There will be a more detailed description of scattering experiments in Chapter 9.)* The protons were

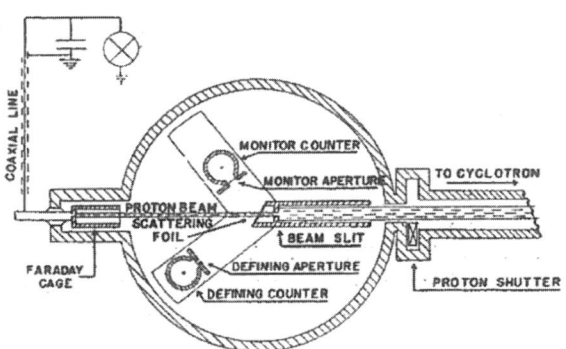

Fig. 6-8. Proton-proton scattering at Berkeley accelerator.

incident on a thin target of nylon, and the two outgoing protons were detected by proportional counters as indicated in the sketch in **Fig. 6-8**. Shankland's main contribution was his expertise on the use of coincidence counters. (*Phys. Rev.* **72** 1131 1947).

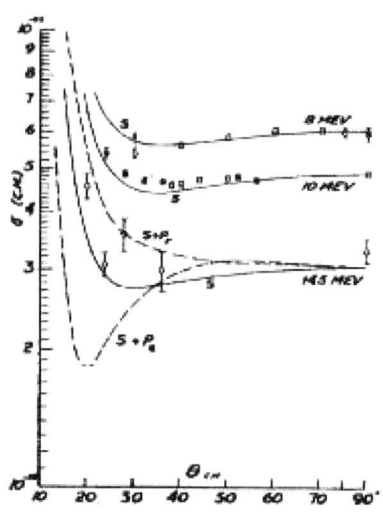

Fig. 6-9. Angular distributions for proton proton scattering.

At these energies, the two equal mass protons leave the target at 90° from one another, so the counters were placed at several pairs of orthogonal angles. The results are shown by the 14.5 MeV data set of **Fig. 6-9** (beneath two sets at lower energies measured by other experimenters). The three curves are calculations by Shankland's Case colleague, Leslie L. Foldy, based on a square-well potential with range 2.8×10^{-13} cm and a depth of 10.5 MeV. They include S-wave scattering (solid), S plus P-wave attractive (lower dashed) and S plus P-wave repulsive (upper dashed). If one just counts the number of standard deviations separating the points from the curves, the S and the S+P repulsive are about equally good. This experiment was one of the earliest attempts to get at the nucleon-nucleon force by scattering experiments. Foldy's work on this and many other theoretical topics will be described in Chapter 9.

Aside on S waves, etc. When two particles collide, they can have orbital angular momentum, like two ballroom dancers. Because of quantum mechanics, the angular momentum can take on only certain values. The lowest value, zero, is called S-wave scattering, the next value, $h/2\pi$ (h is Planck's constant), is called P-wave scattering. (The S,P,D,F letters come from archaic spectroscopic notation – sharp, principal, diffuse, and fine.) If multiple values of the angular momentum contribute to the interaction, they can interfere with one-another (quantum-mechanically), and produce a characteristic pattern in the differential cross-section, like the big dips in the curves in Fig. 6-9.

In the 1950's, Shankland was a physics consultant at the Phillips Petroleum Company's Materials Testing Reactor in Idaho Falls, where he produced a series of papers on neutron cross-sections. Between 1956 and 1958 he participated in experiments at that reactor. Total cross-sections for neutron interactions in carbon and chlorine were measured using a fast chopper to define the neutron energies up to 15 keV.

A chopper is a set of two co-axial spinning disks with holes in each disk. The positions of the holes are offset from one another so that they permit the passage only of particles within a certain range of speeds. These devices were used to create beams of neutrons with a fixed energy, much as bending magnets are used to create monoenergetic beams of charged particles.

Shankland and Miller's Ether Drift

In the 1974 AIP interview, Shankland also discussed his connection with the Miller ether-drift experiments (which we described at length in Chapter 4). The cover page of the Miller article in the Reviews of Modern Physics, including the dedication by Miller to Shankland, is shown in **Fig. 6-10**. The following italicized paragraph is a condensed version of part of the AIP interview.

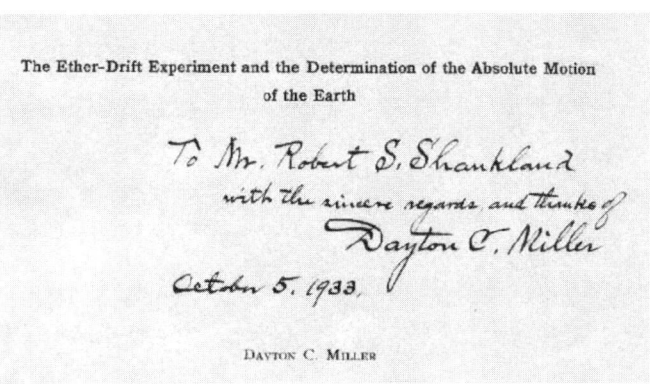

Fig. 6-10. Cover page of Miller's "final" ether-drift paper.

I never worked with Miller when he was making measurements with his interferometer. Those were made at Mt. Wilson and finished during my freshman year. I did help him with some aspects of his final write-up of the paper, but I must say that I never personally could accept Miller's view that it disagreed with relativity. I was absolutely sure that Miller was honest in everything he did, and it was a real puzzle in my mind for many, many years, as to why it was that this periodic effect was there. And I finally decided to spend some time seeing if I could straighten it out, with the help of McCuskey and Kuerti, and with the encouragement of Einstein. We worked away at it, and this was in accord with what Dr. Miller wanted, because shortly before he died, he gave me a great pile of data sheets which I still have, and he kind of pushed them at me and said, "Well, there are the Mt. Wilson observations. You keep them. And you can either burn them up or study them, whichever you think best." This must have been either in January of '41 or shortly before. For years I just had them locked in a closet, but then over a period of nearly 15 years, I would get letters from very distinguished physicists asking me what I thought about Miller's work. So instead of burning them up, we studied them. Einstein wrote me a very nice letter when we got through. He didn't see the final printed version; he died a month before, but he saw every draft.

The resulting paper, "New Analysis of the Interferometer Observations of Dayton C. Miller", authored by Shankland, S. W. McCuskey (chair of the Case astronomy department), F.C. Leone (professor of statistics and mathematics), and Gustav Kuerti (professor of aeronautical engineering and mathematics) appeared in the *Reviews of Modern Physics* (*Rev. Mod. Phys.* **27** 167 1955). These four authors were the most qualified members of Case Institute's faculty to undertake this project. The authors first point out that the Miller measurements made in Cleveland showed very little effect, and that the primary evidence for ether-drift was in the Mt. Wilson data. Miller's raw data consist of many pages, each with twenty rows of sixteen numbers corresponding to the sixteen fringe-shift observations in each of 20 turns of the interferometer. There are data taken at all times of day and at the four seasons of the year. The pages are frequently annotated

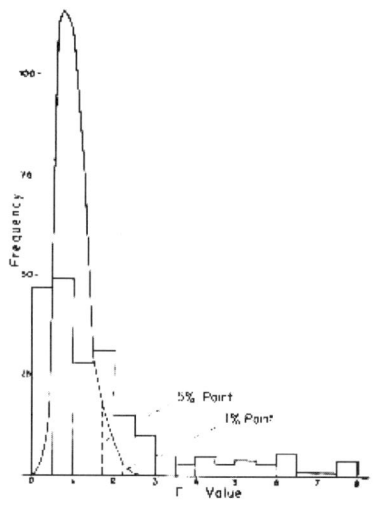

Fig. 6-11. Statistical test for randomness in Miller data.

with "viewing conditions", weather, temperature, etc. Miller's analyses and his interpretation of them were described above in Chapter 4.

Shankland *et al.* first determined the degree of randomness of the entries. For each sheet, they calculated a statistical factor proportional to the sum of the squares of the deviations from the mean. The data were rewritten in units of one twentieth of a fringe, this being assumed to be a reasonable least count. Statistical theory gives the expected distribution of this "randomness factor". In **Fig. 6-11** are shown the randomness for 216 sets of data and the theoretical curve for a random distribution. The conclusion is that structure in the data comes from something other than statistical fluctuations, since a full 36% of the data-points fall in the region on the right where only 5% should lie if the variations were random.

Another demonstration of a non-random signal in the second harmonic (i.e. the presence of a full 360° sine-wave in the fringe shift for a complete turn of the interferometer), is shown in **Fig. 6-12**. Here, the average values of the fringe shift for each of the 16 positions of the interferometer (averaged over a complete sheet or 20 turns) are shown for twenty sheets from the July set. While there is very large scatter in the data, the average of the averages (the large black circles) does show a sine-wave form with amplitude about 0.03 fringe. Similar plots for the other three "seasons" are constructed, and, folded about 180°, are shown in **Fig. 6-13**. Neither Miller nor the four authors could find any argument for the variation in phase among these four curves (i.e., why are they shifted horizontally with respect to one another?) As Miller had done, the authors sought a way that a "cosmic solution", e.g. motion through the ether of the whole solar

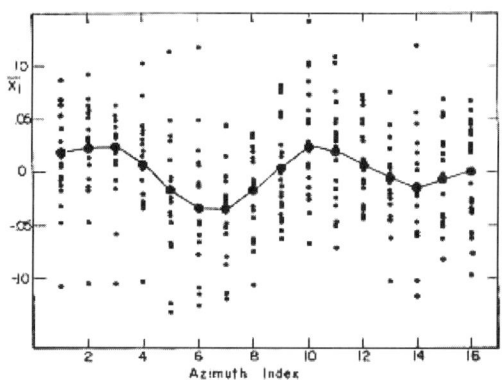

Fig. 6-12. Compilation of second harmonic fringe shifts.

system, might fit the data. They conclude that no consistent solution of this type can be extracted from the data.

The goal, now, is to determine the cause of this residual non-random structure. They turn therefore to a search for systematic mechanical or thermal causes. A mechanical analysis was made of the possible oscillatory motion of the heavy steel cross-beams which

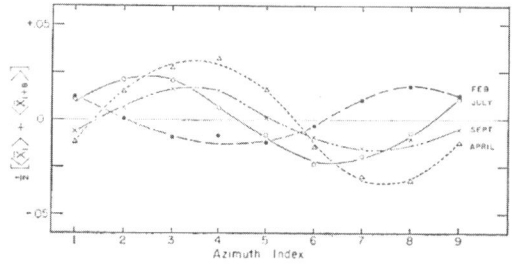

Fig. 6-13. Second harmonic signal for four seasons.

rested on a wooden box floating in mercury. They wanted to determine if some sort of periodic "rocking of the boat" might affect the distances between the mirrors (tilting of a support beam might cause its two arms to sag differently). Knowing the geometry and mass distribution of the system and the restoring forces, they calculated the natural frequency of the rocking motion. This came out as 1.2 s to 1.4 s, so any such motion would have been averaged out over the typically 50-second rotation of the interferometer arms.

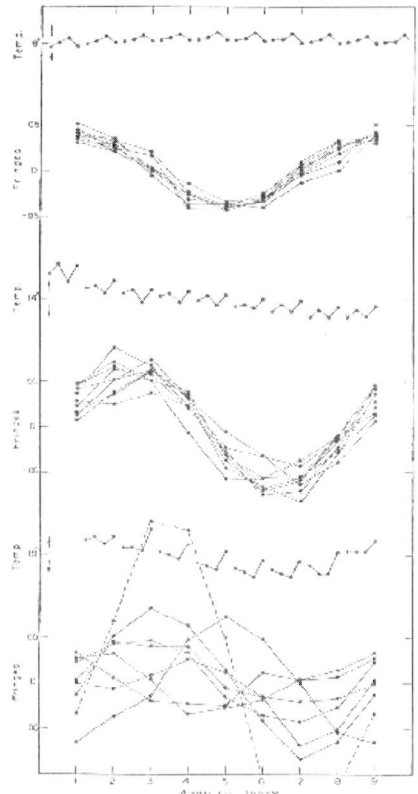

Fig. 6-14. Temperature variations shown with corresponding fringe shift signals.

Next, they searched for possible systematic thermal effects, and here, patient reader, is where they find the answer. Miller himself was very much aware of the importance of avoiding thermal expansions and contractions of the steel arms of the interferometer and changes in the index of refraction of the air between the mirrors. Miller had even purposely introduced electric heaters to establish a temperature gradient across the device and observed a resulting fringe-shift signal of up to twenty-times larger than the 0.03 fringe "ether signal".

Miller provided on each sheet the temperature of the air at each of the four walls of the hut. The authors searched for those sets of data for which the temperature was relatively uniform and constant through the viewing period. Readings taken between midnight and dawn provided the best conditions. **Fig. 6-14** shows the correlation between temperature stability and the amount of fringe-shift observed in three late night sessions. The figure is a bit confusing. The little zig-zag sets, each with four data points, are the temperature readings on the four walls of the hut, taken at fixed intervals during the five-hour period of the runs. The little arrows to the left indicate plus and minus one degree Celsius. The top set (from Cleveland measurements) shows the least variation in temperature; the two lower sets (from Mt. Wilson) show larger variations. Beneath each of these three sets of temperature data is a plot of the fringe shifts observed in the corresponding set of runs. The ordinate scale markers are at plus and minus 0.05 of a fringe. It is clear that the more constant the temperature, the smaller the fringe-shifts. The dashed curve in the bottom set shows how bad things can get when sunlight falls on the interferometer. Calculations were made to determine by how much the temperature should change within the glass-enclosed interferometer arms as a result of a given temperature gradient across the hut. The numbers are consistent with the observed shifts. The authors would like to have established a clear, one-to-one, relationship between measured temperatures and both phase and amplitude observations, but they have to settle for a more general conclusion: *i.e.* that the actual thermal conditions were *sufficient* to cause the observed effects. I quote an entire paragraph in this regard:

"Thus Miller's experiments in 1923 do not rule out the possibility of attributing the remaining systematic effects in the Mount Wilson data, which are most prominent in the second harmonic A_2 and to a lesser degree in the first harmonic A_1, to temperature causes. In what follows, we shall interpret the systematic effects on this basis but must admit that a direct and general quantitative correlation between amplitude and phase of the observed second harmonic, on the one hand and the thermal conditions in the observation hut on the other hand could not be established. The reason for this failure lies in the inherent inadequacy, for our purpose, of the temperature data available." In short: not enough information to tie it down quantitatively!

This careful reanalysis of the Miller data satisfied the authors and the physics community-at-large that Miller's effect was real, but thermal in origin. Thus the Miller experiment joins the many others, performed in the twentieth century and with increasingly sophisticated techniques, in which no evidence for ether-drift is observed. Included in these later ether-drift searches were those done in the 1920's by Michelson himself.

Shankland Discusses Michelson-Morley and Miller with Einstein

Shankland visited Albert Einstein in his Princeton home on five occasions, the first being in February of 1950, the last in 1954. In 1963, he published an article in the *American Journal of Physics* describing these meetings. There were three major topics in their discussions, the Michelson Morley experiment and its influence on the development of the special theory of relativity, the subsequent observation by Dayton Miller of a systematic fringe-shift in his repetition of the ether drift experiment, and some general remarks by Einstein on quantum theory. Ten years later, Shankland sent a second paper to that journal. In his introduction he says that in the earlier paper the discussions "were published almost verbatim with but little comment by me. They have since been referred to in several articles on the history of physics, so it now seems appropriate to supplement the first publication by a more complete discussion of certain statements made to me by Prof. Einstein."

Shankland says that at the time of the first visit, Einstein "was unacquainted with me and may possibly have wondered if I had not come as the successor of Prof. Dayton C. Miller at Case to talk about Miller's 'aether drift' experiments at Mount Wilson." Recall that Shankland was only the fourth in the line of Case chairmen, after Michelson, Reid, and Miller. "When Prof. Einstein realized that I had not come as an advocate for Miller's results his attitude became less formal and our conversations were much more relaxed." Nevertheless, Shankland adds that "when he (Einstein) learned of Miller's result he traveled to Cleveland to see him and they had a long discussion about the Mount Wilson experiments". When I looked through the archives of the *Cleveland Plain Dealer* for references to Einstein, I found a front page article on Einstein's visit. He was in town on a Zionist fund-raising tour in the company of Chaim Weizman, the physicist who would become Israel's first president. This was the same time as the meeting with Miller. The department still has Miller's guestbook with Einstein's signature, dated 1922. (It was also Case graduation day, so Miller must have been very busy.)

Shankland remarks "when we *(i.e. he, McCuskey, Leone and Kuerti)* finally found the cause of Miller's periodic fringe shifts to be temperature gradients across the interferometer, Einstein was genuinely pleased, in fact, wrote me a fine letter on the subject." The letter was reproduced in an article by Shankland, "Michelson's Role in the Development of Relativity", which appeared in October of 1973. (This article in *Applied Optics* was based on the "Naval Academy Lecture" given by Shankland at Annapolis in May of that year. The lecture was given as part of the celebration of the 100th anniversary of Michelson's graduation from Annapolis. *Applied Optics* **12** 2280 1973.)

> *August 31, 1954*
> *Dear Dr. Shankland,*
>
> *I thank you very much for sending me your careful study about the Miller experiments. Those experiments, conducted with so much care, merit, of course, a very careful statistical investigation. This is more so as the existence of a not trivial positive effect would affect very deeply the fundament of theoretical physics as it is presently accepted.*
>
> *You have shown convincingly that the observed effect is outside the range of accidental deviations and must, therefore, have a systematic cause. You made it quite probable that this systematic cause has nothing to do with 'ether-wind", but has to do with differences of temperature of the air traversed by the two light bundles which produced the bands of interference. Such an effect is indeed practically inevitable if the walls of the laboratory room have a not negligible difference in temperature.*
>
> *It is one of the cases where the systematic errors are increasing quickly with the dimension of the apparatus.*
>
> *Congratulating you and your colleagues on your valuable contribution to our knowledge, I am*
> *With kind regards,*
>
> *A. Einstein (signed)*

The letter came into the possession of Shankland's second wife, Eleanor, who in 2000 generously donated it and three other Einstein artifacts to the University. It was sold at auction and the proceeds were made available to the physics department for the support of graduate students .

Concerning the Michelson-Morley result, Shankland states that Einstein's comments over the course of their several discussions were "not entirely consistent", at least as to when and how he had learned of the null result, and what role it played during the period when he was thinking about special relativity.

Finally, concerning Einstein's opinions on quantum mechanics, Shankland lists a series of one-sentence quotations, which he says are as accurate as he could make them, having carefully written them down immediately after his chats. Three short examples which make his position clear: "On quantum theory I am in the opposition." "The ψ

functions do not represent reality." "Bohr always speaks *ex cathedra*." "Conversations with Albert Einstein" *Amer. J. Phys.* **31** 47 1963. "Conversations with Albert Einstein, II" *Amer. J. Phys.* **41** 895 1973. Included in Shankland's personal journals are eleven pages describing his 1950 visit with Einstein, presumably written on the same day. The text is much more detailed and less polished than the papers published over a decade later.

The "Return" of Miller's ether drift, a digression

Sometime in 1998, I was telephoned by the editor of a magazine called *Twenty-first Century Science and Technology*. She had been in contact with the CWRU Archives, and was referred to me. They were planning an article about Dayton C. Miller and were looking for some photos to illustrate it. During the renovation in 1995 of our (i.e. Miller's) building, my colleague Bill Gordon and I gathered up quite a collection of old apparatus from the attic, as well as several drawers of files. I had randomly glanced through some of the documents, and I knew that we did indeed have a few gems, such as Miller's accounts books, his guest book with Einstein's signature, hundreds of lantern slides, etc. Up to that time, my knowledge of Miller was little more than a portrait on the wall.

I sent a few photos of Miller and his interferometer. The CWRU archives did likewise. I was thanked by the editor, Marjorie Hecht, who promised to send a copy of the magazine. The handsome, glossy product arrived a few months later. There was Miller on the cover; and the headline read "Michelson-Morley-Miller: The Coverup" The 25-page article described the Miller experiments in detail, presenting them as proof that the ether does exist and that Einstein's relativity theory is wrong. They mention, in a section titled "The Debunkers", the Shankland, *et al*. 1955 paper and the correspondence between Shankland and Einstein, implying that these two fellows conspired to cover up the Miller results. The magazine, it turns out, was published by the organization of Lyndon Larouche, the fellow who has built his several presidential campaigns on the exposure of a wide array of evil plots against the American people, including, it seems, special relativity. The editor with whom I had corresponded is Mr. Larouche's wife.

This magazine incident was a useful learning experience. In 2000 I received an enquiry by a gentleman from Oregon concerning Miller. The fellow was planning a trip to Cleveland and asked if any of the original Miller data were available. I did an internet search for references to the prospective visitor, and found that, as the director of an institute specializing in orgonics research, he had lectured and published on Miller, the ether, and relativity. Orgonics, the study of the properties of an energy field which surrounds the human brain, apparently needs a medium of some sort.

When he came to Cleveland, I asked him why he was interested in the Miller data. He said that he would reanalyze it using modern techniques. He was aware of the Shankland reanalysis paper, but dismissed it as biased. After he left, I searched through the dozen large file drawers of Shankland papers. The Miller data sheets were there all along, hidden among some architectural acoustics papers. I have since turned them over to the

CWRU Archives so that anyone interested may access them. Anyone wanting to tease out, once again, the tiny and random-phased signal from those numbers had best study the Shankland paper first. It wouldn't hurt to check out recent results from repetitions of the MM experiment using resonant microwave cavities. These report no effect at the one part in 10^{13} level. (Lipa *et al. Phys. Rev. Lett.* **90** 060403 2003). Even Miller's small ether-drift would raise havoc with today's global positioning system.

Shankland edits the Compton papers

In 1973, Shankland edited a collection of selected papers by his mentor: *Scientific Papers of Arthur Holly Compton: X-Ray and Other Studies* (University of Chicago Press). This 777-page compilation, including an extensive introduction, appendices, and bibliography, was received by the history of physics community as a valuable contribution. Roger H. Stuewer, historian of science at the University of Minnesota, reviewed the work and congratulated Shankland on a compilation from which the reader "can learn a great deal about the way research in physics is actually pursued." For those who may be interested in this project, an extensive file of Shankland's notes and correspondence on the subject may be found in the CWRU Physics Archives.

Fig. 6-15. Reines, Glennan, Shankland, Foldy.

Fig. 6-15 is a photograph taken on the day a new portrait of Shankland was dedicated. Participating are Chairman Frederick Reines, Case President T. Keith Glennan, Shankland, and theorist Leslie L. Foldy. It hangs today in Miller's big lecture hall.

Robert Shankland retired in 1976, but continued to work on his papers until his death in 1982. His wife, Eleanor, is well known as an artist and, as mentioned above, a loyal friend of the university. She has done a series of drawings of the buildings on the campus, among which is this one of the Rockefeller Building. It is easy to see the artist as she sits at her easel, but you have to look closely to find Bob Shankland at the window of his office.

Fig. 6-16. Eleanor Shankland's drawing of Rockefeller.

Chapter 7   Case Experiment Takes Off

Olsen, Crittenden, Shrader, Gregg, Smith, Scharenberg, Silverstein
37-60   38-56   40-69   45-61   42-68   61-65   64-69

In 1935 the Case department consisted of Miller, Albright, Nusbaum, Hodgman, Wallace and Shankland. Only Shankland was doing work that could be described as "modern" experimental physics. By 1945, the department had added five young experimentalists: **Olsen, Crittenden, Shrader, Gregg and Smith.** A full research program had come into being, and the directions to be taken by physics research at Case were beginning to be defined. **Fig. 7-1** is a photograph of the members of the department in 1945.

Research costs money

While both Case and Western Reserve had modest research programs in technology and the sciences from their beginnings, things really started to take off during and after World War 2. This was true of course at all American research universities. Both industry and the government took advantage of the concentration of knowledgeable faculty and they were willing to pay for it. The universities were happy to accept large sums in overhead charges which they would invest in expanding their faculties and facilities.

Fig. 7-1. seated: Crittenden, Smith, Shankland, Shrader, Gregg.

The chemistry departments of both Case and WRU and the engineering departments at Case had established relationships with corporations such as Dow Chemical, General Electric, Sherwin Williams, Lubrizol and Standard Oil. The physics departments had rather modest industrial funding before 1942. Their research was in general too "basic" for commercial applications. The war changed this. Industry and government came to the physics community asking for help in electronics for communications and radar, in sonar, optics, materials and basic atomic and nuclear physics.

The research programs of the two physics departments would be funded almost exclusively by the Department of Defense, the Atomic Energy Commission, and later by NASA and the National Science Foundation. The critical role of science in determining the outcome of the war would open the minds and purses of the federal government, even for the rather basic research pursued by the CIT and WRU physicists. The resulting increases in funding would allow the expansion of each department from 5-6 in the 1930's to 25-30 in the 1960's.

Darwin H. Stapleton, former professor in the CWRU department of history, discusses the funding situation at CIT and WRU and some of the problems it entailed in a 1993 paper. "The Faustian Dilemmas of Funded Research at Case Institute and Western Reserve, 1945-1965". *Sci. Tech. & Human Values* **18** 303 1993. His main point is that, while the money was welcome, the control of the research (and associated teaching) would be taken out of the hands of the faculty, in violation of the nationally accepted standards of academic freedom and faculty governance. Most of the faculty and many in the administration were uncomfortable with the pursuit on campus of classified defense research.

As an example of classified work, Stapleton describes a program which accounted for half the CIT federal funding between 1951 and 1958. The "Doan Brook Project" involved scientists, engineers, and social scientists, had a staff of over eighty, and supported about 40 grad students. This "systems analysis" research dealt with aerial seeding of antitank and antipersonnel mines. In 1962, Case's Faculty Committee on Research declared that the publication of research done on campus should not be restricted. Classified projects would no longer be undertaken.

Another important impact of the quest for external funding from government, industry, and even from charitable foundations, was considerable pressure on the two institutions to federate. The inevitable union took place on 1 July 1967. I quote an entire paragraph from Prof. Stapleton's paper because it so much relates to our story.

"The initial years of federation were difficult ones, especially for the science departments. Resistance by Case alumni to the fund-raising effort called for by the Heald Commission (author's note: this was the outside advisory group which recommended federation) exacerbated the effect of the general decline in federal funds in the latter 1960's. Moreover, there were serious problems with faculty morale as the administration pressed for unified departments of physics and mathematics in addition to chemistry and had to overcome delaying tactics and in some instances outright refusals to cooperate with the federation process. Faculty found that the merged departments were expected to be smaller and many left in anticipation of a future termination. It was an agonizing period for much of the university community."

Olsen and the GE connection

We return to the Case physics research story and the new practitioners. **Leonard O. Olsen**, with a PhD from the University of Iowa, was hired in 1937. He studied the interaction between light waves and the atoms of rare gases. This work led to a connection between the department and the General Electric Company's Cleveland-based lighting research group at Nela Park (acronym for National Electric Lamp Association), a connection which was to flourish for at least four decades. In two 1941 papers, Olsen and Iowa colleagues present theoretical and experimental results on the quenching and depolarization of ultraviolet light (the 2537 Å mercury line) in He, Ne, Ar and Kr gases as a function of pressure and applied magnetic field. Olsen would return to this research after the war, publishing a paper based on work done with his MS student, George Kerr. Kerr later joined the research group at General Electric. This work was similar to that done in 1941, but this time nitrogen and oxygen were studied. An expanded experiment involving eleven different gases was published in 1960. The techniques were brought up to date with the use of photomultiplier tubes as detectors, and with the data analysis done on the department's new state-of-the-art IBM 610 computer. "Collision Processes in Mixtures of Mercury Vapor and Foreign Gases" *Phys. Rev.* **119** 691 1960.

Toward the beginning of World War II Olsen and several colleagues at Case put together a crash program to train students in acoustics, in an effort to provide new manpower for war-related research. In 1942, thirty-two students from Case, from industry, and from nearby liberal arts colleges participated in a program which introduced them to electro-acoustics and supersonics. Among the instructors in the program was MA student Earle Gregg (to be introduced shortly) who discussed piezoelectric and magnetostrictive transducers. Most of the students went on to such places as the labs of MIT or directly into military research groups. ("Training Men in Acoustics and Supersonics for War Research" *Amer. J. Phys.* **10** 262 1942.) With Shankland's (Chapter 6) and Foldy's (Chapter 9) wartime research in New York and this related training effort back in Cleveland, Case contributed valuable expertise in acoustics to the war effort.

Crittenden and plans for the Case betatron

Eugene C. Crittenden, Jr., hired in 1938, had published two papers on beta decay in connection with his doctoral research at Cornell. These concerned cloud chamber studies of beta-ray spectra of radioactive nuclei formed at the Cornell cyclotron. His early publications at Case include a paper on a teaching-lab experiment on forced damped oscillations (*Amer. J. Phys.* **17** 282 1943); a paper on the uniformity of low magnetic fields produced by Helmholtz coils (*Rev. Sci. Instr.* **15** 270 1944); and a paper on an electronic flow-meter designed for biological applications (*Rev. Sci. Instr.* **15** 343 1944).

In 1946, Crittenden spent time at the Radiation Laboratory at Berkeley where he studied the theory and operation of particle accelerators. This was at about the same time his later colleague, Leslie Foldy, was at the "Rad Lab", as we shall describe in Chapter 9. Crittenden wrote two papers for the *Journal of Applied Physics*, one on "Methods for Betatron or Synchrotron Beam Removal", the other on "A Graphical Method for Determining Particle Trajectories". (*J. Appl. Phys.* **17** 444 and 447 1946) He remarks that until that time, betatrons were used to create beams of X-rays produced when the internal electron beam hits a target. Since the electrons themselves were not extracted from the machine, Crittenden's studies on the production of beams of electrons would help to open up a new era in accelerator experiments.

Both Crittenden and Foldy subsequently played a role in the design of the betatron to be built at Case. This machine, with radius about two meters and a seven ton electromagnet, was designed to produce a beam of 25 MeV electrons. The proposal presented to the Atomic Energy Commission stated that the research done with the betatron would include the study of fundamental particles at energies available only in cosmic rays, the study of nuclear structure, the production of radioactive isotopes and industrial X-ray radiography. The design called for a toroidal ceramic vacuum chamber placed between the pole pieces of the magnet. **Fig. 7-2**. An alternating current of order 100 Amperes with frequency 180 Hz would energize the electromagnet. During the half-cycle when the magnetic field was rising, the induced electric field would accelerate the electrons. During that one 360th of a second, the electrons would reach almost the speed of light as they spiraled outward toward the edge of the vacuum chamber. At this point, the device, described in a 1950 paper by Crittenden and grad student Sherwood Fawcett, (*Rev. Sci. Instr.* **21** 935 1950) would superpose a 20 microsecond bump in the magnetic field which would nudge the electron beam out through the glass wall of the machine. The cost for the "Case Betatron" and its basement housing was estimated at $39,000 (~$300,000 in 2005 dollars). We shall return to Crittenden later, when he becomes interested in thin metallic films.

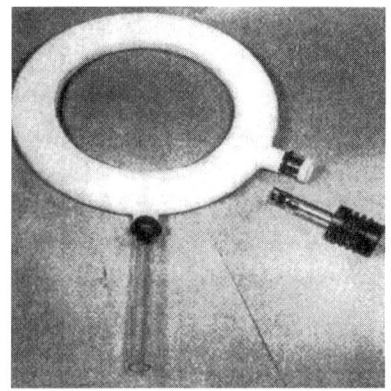

Fig. 7-2. Ceramic vacuum chamber for betatron.

Shrader, Gregg and the betatron

Erwin F. Shrader joined the department in 1940, a year before receiving his PhD from Yale. That year, he sent a short letter to the *Physical Review* describing his determination of the nuclear spin of the isotope ^{37}Cl (*Phys. Rev.* **58** 475 1940). He had used thermal diffusion of HCl gas to obtain a sample of chlorine enriched in the ^{37}Cl isotope. By measuring the absorption spectrum of this sample, and specifically the intensity ratios of several lines in the vibrational spectrum, he concluded that the spin of ^{37}Cl is 5/2, the same as that of ^{35}Cl. For our story, it is interesting that Polycarp Kusch at Columbia took note of the young Shrader's work, and soon published a letter with him (*Phys. Rev.* **58** 925 1940). Kusch worked with molecular beams and used Shrader's ^{37}Cl spin to extract the magnetic moment of that nucleus. Kusch (a 1955 Nobel prize winner) had done his undergraduate work at Case (BS 1931), and it is conceivable that he put the Yale student in contact with Case.

Earle C. Gregg earned a Case MS under Crittenden in 1942, joined the faculty in 1945, and completed his PhD with Erwin Shrader and Shankland in 1949. Gregg and Shrader would devote five years to building and testing the betatron. As early as 1946 Gregg wrote short papers on devices to measure the magnetic fields in the machine. A complete description of the machine (*Rev. Sci. Instr.* **22** 176 1951), includes the photograph of the ceramic vacuum chamber shown in **Fig. 7-2**.

The machine was placed in an underground bunker, adjacent to the south end of the Rockefeller Building. The control room was within the building, and protected from radiation by thick concrete walls. (As in many other experimental papers from Case for the next decade or two, departmental engineer Mr. August (Gus) Hruschka is thanked, in this case for "most of the construction".) The photograph of the completed machine (**Fig. 7-3**) was taken when the president of Case, T. Keith Glennan, was showing the machine to the head of the Atomic Energy Commission, Lewis Strauss. (Glennan would later take over the AEC leadership.) **Fig. 7-4** is a photo of Gregg and Shrader in jackets and ties checking out the betatron magnet cooling system.

Fig. 7-3. Glennan shows betatron to AEC Chief Strauss.

By 1952, Shrader completed probably the first experiment done with the gamma rays produced by the Case betatron: the measurement from 5 to 13 MeV of the photo-disintegration of the deuteron. This was a very hot topic at the time, as it shed light on the nuclear force which binds the neutron and proton together. Shrader, and his student Victor Krohn, aimed the gamma rays coming from the betatron at deuterons and measured the angular distribution of the outgoing protons. In this case, the deuterons were incorporated in nuclear emulsions impregnated with D_2O (heavy water).

Fig. 7-4. Gregg and Shrader, dressed for research.

*The **deuteron** is the simplest nucleus, consisting of a single proton and a neutron. It is bound together by the nuclear or "strong" force and requires an energy of 2.2 MeV to break it apart. It was the object of study by experimentalists and theorists the world-over as a key to understanding the strong force. **Nuclear emulsions** were a widely used technique for observing the tracks of energetic charged particles. They were basically very thick photographic films (without the celluloid backing). Charged particles passing through would ionize the silver salts, and, on development, grains of silver would mark their paths. Measurement of the distance traveled in the emulsion gives the kinetic energy of the charged particle.*

Fig. 7-5 is a view of the lab in Rockefeller where the detectors were built and tested. It shows a nice assortment of the bulky vacuum tube electronics available in the

1940's. **Fig. 7-6** is a full view of the betatron showing the heavy-duty crane above the magnet coils and iron yoke and the vacuum pumps and cold-traps beneath.

Fig. 7-5. Betatron control room.

The gamma ray beam was collimated to about a two centimeter diameter. It just grazed the 200 μ thick emulsion plates so the forward traveling protons would remain in the emulsion. The developed plates are then examined by using a microscope. One sees little strings of silver grains, the tracks ranging from a few microns in length up to a millimeter or two. From known range-energy relations, one gets the energy of the proton which was knocked out of the deuteron. The angle between the proton direction and the direction of the incoming gamma is measured (allowing for the fact that the emulsion thickness after development and drying has shrunk to only one sixth of its thickness during exposure). The energy of the incoming gamma could be calculated from the energy of the proton and its direction. The resulting angular distributions were tabulated for several energy ranges and these were compared with available theoretical predictions. This experiment provided the research topic for Shrader's grad-student, Harold Fleisher. "Photodisintegration of the Deuteron" *Phys. Rev.* **86** 391 1952.

Fig. 7-6. The assembled betatron.

Shrader and Crittenden's graduate student, Robert Strough, published an interesting paper describing a technique for generating quick bursts of electric current. These could be used to produce the pulsed high magnetic fields needed in a betatron. The idea is to use compressed air to mechanically spin-up an 11 lb beryllium-copper rotor to 20,000 rpm. A current is supplied by a battery to a nearby coil, creating a magnetic field in the spinning rotor. This field, according to Faraday's law, produces a high voltage between the center and the outer edge of the rotor. This voltage drives a current, via liquid mercury "brushes", through the betatron coil. In their experiment they were able to produce a current of 56,000 amps in three tenths of a second. The current drops off equally quickly as the 30,000 Joules of mechanical energy is drained from the rotor. In their conclusions, the authors suggest that one could build a 1 GeV betatron with two 700 lb rotating disks. They thank Richard L. Garwin who proposed the scheme in the 1947 BS thesis he wrote when he was an undergraduate physics major at Case. "Pulsed air-core series disk generator for production of high magnetic fields" *Rev. Sci Instr.* **22** 578 1951.

Storing energy in a flywheel became a reality. When I was a grad student working at Brookhaven Lab in the 1960's, I could hear the groaning sound all night long of the big flywheels which every few seconds traded energy with the magnetic fields in the AGS accelerator.

Gregg and early biophysics

Gregg continued physics experiments using the betatron over the following 10 years. During the same period, however, he became interested in physiology and the applications of ultrasound in medicine. In a paper entitled "Ultrasonics: Biologic Effects" (*Medical Physics* **II** 1132 1950), he described how ultrasound is produced and discussed in detail its effects on small beasties like tadpoles and bacteria and its possible therapeutic uses. (Forty years later, William Tobocman (Chapter 9) would be applying nuclear physics scattering theory to the analysis of the passage of ultrasound waves through biological tissue in medical imaging devices.) Another paper described a device Gregg designed to measure physiological response to vibratory motion, in essence replacing the old-fashioned tuning fork as the tool commonly used by physiologists. This interest in quantitative physiology continued with work on the measurement of pain thresholds and the development of instruments which could provide a measure of neural responses to heat. (*Jour. of Appl. Physiology* **4** 351 1951) A natural follow-up on the measurement of pain was a study of the efficacy of pain-killing medications. (*Jour. of Pharmacology and Expt'l Therapeutics* **106** 1 1952)

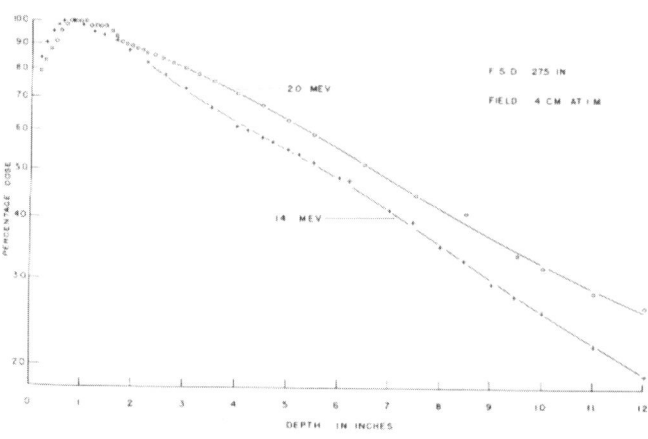

In work which combined his interest in things medical with his access to the betatron, Gregg designed a "Roentgen Ray Dosimeter". (*Amer. Jour. of Roentgenology* **76** 979 1956) This small anthracene crystal detector was used to measure the rate of energy deposition at various depths in a water-filled tank, water being a good approximation to living tissue. **Fig. 7-7** shows a typical plot of "dose" versus depth for two photon energies. While usual medical diagnostic x-rays are in the 50 keV range, this work was done at energies available at the CIT betatron: 14 and 20 MeV; it was no doubt related to therapeutic irradiation.

Fig. 7-7. Data from x-ray dosimeter: dose vs. depth in water.

Nuclear physics at the betatron

Shrader continued the nuclear physics program at the betatron. In one experiment beams of gammas at various energies were incident on a ^{63}Cu target. The gamma knocked out a neutron, leaving a radioactive ^{62}Cu nucleus. "The $Cu^{63}(\gamma,n)Cu^{62}$ Cross Section" (*Phys. Rev.* **87** 685 1952). The ^{62}Cu decays with a half-life of 9.7 min by electron capture and γ emission. The cross section for the reaction was determined by detecting these gammas. **Fig. 7-8** shows the results with the cross section peaking at about 0.1 barn at 18 MeV.

*A **barn** is a unit of area 10^{-28} m^2; the name comes from "you couldn't hit the side of a ..." or maybe "as big as a ...". The probability that a reaction will take place (the "**cross-section**") is given as an area because it is related to the effective size of the target particle which is "presented" to the projectile. It can be much larger or smaller than the actual size of the target particle. During the 1940's and later, cross sections like this were measured for thousands of nuclear reactions, resulting in enormous tables of data (the most famous of which had a picture of a big cow-barn on the cover). These data were essential in the design, for example, of nuclear reactors and weapons, and for the development of models for nuclear structure and even for the syntheses of nuclei in the hot centers of stars.*

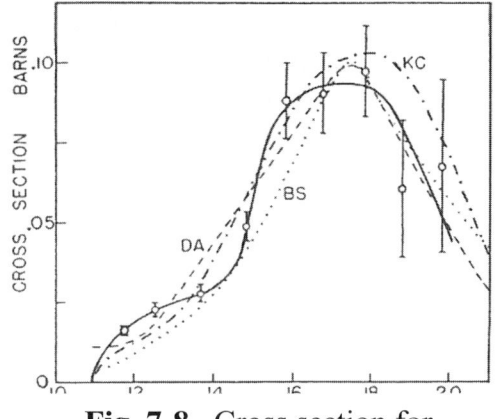

Fig. 7-8. Cross section for $Cu^{63}(\gamma,n)Cu^{62}$.

Shrader's next betatron experiment, the measurement of the absorption of gamma rays by heavy metals (Cu, Sn, Pb and U), used a magnetic *pair spectrometer*. All accelerator labs need detectors, and this device was a major addition to the Case Betatron Lab. The spectrometer is shown in cross-section at the right side of the sketch in **Fig. 7-9**. In this experiment, the incident gammas from the betatron pass through a sample of the heavy metal where some of them are absorbed. A known fraction of the transmitted gammas strikes a thin piece of material, the *radiator*. There, in the electric field of an atomic nucleus, the gamma converts into an electron-positron pair. These two particles then enter the spectrometer. There, they travel in opposite directions along circular paths in the magnetic field. Each enters a crystal, making a flash of light which is registered by the RCA photomultiplier tube. With the betatron supplying a steady flux of gammas, the absorber could be remotely placed into and removed from the beam at regular intervals. The change in the counting rate in the spectrometer measures the rate at which the

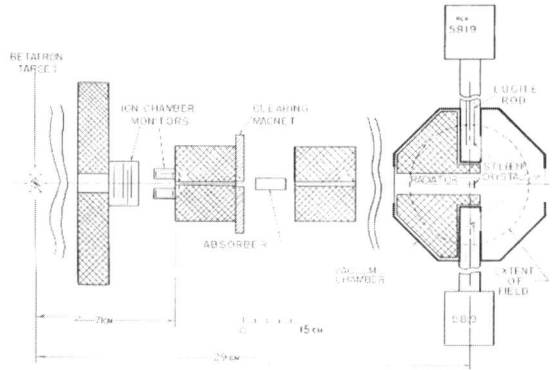

Fig. 7-9. Gamma ray absorption by metals. Pair spectrometer at right.

gammas are absorbed by the metal. The absorption cross-sections were measured for the four materials, each at three gamma energies, roughly to about one percent uncertainty. ("Absorption of 5.3-Mev, 10.3-Mev, and 17.6-Mev Gamma-Rays" *Phys. Rev.* **88** 612 1952, with grad students Earl S. Rosenblum and Raymond M. Warner, Jr.).

Shrader and Shankland both served as thesis advisors to W.H. Voelker (Case PhD 1952) who used the 15 MeV γ's from the betatron to look once again at Shankland's early interest, Compton scattering. The beam, detectors, and coincidence electronics were vastly better than those which Shankland had at his disposal in 1937 (Chapter 6), and the analysis significantly more sophisticated. The resulting differential cross-section compared favorably with the theory of Klein and Nishina. This is a more advanced version of the Compton formula which includes consideration of the electron and photon polarizations.

An interesting betatron experiment, with more of a nuclear-physics flavor, was one published in 1960 by Voelker and D.G. Proctor: the photodisintegration of ^{6}Li. The reactions ^{6}Li(γ,np)^{4}He, ^{6}Li(γ,n)^{5}Li, and ^{6}Li(γ,p)^{5}He were studied by measuring outgoing protons, neutrons, and proton-neutron coincidences. *(The notation: A(a,b)B means a hits nucleus A, and b flies away leaving nucleus B)*. The interest in this fairly simple nucleus stemmed from the possibility that it consists of an alpha particle core with an orbiting deuteron. If the incident γ interacts with the deuteron part, then "there should be a strong 180° correlation between the neutron and proton". The experiment did not show such a correlation. "Photodisintegration of ^{6}Li" *Phys. Rev.* **118** 217 1960; and "Photoneutron crossections of ^{6}Li and ^{7}Li" *Phys. Rev.* **113** 886 1959 by grad students Voelker and Tom Romanowski. Voelker remained in the department as a research associate, responsible for the continued operation of the betatron.

In another betatron experiment published in 1960, new assistant professor **Arthur Benade**, whom we shall meet again in Chapter 12, and his grad student, Robert Chrien, looked at the protons and neutrons ejected from aluminum and copper targets exposed to 20.8 MeV γ's. They found that photoproton and photoneutron rates were about equal at all energies in aluminum. In contrast, in copper the photoneutron rate rises from two to six times the photoproton rate as the γ energy is raised from 12 to 20 MeV. Aluminum has 13 protons and 14 neutrons, while copper has 29 protons and 34 or 36 neutrons. Many of copper's neutrons sit in an outer shell and are more easily knocked out of the nucleus by the gamma. ("Photoproton and Photoneutron Production in Aluminum and Copper" *Phys. Rev.* **119** 748 1960)

On a slightly different tack, it might be interesting to mention a paper by Shrader (*Nucl. Instru. & Meth.* **13** 177 1961) which describes a new electronic photomultiplier circuit which uses a *tunnel diode*. The significance is that up until this time, detector electronics were based on vacuum tubes, and this was an early step toward the use of "solid state" devices. We shall return to Shrader later in this chapter when the next generation accelerator is built.

Earle Gregg, between 1950 and 1960, provided research projects for eleven MS students, and four PhD students, mostly with betatron measurements. Both the detectors employed and the physics studied became progressively more sophisticated. A few examples will illustrate the development of the program:

1. The energies of electrons produced in aluminum by γ's were measured using a new magnetic spectrometer. The angular distributions for four electron energy ranges are shown in **Fig. 7-10**. ("Energy Spectrum of Electrons Produced in Aluminum by 17.8 MeV Bremsstrahlung", *Phys. Rev.* **102** 1 1956)

2. Another measurement was of the scattering of γ's in iron, with sodium iodide detectors; typical angular distributions are shown in **Fig. 7-11**. ("Scattering of High-Energy Gamma Rays", *Jour. Appl. Phys.* **27** 697 1956)

3. Gregg built and tested a magnetic spectrometer which used Compton scattered electrons to determine the energy and relative intensities of the betatron's photon beam. ("Bremsstrahlung Measurements with a Compton Electron Spectrometer" *Phys. Rev.* **105** 619 1957)

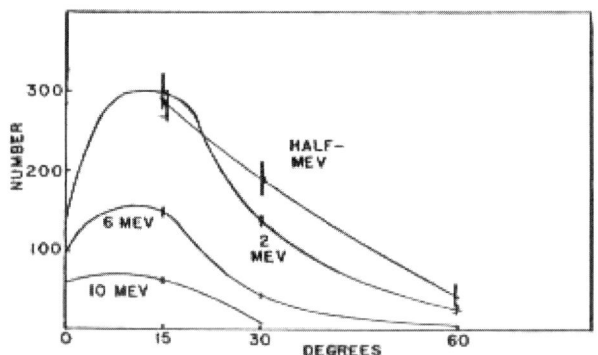

Fig. 7-10. Angular distributions for photoelectrons produced in aluminum.

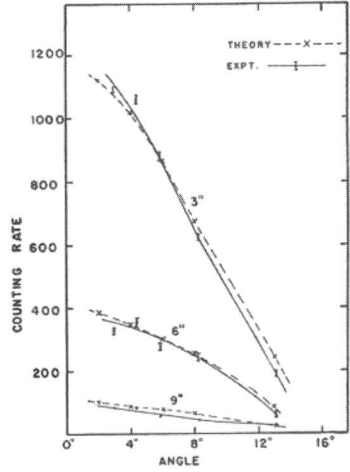

Fig. 7-11. Scattering of gamma rays by iron.

At about this time, chairman **Shankland**, who had become interested in reactor physics during his summers at the Idaho Materials Research Reactor, made a very different use of the betatron: a study of the fragments from the fission of ^{238}U nuclei induced by 17 MeV γ's. Nuclear emulsions were impregnated with the target uranium atoms, exposed to the collimated γ beam, and then scanned for photodisintegrations. About 1100 events were observed, predominantly having two outgoing tracks. The tracks were measured and the directions of the fission fragments (relative to the beam) were recorded. Shrader had used emulsions as detectors a few years earlier (see above), so the techniques of developing, scanning and measuring emulsions were available to Shankland. The resulting technical report states that the angular distribution was about 90% isotropic, with some indication of a $\sin^2\theta$ term, i.e. an excess at 90 degrees. Results like this became part of the vast store of experimental data on fission reaction rates and cross-sections.

Note that up to this point, all of the experiments done with the betatron involved beams of *gamma rays*, and the possibility of producing an external beam of high energy electrons had not yet been realized.

After fourteen years of research, mainly associated with the betatron, Earle Gregg, accompanied by William Voelker, left the department in about 1960 to work on the development of medical instrumentation at the CWRU School of Medicine.

Smith and condensed matter experiment

We back up a few years to describe the beginnings, in the 1940's, of the Case experimental condensed matter physics program, a discipline which would occupy about half of the department's researchers until the present time. This area of research has been variously called *solid state physics*, *condensed matter physics*, and *materials physics*.

Fig. 7-12. Chuck Smith and Polycarp Kusch.

Charles S. Smith, Jr. received his DSc from MIT in 1940. He had earned his BS at Case. His doctoral dissertation was on the determination, by x-ray diffraction techniques, of the crystal structure of iron-tungsten alloys. After a yearlong stay at the University of Pittsburgh, he joined the Case department in 1942 where he would remain until 1968. Smith is shown with alumnus Polycarp Kusch in **Fig. 7-12.** Smith's interest in crystal structure and the mechanical properties of solids laid the groundwork for materials research in the department. A comprehensive paper, written with R. L. Barrett of the Case geology department, appeared in 1947: "Apparatus and Techniques for Practical Chemical Identification by X-Ray Diffraction" (*Jour. Appl. Phys.* **18** 177 1947). The abstract ends with a reader-friendly promise: "The discussion is designed especially for the person who wishes to make use of this important analysis tool but who is not an expert in x-ray diffraction." They covered x-ray sources, cameras, films, sample preparation, measurements and interpretation.

Smith and his student, John R. Neighbours, developed a method for the determination of the elastic constants for cubic crystals by measurements of 10 MHz ultrasound wave velocities in the crystal. (*Jour. Appl. Phys.* **21** 1338 1950) The values of the velocities along the various crystal axes could then be used to determine the elastic constants of nickel (*Jour. Appl. Phys.* **23** 389 1952) and those of a more complex crystal, copper with 4% of its atoms replaced by silicon (*Jour. Appl. Phys.* **24** 15 1953). The work was supported by the Office of Naval Research and the "Case D. C. Miller Research Fund".

Why are such measurements interesting? Determinations of such things as the elasticity, compressibility, sound velocity, crystal structure, as well as the thermal, electric and magnetic properties of solids, provide the building blocks upon which condensed matter physics is based. Ultimately all these properties must be explained theoretically in terms of how the atoms and their electrons are arranged within the material and how they interact with one another. Even more important, new materials might then be de-

signed with whatever properties are desired. In Chapter 17, we shall describe the research of the department's theorists who have worked in parallel with the experimenters.

Smith continued his work with ultrasound techniques for over fifteen years, experimentally determining the elastic constants of a dozen metals and alloys and, in each case, comparing the results with current theoretical models. In the process he provided research projects for eleven doctoral and nineteen MS students, much of whose work was published in *Acta Metallurgica* or the *Journal of Physical Chemistry Solids*. Smith and one of his students are shown in their lab in **Fig. 7-13**. In a paper written with Case theorist John Reitz (who will be introduced in Chapter 17), Smith developed a theoretical description for the elastic shear constants for magnesium and magnesium alloys. (*Phys. Rev.* **104** 1253 1956) The elasticity of this close-packed hexagonal structure was found to depend on changes in the "electrostatic energy of the ion cores in the strained and unstrained geometry", as well as the "short range repulsive interaction of the ion cores". Introduction of small amounts of other metals changes the overall electron density, and consequently the interatomic forces. In addition, changes in temperature affect the electron densities, and thus the elasticities.

Fig. 7-13. Smith (right) and student with 1950's high-vacuum system.

In a 75 page-long review article in *Solid State Physics*, entitled "Macroscopic Symmetry and Properties of Crystals", Smith provides a comprehensive discussion of crystal properties such as resistivity (electric field causes current), pyroelectricity (heating causes electric polarization), piezoelectricity (stress causes electric polarization), elasticity (stress causes volume change), electric susceptibility (electric field causes polarization), and thermal expansivity (heat causes volume change). All these properties were examined along each axis of symmetry of each of several crystal types. (*Sol. St. Phys.* **VI** 175 1958) In Chapter 17 we shall look at these properties from the theorist's point of view.

Smith's papers on ultrasonic measurements of elastic properties of a wide array of crystal materials continue until 1968 when he leaves Case. This line of research was continued by Smith's student, **Donald Schuele**, whom we shall encounter in Chapter 12 as a member of the second generation of Case experimentalists. Smith left the department soon after the federation with Western Reserve, accepting a faculty position at the University of North Carolina at Chapel Hill.

and thin metallic films

A 1945 paper authored by three of our protagonists, **Olsen, Smith and Crittenden**, introduces a line of research which would be a major activity of the department for the rest of the century: the study of thin films. The paper is entitled "Techniques for Evaporation of Metals" (*Jour. Appl. Phys.* **16** 425 1945). The paper begins: "Evaporation techniques have long been used in the preparation of astronomical mirrors and more recently in the preparation of reflecting and non-reflecting surfaces for many uses. Such techniques, particularly those for evaporating metals, have recently assumed considerable importance because of the scientific utility of thin metallic films and because of the scientific interest in the properties of such films." They describe their vacuum system (pressure below 5×10^{-5} mm Hg in a bell jar), the first in a six-decade-long sequence of ever more sophisticated vacuum systems to be used in thin film, surface, and optical materials research in the department. In a paragraph for each of 34 metals, from aluminum to zirconium, the writers describe their observations and make their recommendations for the optimum technique to create metallic vapors and films.

A related paper by the same three authors describes a device for measuring the magnetic properties of very small samples of thin metallic films. (*Rev. Sci. Instr.* **17** 372 1946) (A footnote mentions that the work had been done in 1943 under contract to the government. It was thus one of several Case research papers whose publication was delayed until after the war for security purposes.) Crittenden extended this work on magnetism in thin films (*Rev. Sci. Instr.* **22** 872 1951) in a paper in which he acknowledges help from graduate student **Richard Wagner Hoffman,** who will later become the major player in the thin films and surface physics program. (Chapter 12)

One of Olsen's undergraduate students at this time, a commuter student from Cleveland Heights, was Donald Arthur Glaser who wrote a 1946 BS thesis on some early experiments in surface physics: "Metal Surface Studies by Electron Diffraction". Glaser would receive the Nobel Prize in 1960 for his invention of the bubble chamber.

The use of electron beams as a probe of the structure of thin films would be further developed by Crittenden's student, E. I. Halteman: "Electron Diffraction Evidence for the Existence of Microstress in Evaporated Metal Films" (*Jour. Appl. Phys.* **23** 150 1952).

Crittenden and Hoffman, now his research associate, turned in earnest to the fabrication of smooth, uniform, very thin films: as thin as a few atomic layers. They worked on techniques to minimize impurities at the surface and to spread the material evenly across the glass substrates. From their paper, "Thin Films of Ferromagnetic Materials" (*Rev. Mod. Phys.* **25** 310 1953): "It is recognized that evaporated metal films have acquired a bad reputation as far as being specimens of metal which may be directly compared with pure bulk specimens of the same material." They comment on surface impurities: "Connected with this is the lack of true protection offered by vacuum of the order of 10^{-5} mm of mercury, for which each surface atom is hit approximately once every half-second by a residual gas molecule." The properties of interest are the electrical resistance and magnetization, and their temperature dependence for each type of film. From the theoretical point of view, these properties depend on the disposition of the electrons in the film, a configuration clearly different from that in the bulk material. From the applied physics point of view, these properties determine the applicability of these films to electronic devices and other technologies. In this case, the metal studied was nickel, as pure as they could obtain. **Fig. 7-14** shows, for example, how the extent to which a film can be magnetized drops off rapidly as the film thickness is reduced to a dozen or so atomic layers. Films thinner than this are difficult to magnetize because they apparently break up into disconnected islands of atoms.

Crittenden and Hoffman did further experiments to learn how the atoms in metallic thin films arrange themselves. These involved measurements of electrical resistivity and of the mechanical stress within the films. (*Jour. Appl. Phys.* **24** 231 1953) They looked at the residual stress in the films as a function of the temperature at which the film is laid down, and of the temperature and amount of annealing. (The techniques for the measurement of stress will be described in Chapter 12.) From this, one can learn about the role of vacancies in

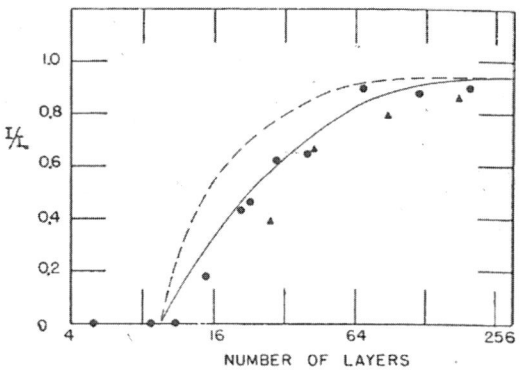

Fig. 7-14. Thin nickel film magnetizability vs. thickness.

the film and the tendency of the atoms to coalesce. More explicitly, the abstract of a paper submitted to the *Journal de Physique et le Radium* provides a compact description of the types of things they were looking at. "Purity of the material, protection from chemical reaction with residual gas after deposition, crystal size and preferred orientation, crystal imperfections, mechanical stress, roughness and agglomeration are discussed. The transition metals are well behaved, the noble metals poor, and Zn, Cd, and Hg very difficult to deal with as regards producing smooth surfaced films down to a few atoms thick." (*Jour. Phys. et Rad.* **17** 179 1956) The goal was to create high quality uniform thin films. Grad students Harold Story and Ned Razor worked on the development of techniques to quantify lattice defects and stress caused by annealing. Hoffman and an army of grad students and post-doc research assistants continued this work on thin films for several decades, as will be described in Chapter 12.

Eugene Crittenden left the department around 1956, taking a faculty position at the Naval Postgraduate (NPS) in Monterey, CA. At Case he had directed the research of nine masters and three doctoral students. In his new position, he would initiate a research program in electro-optics and other areas of special interest to the Navy, such as infrared search and tracking systems and optical atmospheric turbulence studies.

Four years later, **Leonard Olsen** resigned from Case and joined Crittenden in Monterey, where he established a physics baccalaureate program. Olsen devoted his efforts to improving the teaching of physics, including a term as president of the AAPT (American Association of Physics Teachers). Both Olsen and Crittenden were active in faculty governance at the NPS, each serving as president of the AAUP (American Association of University Professors) chapter. Olsen retired from the NPS in 1975 and lived until 1981.

The Case van de Graaff

In 1961, **Erwin Shrader** spent some time at the High Voltage Engineering Corporation's plant in Massachusetts. This company's principal product was electrostatic particle accelerators, specifically van de Graaff accelerators.

*The **van de Graaff** machines were capable of producing external beams of protons or other charged ions with energies of millions of electron volts. They were simply large versions of the popular demonstration machines which thousands of high-school kids have used to make their hair stand on end. An endless belt wrapped around two pulleys is driven at high speeds. A set of conducting whiskers gently brush the belt as it sweeps by, and electrons jump from the belt to the brushes, leaving the belt positively charged. As the other end of the belt passes near a second set of brushes, electrons jump from the brushes to the belt. That leaves a big positive charge on the large metal dome attached to these brushes. The resulting electric potential difference between the dome and ground is used to accelerate protons (i.e. hydrogen ions) or other ions through an evacuated pipe, whence they are directed out of the machine by a bending magnet.*

Shrader's visit to HVE was no doubt connected with plans by Case to buy one of these machines. While there, he co-authored a paper entitled "Production of High Intensity Ion Pulses of Nanosecond Duration". (*Nucl. Instruments and Meth.* **12** 335 1961). This paper described a rather sophisticated set of dipole bending magnets, quadrupole focusing magnets, and radio-frequency electric fields which operate on the ion beam after it exits the van de Graaff. The arrangement delivered 3 MeV "ion bursts of less than one nanosecond duration at a peak current of several milliamperes." The system could also be modified to produce shorts bursts of *neutrons* from the reaction ^{55}Mn(p,n)^{55}Fe. The pulsed nature of the beam would allow the measurement of the lifetimes of nuclear states produced in beam-target collisions because the experimenter would know both the time at which the state was produced and the time it decayed. **Bob Leskovec**, an engineer hired in 1967 as staff physicist, was responsible for the design of the electronics related to the timing of the beam. He is still (in 2005) a member of the university's technical staff; he describes the project in a recent memo: "My first task in 1967 was to develop a "stop-signal" synchronizing system to get a stable lock to the beam pulse for the time-of-flight technique. This was so exciting because we were just at the transition from vacuum tube technology and I was able to pull it off with an entirely "solid state" design using some new devices called tunnel diodes".

Making room: the new wing

Before the van de Graaff could be installed, suitable space had to be prepared for it. This involved a three story high hall for the machine whose belt ran vertically, and large underground experimental areas enclosed by thick concrete walls. This was the opportune time for the Case department to build a large addition to Miller's Rockefeller Building. A three-story brick structure, 75 by 90 feet, was built south of Rockefeller, with its central hallways on each floor placed end to end with those in Rockefeller. The new building had large research laboratory spaces in the basement and three stories of offices, research labs, teaching labs, and class-room spaces above. The new accelerator space was incorporated on the basement and ground-floor levels. An adjacent 600 seat auditorium (named Strosacker) for use by the whole institute was part of the same construction project, and its "basement" area of 75 by 110 feet was added to the physics laboratory space. **Fig. 7-15** shows three views of the construction, with emphasis on the large concrete shielding surrounding the van de Graaff area. (The new building would cover up seven of the physicists' names which were placed over the windows of the old building by Miller in 1905.)

The Case machine was up and running by 1960, with the expert help of engineer Lawrence Hinkley. He is shown, with hand in pocket, checking out the external beam electronics in **Fig. 7-16**. **Fig. 7-17** shows the beam exit ports and part of one of the larger detectors. Larry would become known as "the only person who could get the van de Graaff to run".

Fig. 7-15. Construction of the south wing of Rockefeller.

An early experiment was designed to study a narrow excited state in the calcium nucleus. Protons from the van de Graaff were incident on a target of ^{39}K nuclei. The K nucleus absorbed the proton, becoming ^{40}Ca in an excited state. These excited nuclei then returned to the ground state by the emission of a photon. These photons were then directed at other Ca nuclei which could absorb them, and become raised to the same excited state. However, these photons in general had a bit less energy than the required excitation energy, because the nucleus which emitted them stole some away as recoil kinetic energy.

Fig. 7-16. Larry Hinckley and the van de Graaff.

Now here is the tricky part: the energy of the photon would be boosted (Doppler shifted) to the required level if the nucleus which emitted it were moving with just the right speed in the same direction as the photon. The excited Ca nucleus was indeed moving, having been struck by the proton. If one chose all the angles correctly, the photon could have exactly the right energy to be reabsorbed by the second Ca nucleus, thus being removed from the beam. This technique allows one to measure precisely the energies and the widths (and lifetimes) of the excited state – important data for checking nuclear models. In this case, the excitation energy was 10.3 MeV and its width, coincidently, one-millionth of that, or 10.3 eV. For a state with such a narrow width, the re-absorption could never have taken place without the Doppler trick. (*Phys. Rev.* **124** 1541 1961 with grad student Alan C. Eckert) This technique of observing resonant absorption by moving the target nucleus or, in this case, the source nucleus would become widely used in the study of very narrow states.

Fig. 7-17. Beam chopper at exit from van de Graaff.

Pulsed MeV beams

"A Nanosecond Pulsed Accelerator Facility" is the title of a paper by Shrader *et al.* in *"Nuclear Electronics II"*, a compilation by the International Atomic Energy Agency, Vienna 1962. The ability to produce very short bursts of protons is essential for experiments in which the protons are used to generate neutrons (as mentioned above in the ^{55}Mn(p,n)^{55}Fe reaction), and where the energies of the neutrons are determined by time-of-flight, i.e. the time it takes them to travel a given distance in the laboratory.

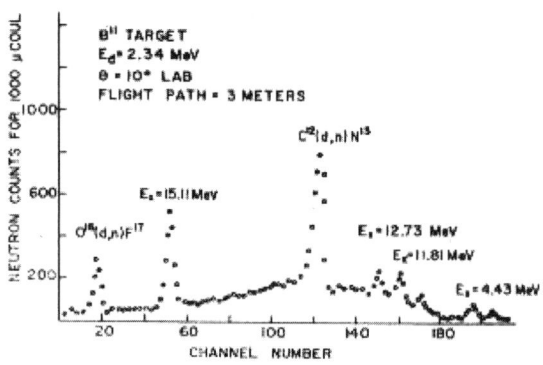

Fig. 7-18. Excited states in carbon 12.

It is amazing how many different ways our players found to use the van de Graaff. As an example, they wanted to study the excited states in the ^{12}C nucleus. Their approach is rather round-about and is not the first one to come to mind. For one thing, they had experience with detecting and measuring the energies of neutrons. For another, the van de Graaff could accelerate deuterons as well as protons. The deuterons were incident on a target of ^{11}B; then, in a "stripping" reaction, the proton was left behind in the B nucleus, making it ^{12}C, and its faithful companion neutron continued on its way alone. The measured energy of this neutron determines the total energy (or mass) of the residual carbon nucleus, thus giving the excitation energies in the carbon. (I have always been fascinated by the fact that when you break up a deuteron like this, you separate the proton and neutron which have been partners for fourteen billion years or so.)

The experimental results could be checked by varying the energy of the incident deuterons (1.6 to 2.7 MeV). In addition to observing well-defined excited states, as illustrated by the sharp peaks in **Fig. 7-18**, they found that many neutrons came from collisions in which the target nucleus was entirely disrupted, with pieces flying out in all directions. This experiment, as earlier ones, relied on the short-pulse capabilities of the accelerator. ("Neutron Producing Reactions by Deuteron Bombardment of B^{11}", *Phys. Rev.* **129** 1275 1963) A follow-up experiment added the detection of the γ's coming from the excited ^{12}C, in coincidence with the arrival of the neutron. ("Angular Correlation in the B^{11}(d, nγ 1511 MeV)C^{12} Reaction" *Phys. Rev.* **131** 2594 1963 with grad student Hee J. Kim.)

Detecting the neutron

In experiments at the van de Graaff in which secondary neutron times-of-flight were measured, the arrival of the neutron was usually detected by the observation of a recoil proton in a scintillator. In all such materials, there are both hydrogen and carbon nuclei, and, in order to determine absolute cross-sections, it is necessary to identify those events in which the neutron scattered off a hydrogen nucleus. Shrader and his students looked into this problem and found that if one measures the intensity of the light produced in the scintillator, the proton events, which produce much more light, can be isolated. **Fig. 7-19** shows the scattered neutron times-of-flight, with (the upper plot) and without (the lower plot) the requirement of a bright flash of light. The peak on the left side is from proton events. The separation is quite good. *Nucl. Inst. and Meth.* **85** 151 1970 "Absolute Normalization of Neutron Scattering Cross Section Data Using Organic Scintillators as Scatterers."

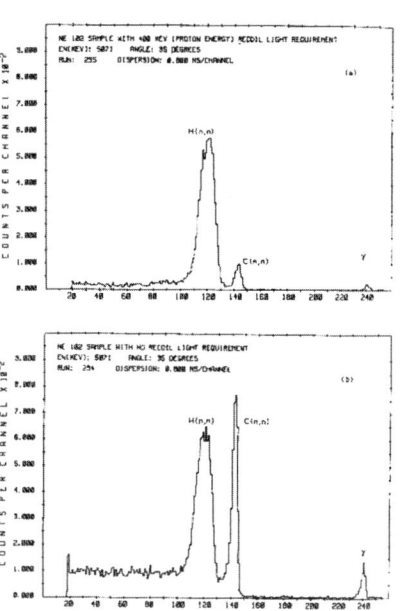

Fig. 7-19. Neutron time-of-flight plots showing separation of H from C events.

Between 1947 and 1970, Shrader advised 12 doctoral students, 18 masters students, and a large number of bachelors degree students, including Richard Garwin who, for the past 40 years, has been one of the country's most prominent physicist advisors to government on nuclear arms issues. Considered in some respects to be the "father of the hydrogen bomb", Garwin is one of Case's most illustrious graduates. Another of Shrader's BS students was Jonathan Reichert whom we shall meet in Chapter 14 when he joins the CWRU faculty. Erwin Shrader left the department in 1970 for a position at Harshaw Chemical Company, a manufacturer of the scintillator materials used in particle detectors.

Scharenberg and Silverstein and the van de Graaff

Fig. 7-20. R. P. Scharenberg.

Rolf Paul Scharenberg joined the department in 1961 having earned his doctorate at Michigan in 1954. **Fig. 7-20.** While he remained at Case for less than five years, he completed a series of measurements of the magnetic moments of nuclear excited states, continuing work initiated by Shrader. Recall that the van de Graaff accelerator delivers 2 MeV protons in nanosecond pulses. As these pass through a tungsten target, some ^{184}W nuclei are excited from the spin 0 ground state to an excited rotational state with spin 2. An external magnetic field of around 40 kilogauss causes these nuclei to precess. As they do so, they decay exponentially (mean life 1.85 10^{-9} s) to the ground state, emitting a 111 keV gamma ray. The outgoing γ's are detected at various angles in the plane of the γ and incident proton. The time at which the nucleus is excited is marked by a signal from the incident proton pulse, and the timing of the hits in the γ counter can be unfolded to give a rate of precession, and therefore, a measure of the magnetic moment of the excited state. Knowledge of the magnetic moment of the excited nucleus allows one to determine the relative contributions of the protons and neutrons in the collective flow within the nucleus. (*Phys. Rev.* **137** B26 1965) The sketch in **Fig. 7-21** is from the PhD thesis of Paul J. Wolfe. In 1969, after five years at Case, and having been promoted to associate professor with tenure, Scharenberg decided nevertheless to pursue his nuclear properties research at the Angular Correlation Laboratory at Purdue University.

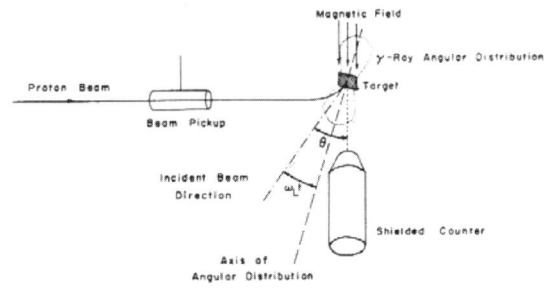

Fig. 7-21. Setup to measure magnetic moments of nuclear excited states.

Fig. 7-22. E. A. Silverstein.

Chairman Reines clearly wanted to get the most out of the van de Graaff program. The Case department added another young nuclear experimentalist in 1964. **Fig. 7-22**. **Edward A. Silverstein** had completed his PhD in 1960 at the University of Wisconsin under H. T. Richards. His thesis was on ^{4}He ^{14}N and ^{3}He ^{16}O elastic scattering. The object was to detect bumps in the cross-section which correspond to excited states in ^{18}F and ^{19}Ne. Subsequently, he spent 2½ years at the University of Padua where he helped install a new van de Graaff accelerator. He would be very useful in the operation and utilization of the Case machine. Silverstein had developed expertise in experiments involving the scattering from gas targets, and he set up an experiment to study proton-proton bremsstrahlung at 3.2 MeV. This required the identification of the small number of events corresponding to pp → ppγ in a sea of elastic scattering events (a few dozen out of 400 million). "Proton-proton bremsstrahlung at 3.2 MeV" *Phys. Rev. Lett.* **21** 922 1968.

Silverstein would later work with Phil Bevington in a search for a three-neutron bound state. This work is described in Chapter 16. In 1969 Silverstein took a position at the radiology department Milwaukee County General Hospital where he would apply his expertise in particle detection.

With the departure of Olsen, Crittenden, Gregg, Smith and Shrader, the first major phase of experimental physics at Case came to an end. While Hoffman and Schuele continued the thin film and the materials programs, the pendulum would begin to swing toward subatomic research.

Chapter 8   Cosmic Rays

Crouch,	Reines,	Frye,	Jenkins,
1952-1987	1959-1966	1960-1993	1960-1995
Woods,	Wang,	Albats,	Koga
1964-1970	1966-1970	1973-1978	1974-1980

In the 1930's and 40's, discoveries in cosmic rays were opening a brand new area of physics: "high energy" or "particle" physics. While there were already accelerators which reached energies of millions of electron volts, high enough to probe nuclei, the particles incoming from space had energies of billions of electron volts. The equivalent accelerators would not be built until the 1950's. In the intervening slice of time, it was the people doing cosmic rays who were making all the exciting discoveries. Flying in balloons and climbing to mountain tops, this athletic group of researchers set up their detectors - Geiger counters, cloud chambers, nuclear emulsion stacks - to catch glimpses of a whole new zoo of particles.

Crouch: cosmic rays and neutrons

Fig. 8-1. Marshall Crouch.

Marshall F. Crouch did his BS at the University of Michigan, put in a three-year stint in the army, and completed his PhD in 1950 at Washington University St. Louis under advisor Robert D. Sard. He did post-docs at Michigan and Harvard before becoming assistant professor at Case in 1952. A photo taken in the 1980's is shown in **Fig. 8-1**.

Among the particles discovered in cosmic rays were the new mesotrons (now called mesons), charged particles with masses between those of the electron and the proton. These were produced by collisions of high energy primary cosmic rays with the atoms of our atmosphere. At Washington U., Crouch *et al.* published two papers on the observation of secondary neutrons caused by incoming cosmic ray mesons. (*Phys. Rev.* **74** 97 1948, and **76** 1134 1949) The setup is shown in **Fig. 8-2**. The experiment made use of new developments in fast electronics which allowed one to count events in which two or more Geiger counters were hit within a well defined time interval. (These were called "coincidence" counters. Earlier versions were used by Shankland in the 1930's, with mixed results as described in Chapter 6.) In this case, the incoming particle must traverse 8 cm of lead (that is, it must be a penetrating µ meson), give a count in both the A and B trays of Geiger counters and plough into a second layer of lead. In addition, *no* hits must be observed in

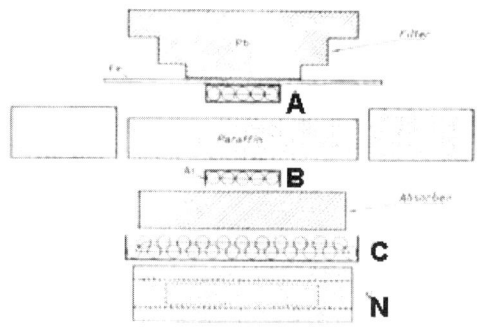

Fig. 8-2. Cosmic muon telescope.

the C tray, indicating that the incident meson stopped in the lead. There it is captured by a positive Pb nucleus, which soon breaks up. Among the fragments are neutrons. The neutrons are detected after they enter a block of paraffin encased special neutron counters (N). The paraffin, which contains many hydrogen atoms, acts as a moderator slowing down the neutrons. The slow neutrons are then absorbed by ^{10}Boron nuclei coated on the counters. These nuclei then fission, giving a characteristic signal in the counter. With the fancy fast electronics, the experimenters could ask for A yes, B yes, C no, N yes.

Cosmic muons as probes of nuclei

Crouch continued this work after he arrived at Case in 1950, with two major improvements. To minimize background counts, part of the running was done in the sub-basement of the St. Louis phone company (and later in a deeper location under a local brewery); and the neutron detectors were replaced by boron-fluoride counters which were twice as sensitive as the earlier counters. The emphasis of this work was the "neutron multiplicity", *i.e.* how many neutrons were emitted when a μ meson breaks up a Pb nucleus. There were theoretical models for this process, one based on a charge exchange with a single proton, the other based on a general explosion of the nucleus. They predicted respectively roughly 1 and 6 neutrons per capture, and, as so often happens, the experimental result fell right in between: 2.16 ± 0.15. (Grad student **Arthur H. Benade**, who would later join Crouch on the Case faculty, is thanked for calibrating some of the counters.) (*Nuovo Cim.* **VIII** 1 1951) and "Distributions of Multiplicities of Neutrons Produced by Cosmic-ray mu-mesons Captured in Lead" (*Phys. Rev.* **85** 120 1952).

Crouch and Washington U. collaborator, Sard, put together a fifty-page review article on the subject of "Nuclear Interactions of Stopped μ-Mesons". It was published in the 1954 issue of *Progress in Cosmic Ray Physics*. (Vol. II. North Holland Publishing, Amsterdam 1954) This paper provided an important summary of work-to-date, with over 100 references, experimental and theoretical. In the first part they report results from experiments similar to their earlier work. As in those experiments, cosmic ray muons are identified by traveling through a slab of lead (only muons could make it through) and then brought to rest in a sample of the material being studied. The neutrons which are then expelled from the affected nuclei are detected, and the delay between the muon time of arrival in the sample and the time of the neutron detection is recorded. From these data, the muon capture rate can be determined. Some examples: the survival time τ for a muon in beryllium (Z=4) is 2.05 ± 0.06 μs, for copper (Z=29) 0.116 ± 0.009 μs, and for lead (Z=82) 0.076 ± 0.004 μs. The data are well fitted by $1/\tau \sim Z^4$, that is, the large positive nuclei snap up the negative muons very quickly.

In the second part of their paper, Crouch and Sard review the data taken from nuclear emulsion and cloud chamber experiments, in which one may observe directly the charged particles emitted by nuclei which have captured a muon. They mention that beams of muons are just becoming available at the new accelerators (which would soon take over muon research from cosmic rays). An interesting feature of their discussion is the hypothesis of the existence of a neutral particle produced in the reaction $\mu^- p \rightarrow n \mu^\circ$. A few years later, the μ° would be identified with the μ-type neutrino, ν_μ. In their conclusions: "…the strength of the coupling between μ-mesons and nucleons is the same order of magnitude as that between electrons and nucleons (β-decay) and that between μ-mesons and electrons (μ → e decay). This agreement is suggestive of some deep relationship that will emerge in a future theory of the fundamental particles." And so indeed it has! This was an early indication of what is known today as the "weak interaction".

Crouch spent the summer of 1956 at Argonne National Laboratory where he participated in a very different type of experiment. It concerned, somewhat indirectly, the interaction between the electron and the neutron, specifically one beyond the expected interaction between the electron's charge and the neutron's magnetic dipole moment. Other experimenters had reported observing such a new interaction. In 1951, Les Foldy (who will be introduced in Chapter 9 as the first theoretical physicist at Case Institute) had published a short note on this subject. (*Phys. Rev.* **83** 688 1951) Foldy pointed out that, in addition to effects which may be caused by the neutron's transformation to virtual π^- - proton states, one could also show "that an electron-neutron interaction of the desired character and magnitude can also be obtained as a direct consequence of attributing to a neutron an anomalous magnetic moment in the manner suggested by Pauli without any further assumption." The determination of the strength of the experimentally observed interaction depended on knowledge of the coherent scattering amplitudes for neutrons on krypton and xenon. It was these quantities which were measured at Argonne by Crouch and his collaborators. Their experiment involved scattering thermal neutrons from the surface of liquid Kr and Xe, both monatomic entities with lots of electrons. Their results, when combined with earlier measurements, showed that the electron-neutron interaction was compatible with that calculated by Foldy. "Coherent Neutron Scattering Amplitudes of Krypton and Xenon, and the Electron-neutron Interaction" (*Phys. Rev.* **102** 1321 1956)

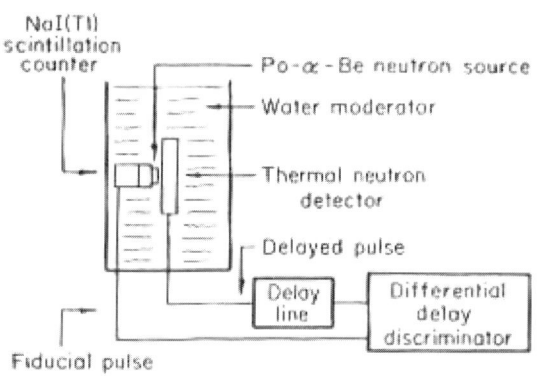

Fig. 8-3. Slowing neutrons in water.

Crouch and his student, Robert Stooksberry, turned their attention to the question of how quickly energetic neutrons would be slowed down in water – a question of particular interest to designers of nuclear reactors. The sketch in **Fig. 8-3** shows the experimental setup. In the "Po-α-Be" neutron source, a polonium nucleus spontaneously emits an alpha, the alpha fissions a beryllium nucleus from which a neutron and simultaneous 4.4 MeV γ are emitted. The γ is detected first, indicating the time of the birth of the free neutron. The neutron bounces around in the water, losing energy until it is traveling slowly enough to interact in the boron fluoride detector. The delay time then is a measure of the time it takes for the neutron to slow down in water. In this experiment, two large, water-filled coaxial cylinders are used to facilitate the corrections needed for neutrons escaping from the sensitive volume. They found that the mean lifetime for neutrons in water is 206.3±5.0 μs, corresponding to a neutron proton capture cross section of

0.330±0.008 10^{-24} cm^2. (*Nucl. Sci. and Engr.* **2** 626 and 631 1957, ibid. **6** 545 1959); "Neutron-Proton Capture Cross Section" (*Phys. Rev.* **114** 1561 1959).

Reines takes over: neutrino physics

Fig. 8-4. Chairman Fred Reines

In 1959, Shankland decided to step down after 15 years as department head, and a committee made up of Erwin Shrader, Martin Klein and Les Foldy was selected to conduct the search for a new chairman. An outside candidate, 41-year-old **Frederick Reines** was selected. (PhD New York University, 1944) He was an experimental particle physicist at Los Alamos National Laboratory. **Fig. 8-4**.

Reines had just published a paper with Clyde L. Cowan on their observation of neutrino induced reactions at a nuclear reactor. (*Phys. Rev.* **113** 273 and 280 1959) *The chargeless, possibly massless, neutrino ("little neutral one") had been hypothesized two decades earlier as the missing player in the decay of the neutron (n → p e⁻ ν), but the neutrino had never been directly observed.* Reines and Cowan devised a way to see the related inverse reaction (vbar p → n e⁺) initiated by an antineutrino from the high flux Savannah River nuclear reactor (*vbar* is the antineutrino). They named their search for the elusive neutrino "Project Poltergeist". The key to the experiment was the observation of both outgoing particles, the neutron and the positron. They detected both γ's from the positron annihilation (e⁺ e⁻ →2γ) and subsequently a third γ coming from an excited cadmium nucleus which had captured the neutron. (The neutron was slowed in a water bath, and the cadmium was in the CdCl$_2$ dissolved in the water.) **Fig. 8-5** shows a schematic of the process. Their result for the cross section for antineutrinos interacting with protons is σ = 11 ± 2.6 10^{-44} cm^2. *This is about 10^{18} times smaller than the cross-sectional area of a proton. With such a small probability for interaction, the antineutrino (and neutrino) can easily fly through the earth, or even through the sun, without interacting.*

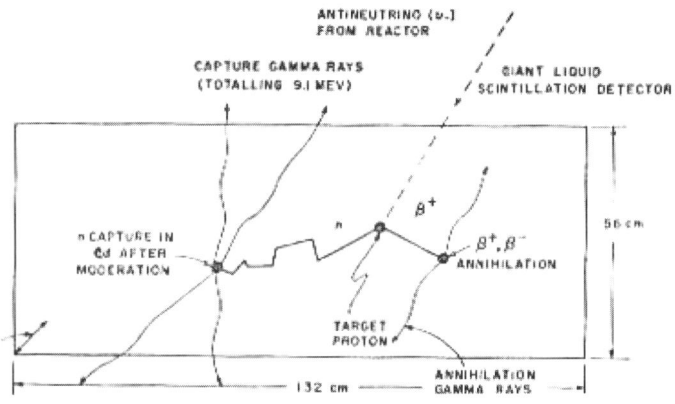

Fig. 8-5. Detecting reactor neutrinos.

A later paper (*Phys. Rev.* **117** 159 1960) reported a refinement of the experiment, including better shielding and comparisons of rates with the reactor turned on and off.

Reines was a great catch for Case, though at the time, who knew that his neutrino work would lead to a Nobel Prize 35 years later? He was to continue his chase after the obscure neutrino while at Case, and for many years afterward. As a, perhaps **the**, recognized expert on experimental neutrino physics, Reines published a review of the field. (*Ann. Rev. of Nucl. Sci.* **10** 1 1960) In it, he describes his earlier results, his plans for an experiment to measure neutrino scattering by electrons, the work by Ray Davis at Brookhaven National Laboratory which showed that the neutrino and antineutrino are two different species, and even some proposals to use nuclear explosions as one-shot "pulsed" neutrino sources. He then discusses what might be done at the new high-energy accelerators, correctly predicting the construction of intense neutrino beams. Neutrinos from space and from interactions of cosmic rays are also proposed as a suitable area for study, something which Reines and his Case colleagues would soon be undertaking. In his discussion of solar neutrinos, he even mentions reactions on the deuteron (currently being exploited at the Sudbury underground neutrino detector in Ontario). In effect, the paper provided a roadmap for the accelerator and cosmic ray neutrino work which would be tackled worldwide over the following forty years.

Reines and proton decay

Reines' interest in hard-to-detect reactions was not limited to those involving neutrinos. As early as 1954 he had participated in an experiment with Maurice Goldhaber in a search for evidence of proton decay. By looking for counts in large detectors placed far underground, shielded from cosmic rays, they were able to set a limit on the lifetime of the proton of greater than 10^{22} years. The idea is that if a proton decays, its positive charge must be carried off by a positron or some other light positive particle, and that this would give a signal in a detector. Therefore, in 1961 at Case, Reines and his first graduate student, Charles C. Giamati, set up a detector 585 meters underground at the Morton Salt Mine under Lake Erie about 30 miles east of Cleveland. (Subsequent similar "low-background" experiments, many involving Reines, would take place at this site for the next thirty-five years.) The 1961 experiment featured a 200-liter liquid scintillation counter inside an iron housing surrounded by a large cylindrical water Čerenkov anti-coincidence counter to eliminate the residual incoming cosmic rays. (A hit in an anti-coincidence counter vetoes the coincidence signal.) A schematic of the setup is shown in **Fig. 8-6**. Knowing the number of protons in the tank, the length of time the detectors were activated, the efficiency of the detectors, and the rate of background counts, they could place a limit on how often one of their protons decays. The new lower limit on proton decay, 10^{26} years, was ten thousand times longer than that reported in the earlier work. (*Phys. Rev.* **137B** 740 1965)

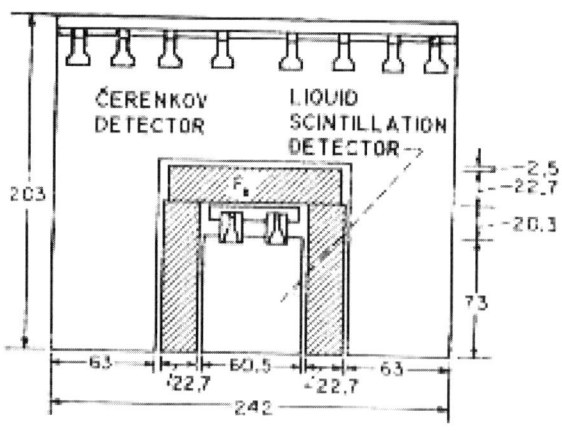

Fig. 8-6. Deep mine proton decay experiment.

Expanding the "rare event" program: Jenkins joins Reines

A year after the arrival of Fred Reines, two new experimental physicists were added to the department. Along with Reines and Crouch, they made up a respectable "particle physics" group. **Thomas L. Jenkins** completed his PhD at Cornell in 1956. He would begin at Case by joining Reines in the study of neutrinos. Jenkins' work will be described both in this chapter and in the chapter on accelerator physics. There is a photo of Tom in **Fig. 16-1**. The second new person was **Glenn M. Frye** whom we shall introduce later in this chapter.

As a graduate student at Cornell, Jenkins worked at the 300 MeV electron synchrotron, writing his dissertation on electron-positron pair production. Subsequently he spent five years at Lawrence Livermore Laboratory working on application of shock hydrodynamics to nuclear weapons design. At Case, in 1960, he soon began work with Reines on a new neutrino experiment at the Savannah River reactor. This time, they were looking at the interaction of neutrinos with the deuteron. While the inverse beta decay reaction $\nu_e \, p \to n \, e^+$ had been studied previously (by Reines and Cowan), $\nu_e \, d \to n \, n \, e^+$ had not. Although both of these reactions turn protons into neutrons, the second has two neutrons in the final state. The Pauli principle requires that these must have opposite spins. Thus the e^+ retains the angular momentum of the incoming ν_e. The proton reaction can involve either of the two orientations of the neutrino angular momentum and so is a mixture of Fermi (e^+ and ν_e spins opposite) and Gamow-Teller (e^+ and ν_e spins parallel) interactions, while the deuteron reaction is pure Gamow-Teller. The reaction was observed in a 97 liter deuterated organic scintillator mixed with a gadolinium compound for the neutron capture. The calculation of the rates was quite complicated as it had to include the efficiencies for picking up photons from the capture of *both* neutrons and from the annihilation of the positron. The resulting cross-section ($3.0 \pm 1.5 \; 10^{-45}$ cm^2/fission antineutrino) agreed reasonably well with the predictions of theory. (*Phys. Rev.* **185** 1599 1969)

Back to the saltmine: underground in Ohio

Back at Case, Jenkins and Reines designed and tested a new liquid scintillator anti-coincidence guard. The plan was to use this setup in a new neutrino experiment at the Ohio salt mine. Neutrino induced events are extremely rare relative to spurious background events, even deep underground. A prototype setup was placed in a hole in the floor of the basement of the physics building under 88 cm of concrete and a two-meter stack of iron ore. **Fig. 8-7** shows the location and the amount of covering material. In the test setup, the main detector consisted of 50 liters of scintillator and two photomultipliers. The anti-coincidence guard which was eventually installed in the mine had

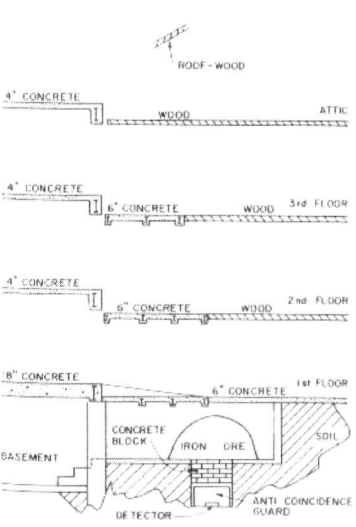

Fig. 8-7. Anticoincidence test setup on campus.

16 photomultipliers looking at 1500 liters of liquid scintillator. **Fig. 8-8** shows engineer Gus Hrushka with the containment tank. The number of counts in the main detector is reduced by a factor of ten thousand when the anti-coincidence guard is switched on. (*Rev. Sci. Instr.* **35** 370 1964)

Fig. 8-8. Engineer Gus Hrushka and containment vessel.

The low-level background available at the nearby Morton salt mine invited a different type of experiment: the search for "double beta decay". The nuclear decay in which two neutrons simultaneously emit electrons, $^Z A \to {}^{Z+2}A + 2e^-$, had been sought by many experimenters. It had been proposed that if the neutrino were its own antiparticle, then this neutrinoless process may very well proceed. In double beta decay, two electrons would be created without any accompanying neutrinos.

Reines had published a paper at Los Alamos in 1956 on an unsuccessful attempt to detect the double beta decay of neodymium, $^{150}Nd^{60} \to {}^{150}Sm^{62}$. At Case, he suggested to Tom Jenkins that he give it a try, this time in the low-background environment of the salt mine. Graduate student Larry V. East assembled his experiment with an anti-coincidence guard similar to the one tested at Case. Because the two electrons in a neutrinoless decay carry away all the kinetic energy, their total energy would be a fixed amount. East used three sets of large scintillators and recorded the pulse-heights associated with two-electron coincidences. He was able to report a lower limit for the lifetime against neutrinoless decay of 5 10^{18} years. (*Phys. Rev.* **149** 913 1966.) In Chapter 10 we'll see that Rolf Winter at WRU had also searched for double beta decay, ten years earlier. Even in 2005, the search goes on in many laboratories.

A second of Jenkins' students, Gary R. Smith improved upon the double β decay experiment. He built a two-arm magnetic spectrometer **(Fig. 8-9)** to direct the electrons from the neodymium source to the detectors, thus eliminating most of the background counts. However, once again only a lower limit could be determined, essentially the same as that reported by East. (*Phys. Rev.* **C4** 1344 1971.) Even in the low-background salt mine, natural radioactivity of the surroundings and the equipment itself place a limit on how small a rate can be observed.

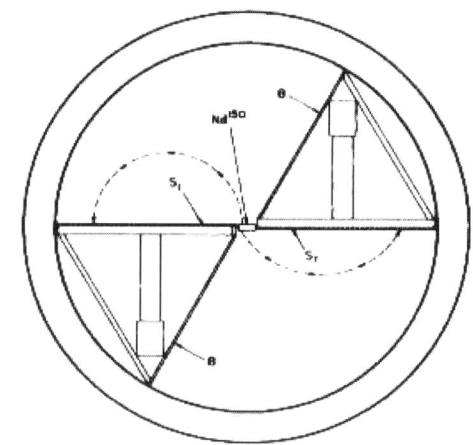

Fig. 8-9. Double beta decay spectrometer.

Neutrinos from the sun

Jenkins, with his students, Fred Dix and Larry Levit, set up another experiment in the saltmine: a search for solar *neutrino* interactions on the deuteron. This was quite different from the earlier experiment at Savannah River where *antineutrinos* from the reactor caused inverse beta decay of the proton in the deuteron. Here, the search was for *neutrinos* from the sun causing inverse beta decay of the neutron in the deuteron: $\nu\, n \rightarrow p\, e^-$. They installed a 2000 liter heavy-water (D_2O) Čerenkov counter with 55 5-inch photomultiplier tubes, surrounded by a foot-thick anti-coincidence shield. This effort, too, was limited by background in the mine. Nevertheless, the idea of sorting out neutrinos and antineutrinos by using a deuteron target is at the heart of today's Sudbury Neutrino Observatory's program, where they are able to sort out the three flavors of neutrinos in the search for neutrino oscillations.

With the arrival in 1966 of **Bill Frisken**, Jenkins returned to accelerator physics, where the beam is a bit more predictable and the backgrounds much more manageable than in double-beta decay or neutrino experiments. We'll describe their accelerator-based work in Chapter 16. Subsequently, in the 1980's, Jenkins teamed up with Glenn Frye in his work on balloon-borne gamma-ray detectors

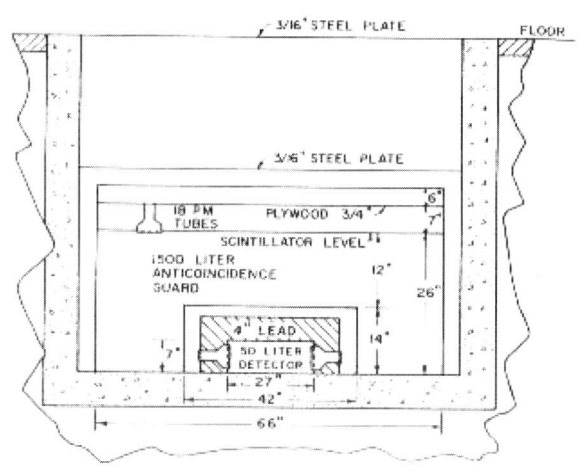

Fig. 8-10. Solar neutrino experiment in Ohio saltmine.

The anti-coincidence shield idea was again put to use in a search for neutrino interactions at the Morton salt-mine site. (*Phys. Rev. Lett.* **12** 457 1964) Reines and Bill Kropp put together a detector which was "surrounded by a large Čerenkov anti-coincidence detector and located 2000 feet underground". **Fig. 8-10.** The goal was to observe recoil electrons from elastic scattering of solar neutrinos. The rate would depend on two quantities, the rate of production of neutrinos in the sun, and the cross-section for scattering in the detector, $\nu\, e^- \rightarrow \nu\, e^-$. The nuclear processes in the sun had long been studied by theorists like Bethe, Bahcall and Fowler, and there were predictions on the rates and energies of the neutrinos produced in them. Reines *et al.* were looking for the 9 to 15 MeV neutrinos in the high energy tail of the distribution from the decay of ^8B. They could set their electronics to select recoil electron signals in this energy range, at the same time requiring no signal in the "shield". "In a counting time of 4500 hours only three events were observed in the energy range 9 to 15 MeV." The rather modest result of this run was "the elastic scattering cross section is ... <35 times the expected value" (assuming the theoretical neutrino production and interaction rates). This experiment marks the beginning of forty years of experiments to detect neutrinos from the sun at underground sites all over the world. The neutrino will turn out to be not only elusive, but to suffer from multiple personalities. In his 1964 letter, Reines turns quickly to a discussion of how large a detector would be required to detect the solar ^8B neutrinos. He estimated 25 m^3 at 2000 feet underground.

"There always remains the possibility of going deeper underground." Reines is looking for a "better hole".

In a paper following-up on the run at the salt-mine (*Phys. Rev.* **137** B740 1965), Reines summarizes his findings, relative to both the solar neutrino flux and the proton decay rate. From the abstract:, "Depending on the assumed decay modes, nucleon lifetime limits in the range 0.6 to 4 times 10^{28} yr were obtained. The upper limits on the neutrino cross-section flux products are $<8.5 \ 10^{-38}$ neutrino per second and $<3.2 \ 10^{-38}$ anti-neutrino per second." (These numbers represent the product of the neutrino flux and the cross section for neutrino interaction – and thus have the simple units of s^{-1}.) The experiment pushes the baryon lifetime upward another factor of 100. The paper includes extremely detailed analyses of the backgrounds and efficiencies.

In the meantime, Reines looks into the possibility of using inverse beta decay to see solar neutrinos. The plan is to use large slabs of lithium or boron, sandwiched between layers of liquid scintillator, to detect the reactions $\nu \ ^{11}B \rightarrow e^- \ ^{11}C$ or $\nu \ ^7Li \rightarrow e^- \ ^7Be$. As an example of the expected rates, for one ton of one-inch thick lithium slabs placed in the salt-mine and scintillators sensitive to 5 MeV electrons, Reines and a new addition to the department, R. M. Woods, calculated a rate of about 180 events per *year*. (*Phys. Rev. Lett.* **14** 20 1965).

Robert M. Woods, Jr. had completed his doctorate at the University of Michigan in 1963. He worked in beta spectroscopy with M. L. Wiedenbeck and had considerable experience in electronic detectors and in the new art of computerized data reduction. Consequently, when he applied to Reines for a position at Case, he was quickly offered an assistant professorship and the opportunity to join in Reines' research. He would work principally on the salt mine experiments until Reines' departure. Finding himself without a research program, and with the post-federation reduction in the department's size, Woods took advantage of an offer of a position with the High Energy Physics Program with the US AEC in Washington.

Reines thought of another use for the low-background environment in the salt mine laboratory: electron decay with charge non-conservation. What if an electron in an inner shell of an atom decides to decay into, say, a neutrino and a photon – thus violating charge conservation? Perhaps one could detect the decay photon in conjunction with a series of x-ray photons coming from the atom as it refills the inner shell. With grad student Michael K. Moe, Reines set up a pair of photomultiplier tubes and a sodium iodide detector in the mine. Result: the lifetime of an electron against such a decay is greater than 2 times 10^{22} years. (*Phys. Rev.* **140** B992 1965)

A later letter, authored by Reines, Kropp and Woods, reported a reexamination of the Fairport Harbor saltmine data. The object was to search for signals coming from muon decays. The idea was to determine whether most or all the observed counts could be attributed to muons. A predictable fraction of the muons would decay in the apparatus, and the resulting decay electron would be detected within a few microseconds. The observed time distribution of such events was compatible with the conclusion that all incoming particles were muons: another disappointing indication of the absence of neutrino signals. (*Phys. Rev. Lett.* **20** 1451 1968)

Cosmic Ray conference at Case - 1965

Having established himself as a world-class experimenter, Reines teamed up with Aihud Pevsner of Johns Hopkins and Larry Jones of Univ. of Michigan to organize a "Conference on the Interaction Between Cosmic Rays & High Energy Physics". It was held at Case in September of 1965. Papers were presented by future Nobelists Reines, Luis Alvarez, and Mel Schwartz, as well as by Jones, Yash Pal of MIT, S. Neddermeyer of Wisconsin, R. W. Thompson of Indiana and theorist J. D. Jackson of Illinois. The conferees had an interesting discussion after the final session, including remarks by representatives of the major funding agencies, NASA and NSF. It had largely to do with identifying those areas of research for which cosmic rays are more suited than accelerator beams. An intriguing remark on funding was made by Reines: "If we weren't meeting here today, then each of us would probably continue trying to push his proposal into various agencies and try to proceed just as in the past. The question we're exploring is, is there any way in which we can combine our effort to make some kind of sensible, all-over program; so increasing the support each one of us receives?" It is not clear that he found an affirmative answer to that question.

In his remarks at the 1978 "Reinesfest" at Irvine, celebrating Reines' 60th birthday, Marshall Crouch described the impressive list of colloquium speakers who visited during the Reines years at Case. "The department had grown a bit when Fred came to Case, and he inaugurated a program of colloquia in which if someone was doing interesting work in physics, that person was jolly-well invited to come to Cleveland and tell us about it himself. In the space of a few years we were inspired by colloquia given by at least a dozen Nobel laureates or laureates-to-be. Names like Luis Alvarez, John Bardeen, Hans Bethe, Owen Chamberlain, Leo Esaki, Richard Feynman, William Fowler, Donald Glaser, Robert Hofstader, Polycarp Kusch, I.I. Rabi, Emilio Segré, William Shockley and Eugene Wigner." Reines' colloquium program seems to have been generously supported by the Case administration.

"Natural" neutrinos

The "better hole" that Reines was seeking turned out to be one reaching 10,500 feet below the earth's surface in a gold mine in South Africa. Reines, Crouch and Jenkins, along with *factotum* technician Gus Hrushka, recent PhD graduate William Kropp, and graduate student Henry Gurr, packed up and crossed the Atlantic with tons of detector equipment. The Case team was joined by collaborators from the University of Witwatersrand in Johannesburg. They set up their detectors in the East Rand Proprietary Mine in Boksburg, east of Johannesburg. The goal was to detect muons produced in the rock surrounding the mine tunnel by high energy neutrinos which were in turn produced in the atmosphere by incoming cosmic rays. To maximize the material through which the neutrinos must pass, the detectors were arranged to catch muons traveling sideways, *i.e.* within 45° of the horizontal.

The two sketches **(Fig. 8-11 and 8-12)** are from a Scientific American article by Reines and the South African physicist J. Friedel Sellschop. (*Sci. Am.* **214** 2 40 1966) They show how the alignment of the two rows of large scintillators would pick up these "daughter" muons, but not the "sister" muons which are produced in the atmosphere. The trick is that the "horizontal" daughter muons must have been formed in the earth, because they could not have made it through the long horizontal trip in the rock. They could only have been produced in the rock by the penetrating neutrinos. The detector consisted of 36 rectangular boxes of scintillator liquid, each roughly 5 m long, 0.5 m high and 12 cm thick. They were arranged in two parallel walls containing 6 "bays" of three elements each.

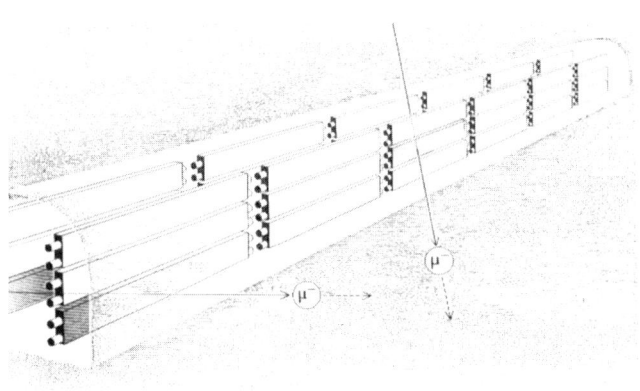

Fig. 8-11. Detecting neutrino induced muons in South African experiment.

Each element had two photomultipliers at each end. The most convincing evidence of a horizontal muon's passage was a signal in all eight photomultiplier tubes on two slabs of a given bay, one in each "wall". The photomultiplier signals were displayed on oscilloscopes and photographed for later scanning. In their letter (*Phys. Rev. Lett.* **15** 429 1965) the authors reported seven such events over the course of about five months of running, the first event being recorded on 23 February 1965. This was the first observation of neutrino induced interactions occurring in nature (as contrasted with Reines' earlier observation of neutrinos from a nuclear reactor.) The experiment heralded a new generation of deep underground neutrino observatories.

Marshall Crouch and Bill Kropp have written an account of the setting up of the experiment and its principal results. Here are several excerpts:

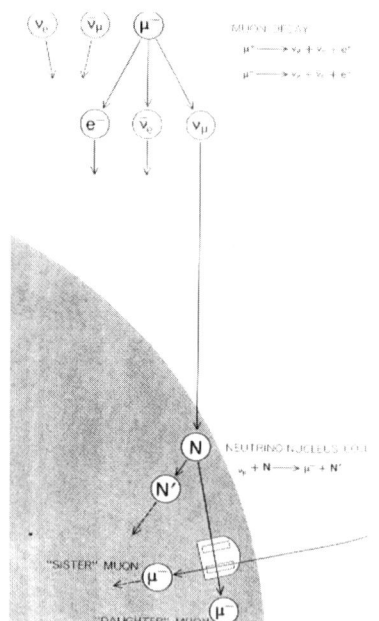

Fig. 8-12. Cosmic neutrinos produce muons deep underground.

"The logistics of moving the many tons of equipment by land, sea, and air, from the US to South Africa was elaborate and complex, and was tangled by both US and South African custom procedures, by maritime strikes, and lost, stolen, and misdirected shipping crates. But these difficulties paled in comparison to the problems associated with transporting the equipment through the labyrinth of the mine to the laboratory site. We will long remember the work chants of the black native miners in Zulu or Xhosa or Basuto as they manhandled the crates in their tortuous route to our crosscut.... The environmental conditions were extremely hostile, the rock temperature was about 135 deg. F, while the air temperature

with the help of the mine's massive air conditioning plant was a cool 95 deg. The humidity was always near saturation. Our engineer, Gus (Hrushka) lost nearly 30 pounds during the six month construction stage……In all, 35 neutrino events were recorded during the first phase of the experiment, with an additional 132 events, with much improved angular resolution, observed with the upgraded array. In addition, new limits were placed on the fluxes of extraterrestrial neutrinos, and on the lifetime of the proton for many possible decay modes. Finally the muon depth-intensity curve was extended many decades, reaching for the first time slant depths so great that only the constant flux of neutrino induced muons is observed." (W. Kropp, M. Crouch, in a memoir dated 7 September 1989)

Of course, a new lower limit on the baryon lifetime was set with the equipment in the South African deep mine laboratory: 2 times 10^{28} years, a hundred times longer than the 1961 Ohio salt mine number. (*Phys. Rev.* **158** 1321 1967)

Even with all the work and excitement associated with the deep-mine neutrino experiments, Reines decided in 1965 to direct a Case grad student, Frank Nezrick, in an improved version of the reactor experiment done seven years earlier. Returning to the Savannah River reactor with large sodium iodide detectors (29 cm diameter 7.6 cm thick cylinders from Harshaw Chemical), the two men studied the inverse beta decay reaction vbar p $\rightarrow$ e^+ n taking place in a liquid-scintillator target. The signal from the positron was quickly followed by signals from the positron annihilation ($e^+ e^- \rightarrow 2\gamma$). Within the following 50 μs the neutron is captured by a gadolinium nucleus which subsequently emits other γ's. As mentioned before, one advantage of looking for reactor-produced neutrinos as opposed to solar neutrinos is that one can turn the reactor on and off. They observed 549 events in 2484 hours (0.22 per hour) with reactor ON and 12 events in 357 hours (.03 per hour) with reactor OFF. In contrast to the 1958 Reines-Cowan effort, this experiment measured the *energy* of the positron, and thus determined the energy spectrum of the antineutrinos coming from the reactor. (*Phys. Rev.* **142** 852 1966)

CWRU comes; Reines goes

By 1966, it was clear that Case and Western Reserve would join to form a new institution, Case Western Reserve University. (Still today, almost 40 years later, people are trying to think of a more manageable name. With a significant number of loyal Case and Reserve alumni still fine-tuning their wills, it may take a few more years for this to happen.) *(Flash: the 2003 administration has adopted "CASE" as the name by which the world should know the institution, having determined that "CWRU" can be pronounced only in Welsh.)* Whether the impending union of the two physics departments had anything to do with Reines' decision to leave, or whether it was the attractiveness of the position he was offered at University of California Irvine, would be interesting to learn. In any event, in 1966 Reines became the founding Dean of the School of Physical Sciences at UCIrvine, where he was to remain for the rest of his career. He took his young colleagues Henry Gurr and Bill Kropp, and the multi-talented Gus Hrushka with him to California. His neutrino and baryon decay work would continue for many years, including new experiments at the Ohio salt mine and the Johannesburg gold mine. He returned to Case on several occasions, including a visit to give a talk at the October 1987 Michelson-Morley Centennial celebration. It just happened, the previous February, on Valentine's Day, that the proton-decay detector in the Ohio salt mine had seen a burst of neutrinos from Supernova 1987A – rather nice timing considering that the expanding shell of neutrinos had traveled 175,000 years to get to Fairport Harbor just in time for the centennial celebration. The signal consisted of seven neutrino events in ten seconds, while the usual background rate was one or two events per *day*!

Reines' colleagues, friends and family were very pleased that he was awarded the Nobel Prize in physics in 1995 while he was still well enough to enjoy it. He died three years later.

In 1966, the theoretical physicist and historian **Martin Klein** (Chapter 9) was appointed interim head of the Case department. **B. S. Chandrasekhar,** whom we shall meet in Chapter 15, was the chairman of the Reserve department, and the two of them became co-chairs of the new, greatly inflated CWRU department. There had been 22 members of the WRU department and 31 in the CIT department. The total number was clearly much too large. It would be reduced by half over the following ten years, roughly where it remains in 2005. (Other departments did not suffer the same fate because they did not have large numbers on both sides.)

Glenn Frye

Glenn M. Frye completed his doctorate at the University of Michigan in 1950. Before his arrival at Case in 1960, he had spent seven years at the Los Alamos Scientific Laboratory where he studied charged particles coming from neutron induced nuclear reactions in light nuclei. The neutron beam was produced in collisions of deuterons with tritium nuclei. (The deuterons were accelerated in the Los Alamos Cockcroft-Walton 250 keV machine.) The outgoing charged particles - protons, deuterons, and tritons - were identified in nuclear emulsions.

In one of these experiments, Frye looked at neutrons scattered from ^{6}Li and ^{7}Li nuclei, identifying those scattered elastically (i.e. the neutron just bounces off the nucleus without exciting it or breaking it up), and those in which the nucleus is left in a well-defined excited state. The differences associated with the extra neutron in the ^{7}Li nucleus cast light on the structure of these two isotopes. (*Phys. Rev.* **93** 1086 1954 and *Nucl. Phys.* **52** 505 1964) I recently asked Frye why there was a ten-year delay in publication of the second paper on the lithium work. He said that it had been classified for a long while. The government wanted to keep to itself any information on the production of tritium, a key component in thermonuclear devices.

Frye then moved on to experiments at the newly operating Bevatron at Berkeley. This 6 GeV proton accelerator was built largely to search for the antiproton, having just enough energy to reach the threshold for p p → p p p pbar. The discovery was made promptly, and Frye was among those who hurried to study these new examples of antimatter. He and his collaborator, Alice Armstrong, exposed a stack of nuclear emulsions to a magnetically analyzed 700 MeV/c negative beam coming from a Cu target in the machine. Among the many, mostly π^--induced, collisions were sixteen antiproton induced events. These were measured and the charge and momenta of the annihilation products, all pions, were determined. (*Phys. Rev.* **110** 170 1958, *Nuovo Cim.* **13** 77 1959) There is a photo of Glenn Frye later in this chapter. **Fig. 8-23.**

Physics by balloon, cosmic gammas and electrons

While his colleagues, Reines, Crouch and Jenkins, were putting their detectors as far below the earth's surface as possible, Frye was doing just the opposite. In two papers published in the same week, Frye and his grad student, Lawrence H. Smith, describe data taken with spark chambers flown at an "altitude" of 3.5 millibar pressure on balloons launched from the National Center for Atmospheric Research in Palestine, Texas. (*Phys. Rev. Lett.* **17** 733 1966 and *Phys. Rev.* **149** 1013 1966) *(The atmospheric pressure drops off logarithmically as the altitude increases, so 3.5 millibars is about 40 Km.)* The first was an attempt to observe high energy gamma rays from discrete astronomical sources. The second measured the high energy electron flux.

There had been proposals that powerful discrete radio sources might be fueled by antimatter annihilations, producing copious neutral π mesons, which in turn would decay into energetic γ's (in the hundreds of MeV range). The flying spark chamber was triggered when there was no incoming charged particle, but there were two outgoing charged particles, presumably an electron-positron pair coming from the materialization of an energetic γ. The spark chambers were photographed, along with digitized information on the orientation of the detector, so that the direction of the incoming γ could be determined. (The cameras had to be recovered after the balloon returned to earth, sometimes after a wild chase across Texas.) Special effort was made to look toward such objects as the Crab nebula and Cygnus A, but only upper limits (order of 10^{-4} γ per square cm per sec) could be set.

The work done by Frye did not go unnoticed. The following is a paragraph from a letter in Bob Shankland's files (in the departmental archives). It is from Luis Alvarez to Shankland, dated March 1965.

"Jan and I spent the week end with friends in Palm Springs, and Edward Teller was also there. Edward told me that he had figured out where the quasars found their tremendous energies. He said he felt sure that they must be collisions between galaxies and anti-galaxies. He had calculated the number of gamma rays that would be incident on the top of the atmosphere, and he felt sure that the experiments to look for them would be very difficult, but he hoped that somebody might undertake them in the future. I told him of Glenn Frye's beautiful spark chamber experiments, and from then on Edward was disconnected from the rest of the guests. He sat and thought, and didn't pay a speck of attention to anyone else at the party. Finally he disappeared from the room, and after quite some time, he came back with a big smile on his face, and said he had just been talking with Glenn Frye on the phone. He said his theory was on the ropes and probably wouldn't survive, but at least he was very happy to have heard the experimental results when he did."

Frye's next paper described the flights of a lead-plate spark chamber in which energetic *electrons* would produce recognizable showers. Two separate flights were made from the Texas launch facility so that the electron flux could be measured at two very different altitudes. The object was to determine the increase in the number of electrons as one looks deeper in the atmosphere. The altitude was given in g/cm^2 of residual pressure. At the higher level (2.16 g/cm^2) the flux was 4.8 ± 5.4 electrons/steradian m^2 sec and at the lower level (4.35 g/cm^2) it was found to be 16.3 ± 5.3 electrons/steradian m^2 sec. The conclusion was that no more than 5 electrons/steradian m^2 sec arrive vertically at the top of the atmosphere; this is less than one percent of the primary *proton* cosmic ray flux. The higher flight was made with a 6-million cubic foot ½ mil polyethylene balloon; it flew for 8 hours and ended up near Montgomery, Alabama.

In 1966, Frye was joined in the balloon work by a newly appointed associate professor, **Chia Ping Wang**. Wang had done his doctorate at the University of Singapore in 1953 and had held faculty positions in Chinese and Hong Kong universities, followed by two years at Catholic University in Washington. At CWRU, in addition to his work with Frye, Wang published a study of particle multiplicities in high energy collisions. He compiled pion-nucleon and nucleon-nucleon data from over fifty accelerator experiments, with energies up to 27 GeV. He proposed a model for the sub-structure of the target nucleon which appeared to follow from the observed multiplicities. (*Phys. Rev.* **180** 1463 1969.) This work was extended to 60 GeV data on pion-nucleon and antiproton-nucleon from the Serpukhov accelerator in the USSR. (*Phys. Lett. B* **30** 115 1969.) Wang left the CWRU department in 1970. He continued to publish on this approach to nucleonic structure while associated with several university departments, including MIT and Cambridge.

Frye and Wang returned to the search for high energy γ's from discrete astronomical sources, but again came up empty handed. After a group at Rochester reported a signal from somewhere in the constellation Cygnus, the Case group mounted three balloon flights with the electron-positron-pair-detecting spark chambers. Their maximum flux turned out to be ten times *less* than what had been reported by the Rochester group. (*Phys. Rev. Lett.* **18** 132 1967 and *Canadian Jour. Phys.* **46** S448 1968)

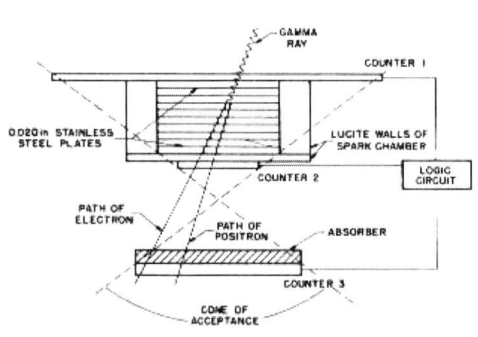

Fig. 8-13. Balloon borne γ-ray telescope.

An even more elaborate balloon-flight experiment was undertaken in 1967, again from the Palestine, Texas launch facility. The object was to make a more detailed measurement of the electron component of the primary cosmic rays. To better separate the electrons from protons and other heavier particles, a gas Čerenkov counter was placed above four aluminum and fifteen lead plates in the spark chambers. To reduce the power requirements, glass sheets were placed over all the metal plates to reduce the amount of charge in each spark. One must appreciate the experimental challenges of designing self-contained detectors like these, along with the necessary power supplies and data recording systems. Two flights were made, one at 5.30 g/cm² residual atmosphere took data for 5 hours and ended up in Arkansas, the other at 2.72 g/cm² for 2 hours on its way to Louisiana. It was determined that the electron flux more energetic than 3 GeV (4.3 ± 0.97 electrons/sterad m² s) was independent of altitude and may be attributed to primary cosmic rays. (*J. of Geophys. Research* **74** 53 1969)

Data on cosmic gamma rays from a flight in August 1968 out of the Texas launch site were published several years later. The object was to see if the flux at 10 MeV were consistent with power-law extrapolations from lower energies. The data consisted of 149 electron pairs which satisfied all the selection criteria. The use of the spark-chamber telescope allowed the determination of the direction and the energy of the incident gamma. The telescope package is sketched in **Fig. 8-13**. The authors conclude that "the measured value is an order of magnitude *above* the extrapolation of the power-law which holds below 1 MeV." They propose that the observed excess is due to a diffuse gamma flux of cosmic origin, superimposed on gammas which are secondary atmospheric radiation. "The Cosmic Diffuse Gamma-ray Flux at 10 MeV" (*Astrophys. J.* **182** L51 1973.)

Discovering discrete cosmic gamma sources

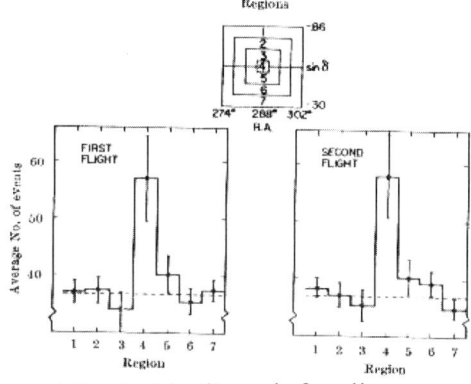

In February 1969, Frye and his grad students Alan Zych and Jon Staib teamed up with a group from the University of Melbourne on two balloon flights from New South Wales. This time, they identified a point source of high energy gammas. Using two different sets of spark-chamber and Čerenkov detectors and two different gondolas, the collaboration found a remarkable excess of events above 50 MeV in a particular 4° half angle cone in the direction of Sagittarius. (*Nature* **223** 1320 1969) **Fig. 8-14**. (The little square shows the seven regions on the sky which are plotted beneath.) A follow-up paper detailed the results of the two February flights, and a third flight in November of 1969. Data from the three flights were combined, and three gamma-ray sources were observed. The measured intensity for each of the three sources was about 1.5 10^{-5} γ cm^{-2} s^{-1}. (*Nature* **231** 372 1971) The third flight ultimately yielded a fourth gamma source. **Fig. 8-15** shows the regions on the sky which were searched by each of the three flights, and the positions of the four γ-ray sources observed. "New Point γ-source Lib γ-1: Evidence for Time Variation and Possible Identification with PKS 1514-24" (*Nature* **233** 466 1971).

Fig. 8-14. Search for discrete gamma sources.

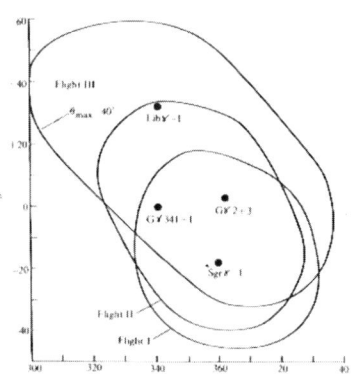

Fig. 8-15. Four γ-ray sources discovered.

Returning to the northern hemisphere later in 1969, Frye and Wang completed four flights with a 30-plate spark chamber with angular resolution 2.6°. They did an extensive survey of the northern sky and found no gamma point sources. They were able, however, to set limits on the γ's coming from a variety of candidates (31 in number), e.g. the crab nebula, Cygnus A, the sun, 3C 273, etc. In so doing, the authors were able to test several proposed theories for energy generation in these systems. (*Astrophys. J.* **158** 925 1969) **Fig. 8-16** shows one of the four "equal solid angle" plots, There are three integers in each solid angle bin: respectively >100MeV events, <100MeV events, and events with single tracks. It's a nice way to look for hot spots in the sky.

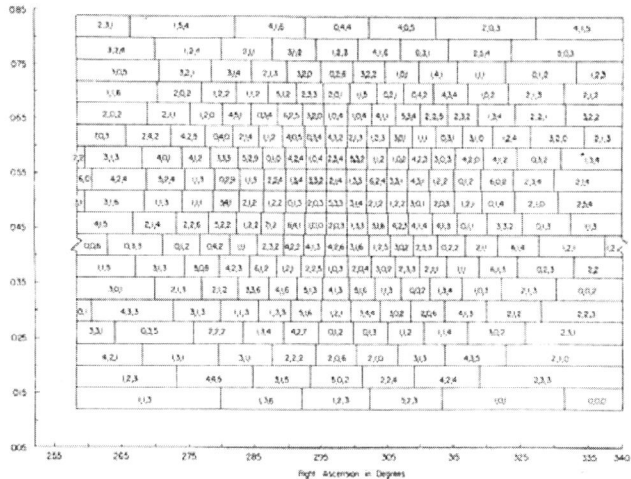

Fig. 8-16. "Equal solid angle" plot showing three categories of γ-ray counts.

Another Texas balloon launch was carried out in the summer of 1969, this time to measure the flux of cosmic ray *neutrons* as a function of altitude. The neutron detector included polyethylene absorbers in which neutrons could hit protons. The recoil protons were observed in spark chambers and the tracks recorded photographically. The balloon was slowly deflated to allow sampling at a series of altitudes. A total of 45,000 pictures was taken. The proton recoil angle and range helped identify the neutron events and their incident directions. The experimenters had considerable difficulty subtracting the large background of electrons produced by γ's. "Altitude Variation of High Energy Neutrons Near the Top of the Atmosphere" (*Acta Phys. Acad. Scien. Hungarica* **29** 709 1970.)

Fig. 8-17. Paul Albats

In 1969, Frye hired 27-year-old **Paul Albats** as a research associate. Latvian-born Albats had completed his BS at the University of Chicago and his PhD research at Cornell. **Fig. 8-17.** In 1973 he was appointed assistant professor. In addition to his work on high altitude balloon-borne experiments, he was instrumental in creating a digital electronics laboratory for the physics-majors.

Frye, Albats and Zych teamed up once again with the Melbourne group for another balloon flight from Palestine Texas. This time, the spark chamber system was designed to detect gamma rays in the 10 to 30 MeV range. The goal was to look for gammas coming from the Crab Nebula Pulsar, NP 0532. This object had been observed to emit electromagnetic radiation from 0.1 to 10 MeV, in pulses with period of about one thirtieth of a second. Frye *et al.* wanted to extend the measurements to higher energies.

The thirty million cubic foot balloon remained for nine hours at an altitude of 46.7 km (the highest flight to that date, according to their paper). An essential component of the apparatus was an onboard digital clock which recorded the time of the events to within a millisecond, and which was synchronized once each second with a ground based clock. Because the pulse from the pulsar had a width of less than 3 ms, a signal from NP 0532 must appear as an excess of events in the same absolute-time 3 ms bin as the lower energy radio frequency, visible, and X-ray signals. **Fig. 8-18** shows the number of events in one millisecond bins during a four hour exposure to the sky within 15° of the Crab. A more than 5 standard deviation peak appears at the correct time. That this was indeed associated with NP 0532 was based on several arguments: the time and the width of the signal were correct; the peak disappeared when a slightly different period was used or when neighboring areas of the sky were observed. The resulting observed energy flux fit rather well with lower energy data on the $E^{-1.2}$ exponential curve.

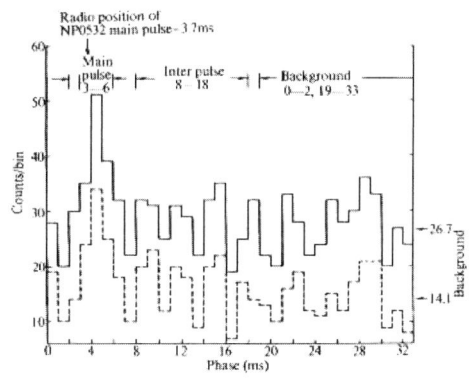

Fig. 8-18. Signal from the Crab Nebula 3.7 msec pulsar.

Fig. 8-19 "Two New Sources of High Energy Cosmic Gamma Rays" *Nature* **231** 372 1971. "New Point Gamma Ray Source Lib γ-1: Evidence for Time Variation and Possible Identification with PKS 1514-24" *Nature* **233** 466 1971. "Detection of 10-100 MeV γ-rays from the Crab Nebula Pulsar NP 0532" *Nature* **240** 221 1972.

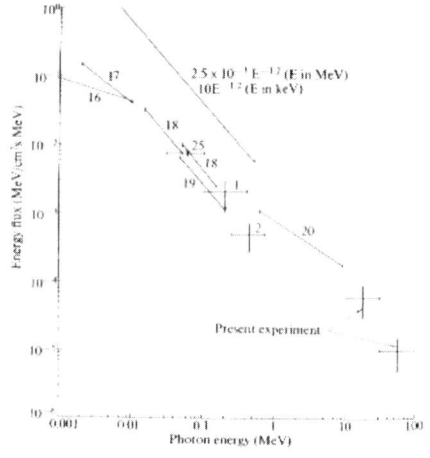

Fig. 8-19. Exponential drop-off of γ-ray flux with energy.

The whole detection apparatus of the preceding experiment was then transported to Australia for a look at another pulsar, the one in Vela: PSR 0833-45. The result, based on 21 events where the background is 8 events, was quite convincing. An interesting feature of the paper is a brief discussion of the relation between intrinsic luminosity and period which, in a certain theoretical model, was predicted to be $L \propto T^{-4.25}$. Given the observed periods and the known distances to the two pulsars, the ratio of Crab to Vela intensities should be 68, compared to the experimental 64 ± 32 – not very conclusive, but indicative of the type of measurement which may be made. "Pulsed 10-30 MeV Gamma Rays from PSR0833-45" *Nature* **251** 400 1974.

Frye and Zych combined data from two balloon flights (one in Panama and one in Australia) and a ground-based run (on the CWRU campus) to determine the flux of γ's produced in the atmosphere by incoming cosmic rays. They used the same spark chamber arrangement which was used in the search for discrete γ sources. **Figs. 8-20 and 8-21** show the measured high-altitude γ intensity as a function of energy (50 MeV to 1 GeV) and the angular distribution of the γ's relative to the zenith. The intensity is given in γ's per (sec g sterad MeV). (*J. Geophys. Res.* **79** 929 1974.)

Back to cosmic neutrons

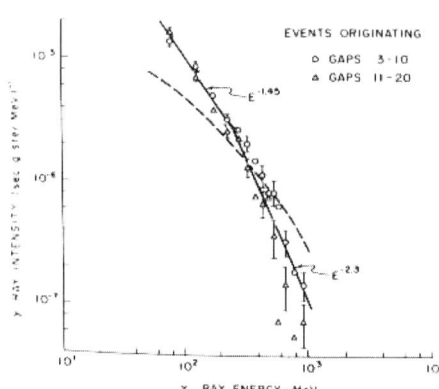

Fig. 8-20. Energy distribution of γ-rays produced in the atmosphere.

Throughout the 1970's Frye, Zych, Albats and their students continued the development of particle detectors which would give detailed information on incident cosmic ray particles and at the same time be self-contained enough to operate while hanging from a balloon dozens of kilometers above the earth. The abstract of a paper in 1977 begins with a compact description of their latest efforts: "A new detection method for 15 to 150 MeV neutrons from an extended source utilizes a spark chamber, time-of-flight technique. Both secondary particles in an n-p scatter are measured – the recoil proton in a spark chamber and the scattered neutron in a time-of-flight hodoscope, enabling the direction and energy of the incident neutron to be determined uniquely. ... The detection efficiency is 10^{-4} at 70 MeV, the acceptance cone 40°, the angular resolution 3.0° and the energy resolution 10%. The detector has been used on high altitude balloon flights to measure the flux of atmospheric neutrons and to search for solar neutrons."

Figs. 8-22 shows a schematic of this detector. An incoming neutron coming from the upper-left hits a proton in the polyethylene stack and the outgoing proton travels downward. The proton first passes through two planes of scintillator and the time interval between the two signals is recorded. The proton then enters a sandwich of spark chambers where the path of the particle is recorded on film; in addition the distance it travels through the chambers is recorded. Meanwhile, the scattered neutron travels off at 90° to the proton, hopefully entering one of the elements of the neutron detecting hodoscope (the 18 cylinders in the schematic).

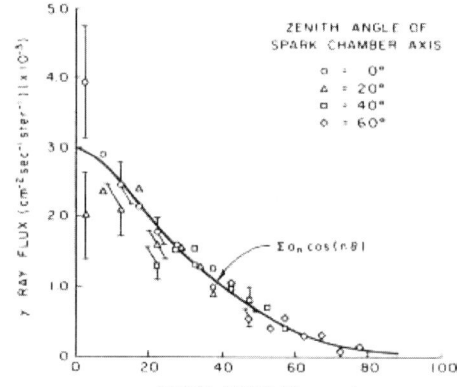

Fig. 8-21. Direction of atmospheric γ-rays.

The time interval between the hit in the first proton scintillator and the arrival of the neutron in the hodoscope gives the speed of the neutron. The position of the struck hodoscope element gives a rough measure of the neutron's detection. All the recorded information appears on the spark-chamber photographs in the form of little images of lights. With all these required signals, one can see why the efficiency for this device in detecting a neutron is only one in ten-thousand – but

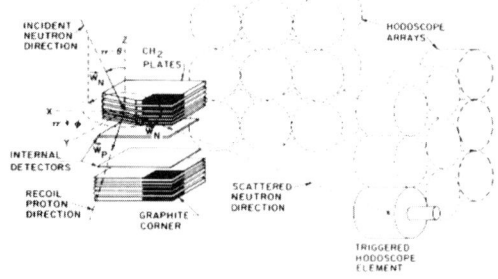

Fig. 8-22. Schematic of balloon-borne neutron detector.

there are a lot of neutrons out there. (*Nucl. Instr. and Methods* **144** 183 1977.) A photo of Glenn Frye, taken in the 1960's, appears in **Fig. 8-23**.

In the spring of 1978, the department voted to promote Albats to associate professor, but this was turned down by the administration. He subsequently accepted a position with the international Schlumberger organization, where he would work on neutron and γ-ray instruments for oil well logging. His departure dealt a serious blow to CWRU's cosmic-ray research program. Nevertheless, Frye would be joined by Jenkins in a continuation of balloon-borne experiments.

In the 1980's, Frye and post-doc Rokutaro (Rocky) Koga built a large multi-wire gamma-ray telescope with arc-minute resolution for flights launched from Australia. In 1987, Frye, Jenkins, Koga and Albats collaborated with groups from Imperial College, Frascati and New South Wales in a stratospheric balloon flight in search of 50 to 500 MeV γ-rays emitted by Supernova 1987A. (*AIP Conference Proceedings* **170** 80 1988).

Fig. 8-23. Glenn Frye.

Under Tom Jenkins' supervision, grad student Ken DelSignore completed his doctoral research by participating in the study of gammas and neutrons emitted in an unusually powerful solar flare in June 1991. The data were collected by the Oriented Scintillation Spectrometer Experiment (OSSE) which was aboard the Compton Gamma-ray Observatory Satellite. The balloon era was ending, and the exciting new satellite-based cosmic ray experiments would soon take over. According to DelSignore's dissertation, this was the most intense gamma ray flare observed to that date.

Glenn Frye retired to emeritus status in 1993, and Tom Jenkins did likewise two years later. Jenkins, a long-time member of the Sierra Club, continues to be active in the environmental protection movement.

Mining more results from the South African experiment

Marshall Crouch participated with the Reines UC Irvine group in a further analysis of the South African gold-mine data, this time to extract the *intensity* of surviving cosmic ray muons as a function of depth in the rock. The same large scintillator hodoscope described above (page 98) was used. The incident muon flux was determined as a function of angle relative to the vertical, thus allowing the experimenters to select different thicknesses of rock overlay. The resulting muon flux could then be parameterized $I_\mu(h) = a_\mu e^{-h/\lambda}$ where $a_\mu = 1.04 \; 10^{-6}$ cm^{-2} sec^{-1} sr^{-1} and $\lambda = 8.04 \; 10^4$ g cm^{-2} and h is the depth in g cm^{-2}. (The uncertainties are about 15% in a_μ and about 5% in λ.) These results could then be compared with the measured energy distribution of muons at sea level combined with the measured rate of energy loss and interaction cross sections for muons.

"Cosmic-Ray Muon Intensity Deep Underground versus Depth" (*Phys. Rev.* **D1** 2229 1970)

Since the neutrino data from the South African experiment included the *time* of arrival of muon neutrinos, the Reines group decided to look to see if there were any coincidences between the arrival of neutrinos and of the *gravitational* wave signals reported by Joseph Weber of the University of Maryland. Weber provided the authors with the times of his signals during the first nine months of 1970, the period during which the neutrino experiment was running. There were two events in which the Weber pulse and the neutrino hit were within two minutes of one another. "we assume that the observed coincidences are consistent with accidentals and take the upper limit on neutrino pulses associated with Weber waves to be 2 yr^{-1}." "Upper Limit on High-Energy Neutrinos from Weber Pulses" (*Phys. Rev. Lett.* **26** 1451 1971.) *(While it is unlikely that Weber had really detected gravitational radiation, the search continues today with some very sophisticated and expensive long-baseline interferometry experiments. Reines was always on the look-out for opportunities to exploit the gold-mine of data from the gold-mine experiment.)*

The "definitive paper" on the South African neutrino experiment was published in two parts, "Experiment" and "Analysis", in the *Physical Review* in 1971. Crouch was included as an author of the first part, along with members of the Irvine and Witwatersrand groups. The emphasis was on the events in which a neutrino coming laterally through a large thickness of rock produced a muon. Much of the material had appeared in the earlier papers, but this longer format allowed the inclusion of a table giving the details for each neutrino event "Muons Produced by Atmospheric Neutrinos: Experiment (and Analysis)" (*Phys. Rev.* **D4** 80 and 99 1971.)

In a two-author letter published three years later, Crouch and Reines present a new analysis of their search for baryon decay in the South Africa mine. At question were baryon decay modes which resulted in μ → e decays. Their new result pushed up the lower limit of the baryon lifetime by a factor of three. Their earlier paper had described a search for muon signals which were in excess of those expected from cosmic neutrinos. In this paper, that search was refined, in that the measured delay between the detection of the muon and of its electron decay-product is included in the selection process. As often seen in Reines' papers, the authors discuss briefly how one might improve the experiment, including increasing the size of the detector from 20 to 100 tons. "Baryon-Conservation Limit" (*Phys. Rev. Lett.* **32** 493 1974.)

Four years later, in 1978, one more paper based on the 1967-1971 gold-mine data was published by Reines and his group at Irvine, along with their Witwatersrand partners and CWRU's Marshall Crouch. "Our basic motivation for this study is to investigate the weak interaction in the high-energy region. Measurement of the muon flux deep underground provides a test of the correctness of the theory of the ν_μ + $\bar{\nu}_\mu$ processes producing muons (*i.e.,* charge-changing ν_μ interactions) which occur in the rock. In addition, the atmospheric neutrino flux represents an important component of the cosmic radiation which should be studied to complete our understanding of cosmic rays as an observed geophysical phenomenon." Recall that the experiment observes muons and the direction in which they move through the underground detector. At large angles, the effective thickness of rock is so great that the muons must be those created when *neutrinos* interact in the rock. As the authors put it: "The depth and array configuration were chosen to permit measurement of the transition from the angular region where penetrating muons produced in the earth's atmosphere predominate (zenith angle < 45°), to the region where the entire muon flux is neutrino-induced."

The measured vertical muon intensity as a function of vertical depth, h, was best fitted by the sum of an exponential drop-off plus a constant term: $I_{v\mu}(h) = A \exp(-h/\lambda) + B_{v\mu}$, with $A = (2.26 \pm 0.16) \, 10^{-6}$ cm^{-2} s^{-1} sr^{-1}, $\lambda = (7.58 \pm 0.09) \, 10^4$ g cm^{-2}, and $B = (2.23 \pm 0.20) \, 10^{-13}$ cm^{-2} s^{-1} sr^{-1}. The value of the attenuation length λ measures the absorption of atmospheric muons in the rock and it can be compared to what one might expect from the strength of the weak and electromagnetic interactions. The value of the depth (and angle) independent term, B, measures the number of muons produced by neutrinos in the rock. **Fig. 8-24** shows the muon intensity as a function of "slant depth", clearly illustrating the exponential drop-off and the constant term. This figure, coming from a 1990 review paper by Marshall Crouch, was supplied by him for inclusion here. It shows the muon intensity as a function of slant-depth below the top of the atmosphere, from the measurements by Millikan in 1925 to those by those made in the South African gold mine. "Cosmic-ray muon fluxes deep underground: Intensity vs depth, and the neutrino-induced component" (*Phys. Rev.* **D18** 2239 1978).

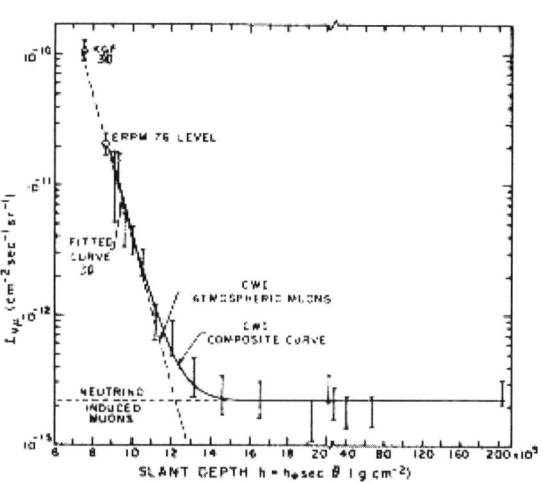

Fig. 8-24. Muon intensity vs. "slant depth" showing neutrino induced µ's.

Crouch goes after quarks (and tachyons)

The success of the quark model in categorizing a large number of sub-nuclear particles and in explaining their properties prompted many groups to search for evidence of fractionally charged *free* quarks. Candidates were sought in everything from meteor fragments to oyster shells. Crouch decided to build a new detector to search for quark signals in the cosmic rays. **Fig. 8-25** shows a sketch of the counter telescope. It consisted of five layers of 135 x 57 x 12 cm scintillator slabs, interspersed with 1000 flash tubes. The required signal was energy deposition in each layer of from 30 to 80% of E_o where E_o is the amount deposited by a single through-going fast singly-charged muon. The energy expected from a fast particle with a charge of only 2/3 e would be 4/9 E_o. The trajectory of the particle through the telescope was determined by the hits in the many layers of flashtubes. Background signals from electron showers are eliminated by single-track discrimination in the flash-tube array. In 1157 hours of running, there were 963 quark candidates. This number was reduced to 115 events with a clear single track and the required energy deposition. These are shown in **Fig. 8-26**, where all events except one pile up at the high end of the energy window. This single event, at 0.57 E_o, is somewhat above the 0.44 E_o expected for a charge 2/3 e particle. The resulting upper limit for 2/3 e quarks is given as 2.2 10^{-6} m^{-2} s^{-1} sr^{-1}. "Search for Relativistic Charge 2/3 e Quarks in the Cosmic Radiation" (*Phys. Rev.* **D5** 2667 1972).

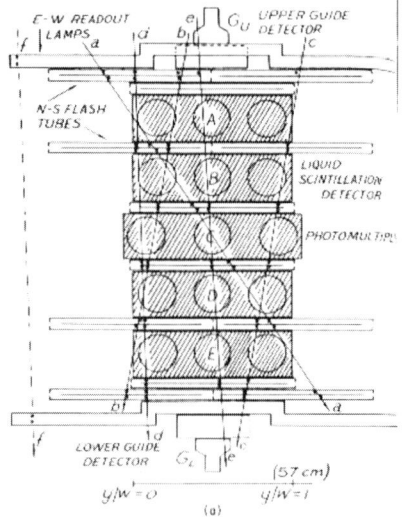

Fig. 8-25. Quark search telescope.

Crouch spent 1974 on sabbatical leave in Japan at the University of Tokyo. In a paper written there with G. Tanahashi, the authors explain: "We have carried out an experiment to search for tachyons produced in interactions of high energy cosmic ray particles which generate extensive air showers, the signature for faster-than-light particles being signals observed in advance of the shower front of relativistic particles." Some earlier experimenters had presented preliminary evidence for tachyons (i.e particles which can travel *only* at speeds greater than the speed of light). The Crouch - Tanahashi experiment was done at the Air Shower Array (ASA) of the Japanese Institute for Nuclear Study. This facility had been in operation for a dozen years; it featured a very large array of detectors for the study of extensive air showers, typically with energies in excess of 10^{15} eV. A trio of one-m^2 scintillation detectors was arranged in a triangle at the center of the ASA. The signals from these three detectors were fed into an electronic delay line where they languished for 96 µs. When a signal of eight or more hits in the ASA was detected, the delay line was read out, providing data of what was happening *before* the shower. The resulting data were compared with similar streams of data taken at randomly selected times. As you may have guessed, tachyons were not discovered in this experiment.

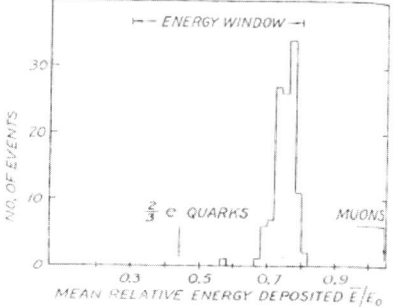

Fig. 8-26. Search for fractionally charged particles.

Marshall Crouch retired in 1987. He lives with his wife in nearby Willoughby, Ohio, and occasionally attends university and department events.

Chapter 9   CIT Theorists

Foldy,	**Klein,**	**Milford,**	**Winterberg,**	**Tobocman,**	**Thaler**
1948-1990	1949-1967	1952-1959	1959-1963	1960-2001	1960-1981

Shankland was away from Cleveland for most of the war, working on underwater sound detection in New York City or England or Florida. When he returned to Case to resume his responsibilities as chair of physics, he appreciated the need to add research in *theoretical* physics to the department's program. His colleagues, Smith, Shrader, Crittenden, and Olsen, were all doing *experimental* physics. While experimentalists are generally familiar with the theory underlying their research, their emphasis is on apparatus and measurement. There were no "pure" physics theorists at Case. Shankland had made many contacts in his travels with young people who were developing the theories stemming from new experimental results in nuclear, particle and condensed matter physics.

Case Institute could expect to expand rapidly, especially given the increased postwar public interest in science and engineering. Many of the troops returning from the war would take advantage of the GI Bill which paid their college tuition. Quoting C.H. Cramer's history of CWRU: "By 1946 every engineering college in the land was operating beyond its normal capacity; some had set up two-shift programs to meet the emergency. At Case there was a ten-to-one ratio between applications and admissions." By 1947, there were almost 3000 students at Case, including 200 in the newly established graduate program. Shankland's student, Earle Gregg, would be awarded the department's first PhD in 1949.

Shankland enjoyed the support of two successive presidents, William E. Wickenden and T. Keith Glennan, and of the trustees, in expanding the department. Between 1948 and 1960, Case would add a half dozen theoretical physicists to its faculty. The Western Reserve department would do likewise, under chairmen Richard Beth and John Major.

With the arrival of the theorists, this history of research comes to a turning point. The earlier chapters describe experiments, including details of the techniques and apparatus employed, most often with quantitative results. Most theoretical papers, however, concern the development of new mathematical techniques or the application of old ones to organize and describe what the experimentalists have seen.

In this and later chapters, we shall give a qualitative overview of the work done by the twenty-odd theorists who become part of the physics faculty. We'll give some background on each person, and list a selection of papers including titles, and comment on a few of the highlights. Hopefully, this will allow the reader to have some appreciation for what the theorists have accomplished.

Foldy – Case's first theorist

Leslie L. Foldy was born in 1919 in Sabinov, Czechoslovakia; he was brought by his parents to America as an infant. In 1937, he graduated from Glenville High School in a European ethnic neighborhood on the east side of Cleveland. He matriculated as a student of Miller and Shankland at Case, where he would later spend most of his fifty-year career as a theoretical physicist. He would become the first theorist to join the Case department. Foldy's extraordinary productivity, his wide interests, and his friendly and cooperative nature had a major impact on the department.

Foldy graduated with a BS from Case Tech in 1941, having written his senior thesis on crystal lattice vibrations. I asked him once what his middle initial L stood for. He said that when he arrived at Case, all the other students seemed to have middle names. He had none, so he opted for a simple L. He went to the University of Wisconsin where he completed a master's degree under the famous French physicist Leon Brillouin. With the outbreak of World War II, Foldy went in 1942 to join his former professor, Shankland, at Columbia University in the Division of War Research, to work on submarine detection. Foldy quickly grasped the theoretical physics implications of the work, publishing three related papers.

In the first, Foldy looks at the scattering of waves from a random distribution of isotropic scatterers, including (and this is the main thrust of the exercise) the interference phenomena present in any wave scattering. The results are applicable to many systems, from sound waves scattered by water droplets in a fog, to electrons scattered by atoms in a crystal. Why was this of interest in submarine detection? Given a source of sound at a certain position and with a certain intensity, what would be the effective signal measured by a detector of a given geometry and size at a second position? It would have been helpful for our purposes if the equations developed in the paper were applied to some measured system, but such is not the case, as with many theory papers. "The Multiple Scattering of Waves" (*Phys. Rev.* **67** 107 1945) The ideas in this paper would be applied not only to problems in the multiple scattering of acoustical waves, but, for decades afterward, would be used in the analysis of nuclear and other scattering processes

While working in 1944 at Columbia, Foldy wrote a brief report on the possible application of Dayton Miller's mechanical integrator, the Henrici analyzer. We described this favorite device of Miller in Chapter 4. Foldy acknowledges that there are other methods to Fourier-analyze waveforms, but this one might have its advantages. He describes a shutterless movie-film camera which captures the signal from an oscilloscope face to produce transparencies which can be placed directly on the Henrici tracing bed. "The Use of the Henrici Harmonic Analyzer to Obtain Frequency Spectra of Pulses". The copy of this report in Foldy's files is stamped "…. information affecting the national defense…"

The two other Columbia papers (*Jour. Acous. Soc. Amer.* **17** 109 1945 and **19** 50 1947), written with Henry Primakoff, developed the theory of electroacoustic transducers. Here Foldy turns his attention to the physical properties of the emitters ("speakers")

and detectors ("microphones"). Once again, he develops integral equations to describe the relation between the pressure and normal velocity at each point on the surface of the transducer (the input) and the voltage and current at its electrical terminals (the output). Input to the calculation are the electric, magnetic, thermal, and mechanical properties of the device. Related to this study is an examination of the validity of the "Reciprocity Theorem", which states (quoting the first paper) "that the ratio of the microphone response of the transducer to its speaker response is a quantity which is independent of the nature and construction of the transducer." Here he is talking about hitting the transducer with an electrical pulse and looking at the resulting mechanical response, and then hitting it with a mechanical pulse and looking at the resulting electrical response. In a 1957 letter about Foldy to the Case dean, Shankland says, "....the most interesting (contribution by Foldy) was his thorough-going proof .... that the principle of reciprocity is the most basic and useful means for the calibration of acoustical transducers. His views on this were accepted only after very vigorous opposition by the Bell Telephone Laboratories and others .... considered to be leaders in acoustical problems."

To Berkeley

In 1945, Foldy was appointed Research Physicist at the Radiation Laboratory at University of California Berkeley. His assignment was to undertake a detailed analysis of just what is going on inside the new particle accelerators. (Ernest O. Lawrence, inventor of the circular machines, was at Berkeley.) In two papers written with young colleague D. Bohm (*Phys. Rev.* **70** 249 1946, and **72** 649 1947), Foldy presents calculations for the motions of charged particles in various types of machine.

The following spring he began graduate studies at UC Berkeley working under J. Robert Oppenheimer on the nature of nuclear forces. His doctoral dissertation had a four-fold title: "Four Studies in Theoretical Physics. I. The Theory of the Synchrotron. II. Theory of the Synchro-Cyclotron. III. On the Meson Theory of Nuclear Forces. IV. The Energy-Momentum Relations for Particles Interacting with Fields." He spent his final year of graduate studies at Princeton because Oppenheimer had moved there from UCB.

An aside on particle accelerators. In a betatron, *electrons are injected into an evacuated space in which an applied magnetic field rises rapidly from zero to some maximum value. The changing B-field induces a tangential electric field which accelerates the electrons while the magnetic field confines them to a circular path. In a* cyclotron, *the charged particles are introduced at the center of a large cylindrical evacuated space in which there is a steady magnetic field. They are then accelerated by an alternating electric field set up between the two halves of the cylindrical space, spiraling outward until they exit the machine. In a* synchrotron, *bunches of particles are introduced into a doughnut-shaped evacuated space where they are accelerated by periodic electric fields at certain points around the ring; they are held in a fixed orbit by a magnetic field which increases in strength as the particles speed up.*

To first approximation, these various accelerators are simply described, but when one looks at details such as off-axis beam particles and their oscillations about a central orbit and the effect of energy loss by the electromagnetic radiation coming from the accelerated particles, the problem becomes a suitable challenge for the mathematical physicist. These two frequently cited papers of Bohm and Foldy present criteria for optimizing machine design so as to capture and accelerate the maximum number of particles.

Foldy and particle physics

To this point, most of Foldy's work may be described as "applied physics", and it was time for him to move on to "basic theoretical research" in nuclear/particle physics. Experiments on neutron-proton and proton-proton scattering and investigations of the deuteron (the bound neutron-proton system) had been reported and theorists were busy trying to deduce from them the nature of the short-range nuclear force. Yukawa had proposed in 1934 that the nuclear force might be due to the exchange of a spinless particle (the meson) with mass about 100 MeV (about 200 times the electron mass). Subsequently, particles with masses in this region were discovered in cosmic rays. Gradstudent Foldy's task was to calculate the phase-shifts (a set of numbers which determine the np and pp scattering angular distributions and cross sections) and the deuteron properties by assuming various combinations of two-nucleon angular momentum states (as described in Chapter 6 in the section on the Shankland experiment) and the spins and masses of the exchanged meson(s). His conclusion was that better agreement with experiment could be found with the exchange of 150 MeV mesons than with 100 MeV mesons. ("On the Meson Theory of Nuclear Forces" *Phys. Rev.* **72** 125 1947) The π meson mass was eventually found to be 139 MeV. However, over the course of the following twenty years, it was became clear that several mesons with various masses and spins participate in the nuclear force.

Fig. 9-1. Bob Shankland and young Les Foldy.

Shankland was very keen on getting Foldy to join the department at Case. In 1948, Foldy accepted his offer, and returned to his hometown and undergrad school to become its first theoretical physicist. **Fig. 9-1** shows the young Foldy and chairman Shankland at the front door of Rockefeller. In a paper written with Robert Marshak (University of Rochester), assistant professor Foldy presented a calculation of the expected cross-section for the production of π mesons in nucleon-nucleon collisions. The authors point out that the new accelerators will be able to produce these particles, seen hitherto only in cosmic rays. The approach is described so: "…we have regarded meson production as a second-order process in which one step consists of the creation of a meson by one of the nucleons, and the other step consists of the scattering of the resulting nucleon by the second nucleon via the nuclear potential between them." The resulting cross-sections are given for two forms for the potential and for three energies. (*Phys.*

Rev. **75** 1493 1949) Later developments in this field, for example the role of excited states of the nucleon, superceded these calculations.

The FW transformation

A second paper was written during Foldy's year at Rochester, this time with the Netherlander Siegfried A. Wouthuysen. It is this paper for which Foldy is best known today. With the advent of quantum mechanics, in which particles are shown to behave like waves, Schrödinger proposed a wave equation which was extremely successful in describing such things as the hydrogen atom, including the prediction of all its energy levels. Later, Dirac proposed a wave equation which is applicable to *relativistic* systems. The astonishing thing about the Dirac equation was that it had solutions for both positive and negative energies and for two values of a new undefined quantum number. It was soon understood that the negative energy states describe anti-particles such as the positron, and that the new quantum number specified the two possible orientations of the spin of the particle. The bad news was that the four components of the predicted wave functions were mathematically entangled so that it was difficult to write a wave function for a given electron in a given spin state. Foldy and Wouthuysen discovered a transformation (i.e. a mathematical operation) which changed the wave function to a form which had separate solutions for positrons and electrons and separate solutions for spin-up and spin-down. The new form makes it easier to interpret the solutions of the Dirac equation, as well as simplifying the analysis of the interactions of Dirac particles with electromagnetic and other fields. Furthermore, it makes it easier to mesh the predictions of the Dirac theory with those of the non-relativistic theories in the energy region where both theories should apply. "On the Dirac Theory of Spin ½ Particles and its Non-relativistic Limit", (*Phys. Rev.* **78** 29 1950.)

Foldy soon followed up on these ideas with a short letter which pointed out that the application of the Foldy-Wouthuysen (FW) transformation to the equation describing a neutron in an external electromagnetic field predicts the value of the electron-neutron interaction (roughly parameterized as a potential well with depth 3.9 keV). He pointed out that this value is consistent with that deduced from shifts of spectral lines associated with the additional neutrons in certain isotopes. According to Foldy's colleague, Phil Taylor, Foldy "thought of this while brushing his teeth one night; worked it out and submitted to *Phys. Rev.* the next day." "The Electron-Neutron Interaction" *Phys. Rev.* **83** 688 1951.

Two back-to-back papers the following year pursued additional applications of the FW transformation. In the first, Foldy further develops the ideas of the FW paper. He explains how the interaction of Dirac particles with external electromagnetic fields, when suitably transformed, can be broken into a sum of terms representing the various moments of the charge and current distributions with the moments of the interacting field. "The Electromagnetic Properties of Dirac Particles" *Phys. Rev.* **87** 688 1952.

The second paper addresses new measurements of the electron-neutron interaction based on the scattering of thermal neutrons by atoms of monatomic gasses. (These atoms have a full shell of electrons on their outside surface which the neutron can sample.) Foldy used the techniques elaborated upon in the preceding paper to isolate the contribution of the neutron's intrinsic magnetic moment. He then ascribed the difference between the experimental value of the strength of the interaction and his calculated value (a rather wide-open 320 ± 400 eV) to a possible contribution from "meson theory". In the latter, the neutron is expected to spend part of the time as a proton and a negative pi meson, so that the electron might see the fleeting virtual charged particles. Clearly, the theory was a bit ahead of experiment, but at least an upper limit on the "meson" contribution was implied. "The Electron-Neutron Interaction" *Phys. Rev.* **87** 693 1952. A later paper returns to the neutron-electron interaction, where Foldy considers the possibility that the neutron has internal structure, with separated electric charges, which the electron will feel. (This was well before the three charged quark picture of the nucleons appeared.) "Neutron-Electron Interaction" *Rev. Mod. Phys.* **30** 471 1958. "Electric polarizability of the neutron" *Phys. Rev. Lett.* **3** 105 1959. In Chapter 8 we mentioned Marshall Crouch's experiments on the neutron-electron interaction and how they related to the Foldy calculation.

Years later, perhaps in the 1980's, Foldy wrote a five page description of the development of the FW transformation: "Origins of the FW Transformation: A Memoir". Because this paper has not been published elsewhere and the FW technique represents a significant advance in quantum mechanics, I have included, with permission from the Foldy family, the entire text in **Appendix G**.

The effective electric charges of elementary particles are modified by the presence (in the vacuum) of particle-antiparticle pairs which pop into existence for a very short time and whose electric charges change the electric fields around charged particles. This is called "vacuum polarization". Foldy pointed out that the observed effects were well explained by virtual electron-positron pairs, and that there was no need (or room) for proposed pairs of some new lighter entity. "Elementary particles and the Lamb-Retherford line shift", *Phys. Rev.* **93** 880 1954. This was followed by calculations of the effects of vacuum polarization on low energy proton-proton scattering, and on Coulomb energies in nuclei, both of which depend on the effective charges of the protons. "Some physical consequences of vacuum polarization" *Phys. Rev.* **95** 1048 1954.

Scattering theory

In the 1930's, experimental physics was largely concerned with the properties of atoms, in the 1940's and 1950's the major interest was the atomic nucleus, and in the 1960's, it was the properties of more fundamental particles, like protons, electrons, pions. This progression toward the study of smaller structures followed the development of more and more energetic beams of projectiles: 10 keV x-rays and electrons to scatter from atoms, 10 MeV gamma rays and protons to aim at nuclei, 1 GeV electrons, protons and pions to probe even more deeply. The common thread is "the scattering process", and the challenge to the theorist is to deduce the properties of the target and the nature of the interaction between projectile and target from the experimental observations. This challenge would be taken up by Foldy, by his fellow theorists Tobocman and Thaler, and later by a half-dozen other members of the Case and WRU departments.

An aside on scattering. What happens when beam particle ***a*** hits a target particle ***A***? Each of these particles can be as simple as an electron or as complex as a large atom. The simplest thing that can happen is ***a A*** → ***a A***, i.e. the incident particle bounces off the target particle, giving it some of its kinetic energy. There is no change in the mass or internal structure of either particle. This is **elastic** scattering. Analysis of the elastic scattering of alpha particles by matter led Rutherford to discover that most of the mass of the target atom resides in a tiny positively charged ball, the nucleus.

But many other things can happen in a scattering experiment. The particle ***a*** might be absorbed and a new particle ***b*** created: ***a A*** → ***b B***; the target ***A*** might become excited, reducing the energy of the incident particle: ***a A*** → ***a A****; new particles might be created or blasted out of the target: ***a A*** → ***b c B***; or ***a*** might interact with some component of the target, chipping off a piece. All these processes take place through forces: electromagnetic, strong or weak; or, in terms of particle exchange, by the emission and absorption of the force carriers: photons, gluons, or weak bosons.

The experimentalist measures the reaction cross sections (i.e. the probability that a given reaction will occur) and the differential cross sections (i.e. the probability that the scattered particles will travel in a given direction). It is the job of the theorist to deduce what is interacting with what, and what forces come into play.

Collisions of small particles at high energies are governed by quantum mechanics, where particles are waves, and systems of particles exist in certain allowed energy states, and collisions take place in certain allowed angular momentum states. The theorist treats the incoming and outgoing particles as waves, and the collision as characterized by a finite number of "partial waves" (i.e. angular momentum states). A quantum mechanical analysis of the scattering process yields a set of coupled partial differential equations for the partial waves. These are too difficult to solve exactly. This is where the theorist uses a bag of mathematical tricks to devise approximate solutions. Many of the papers by Foldy, Tobocman, and Thaler, and, later, of Kowalski, Kisslinger, Shakin and Brown are concerned with this sort of calculation.

One example of Foldy's important contributions to scattering theory is described in a paper he wrote with R. F. Peierls at Brookhaven in the summer of 1962. In this work, rules were derived for interactions which proceed by the exchange of a virtual particle. In particular, the theory sets limits on the values of the isotopic spin of the exchanged "entity". Isotopic spin, then called T, is a quantum number related to the number $(2T+1)$ of particles in a family of particles, e.g. $T = \frac{1}{2}$ for the two member family of neutron and proton or $T = 1$ for the 3 member family of π^+ π^0 π^-. "Isotopic spin of exchanged systems", *Phys. Rev.* **130** 1585 1963.

The Versatile Foldy

Foldy's interests were broad, concerning a wide range of physics theory. For at least four decades, the department benefited from his willingness to discuss, enlighten, and advise on almost any corner of physics research. A lot of physics was accomplished at the daily roundtable lunches in the Rockefeller building. A clear indication of the breadth of Foldy's research interests can be found in Appendix B where the titles of his students' masters and doctoral dissertations are listed. The photo of Foldy in **Fig. 9-2** was taken around 1980.

Fig. 9-2. Les Foldy.

Solids

A paper on a **classical electromagnetism** problem which Foldy wrote with colleague J. R. Reitz will be described in Chapter 17. In a paper related to the analysis of the experiments by W. L. Gordon (Chapter 11), Foldy looked at the theory of electronic energy levels in solids. "The present paper can be considered a 'derivation for experimentalists' of the (theoretical) inversion formula". "Inversion Scheme for Obtaining the Fermi surface from the de Haas-van Alphen Effect" *Phys. Rev.* **170** 670 1968. Later works present calculations of the frequency spectrum of lattice vibrations for bcc and fcc lattices and the phase transitions between them. "Phase Transitions in a Wigner Lattice" *Phys. Rev.* **B3** 3472 1971. "Electrostatic stability of Wigner and Wigner-Dyson lattices" *Phys. Rev.* **B17** 4889 1978.

Atoms

An example of Foldy's work in atomic physics is a comparison of two theoretical methods of calculating the total binding energy of the electrons in an atom as a function of atomic number. "A Note on Atomic Binding Energies" *Phys. Rev.* **83** 397 1951. Another is the work on "anticrossings" with colleague Tom Eck which will be described in Chapter 12.

Nuclei

Foldy did some work on the structure of nuclei, looking at how the magnetic properties of the nucleus could result from the individual magnetic moments of the constituent nucleons. This early paper was co-authored with undergraduate Fred Milford, whom we shall meet shortly. "On the Deviations of Nuclear Magnetic Moments from the Schmidt Limits" *Phys. Rev.* **80** 751 1950. In a paper written during a summer spent at Brookhaven, Foldy looked at the conditions for a nucleus to decay through the *simultaneous* emission of an electron and a photon. "Beta-Gamma Emission through Virtual States" *Phys. Rev.* **128** 1776 1962.

Energy loss

The processes by which particles lose energy as they pass through matter was of special interest to experimenters in their design of detectors. "Diffusion of High Energy Gamma-Rays through Matter. *Phys. Rev.* **81** 395 and 400 1951, **82** 927 1950 (with PdD student Richard K. Osborn). "Energy Degeneration of Cosmic-Ray Primaries" *Phys. Rev.* **81** 13 1951 (Fred Milford's BS thesis).

Accelerators

In 1961, Foldy, who often spent his summer months at Brookhaven Lab, did some thinking about how to improve the beams in particle accelerators. Recall that he had made significant contributions to this area back in 1947, even before he started his graduate studies. "Method for expanding the phase-stable regime in synchronous accelerators" *Nuovo Cim.* **19** 1116 1961.

Many-body problem

Foldy and his colleague Bill Tobocman (to be introduced later in this chapter) wrote a two-page paper pointing out some basic drawbacks in the application of accepted scattering theory to systems of three or more interacting particles. An example they give is scattering of a neutron from a bound state of a carbon nucleus and proton. "Application of Formal Scattering Theory to Many-Body Problems" *Phys. Rev.* **105** 1099 1957.

With his graduate student Richard Krajcik, and twenty years after the publication of the FW transformation paper, Foldy responded to claims by several authors that there were some problems with the application of FW theory to the electromagnetic interactions of relativistic particles. Foldy and his student point out certain omissions in the challengers' calculations and conclude that with their inclusion, the FW form does not violate accepted theorems, since, as they playfully state, " Theorems of such impeccable lineage demand proper respect by electromagnetic interaction Hamiltonians." "Electromagnetic Interactions with an Arbitrary Loosely Bound System" *Phys. Rev. Lett.* **24** 545 1970.

Strong interactions

In a 1978 paper, Foldy examined data coming from the multi-GeV accelerators in which new short-lived resonant states were being discovered. He compared his calculated phase shifts with data from Brookhaven bubble chamber experiments. "A single-term separable potential with a simple analytic form which can be made to fit either experimental πN phase shifts in the $\Delta(1232)$ channel or the experimental $\pi\pi$ phase shifts in the $\rho(767)$ channel over a much larger range in energy than has been possible with most previous single-term separable potentials is described." (Reactions such

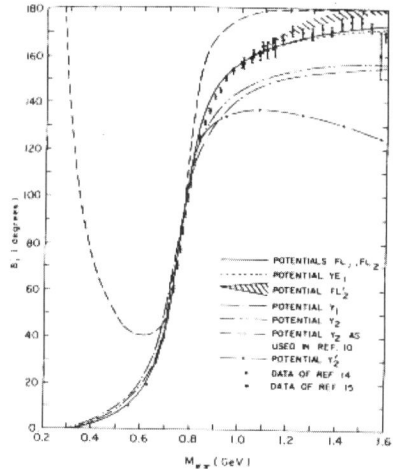

Fig. 9-3. Comparison of calculated phase shifts with experimental data.

as these will be described in Chapter 16.) **Fig. 9-3** shows the experimental $\pi\pi$ phase shifts along with a series of curves corresponding to different choices of potential. "Families of improved separable interactions for πN and $\pi\pi$ scattering for applications to three-body problems" *Phys. Rev.* **D17** 3065 1978. Foldy then sought a more fundamental description of high energy interactions. He concludes: "In summary, as an alternative to one-pion exchange S-matrix calculations, one may learn quite a bit concerning meson exchange, electromagnetic and weak currents by the application of various symmetries and conservation laws." "Symmetries, conservation principles, and the phenomenology of meson exchange currents" *Mesons in Nuclei* North-Holland Publishing Co. 1979.

In 1986, Foldy and his last graduate student, Sam Stansfield, undertook a "classical-quantum" project. The result was an interesting paper on the solution of the Schrödinger equation for a particle moving in a hypothetical potential: the superposition of a $1/r$ (Newton or Coulomb) potential plus a concentric harmonic oscillator potential. *Phys. Rev.* **A35** 1415 1987.

Les Foldy was appointed the first "Case Institute Professor" in 1966. He supervised the work of more than twenty doctoral students, many of whom went on to academic careers. He regularly spent summers at Brookhaven Laboratory and enjoyed productive sabbaticals at the Bohr Institute in Copenhagen and at CERN in Geneva. In a 1953 letter to Foldy, Niels Bohr writes, "you will be most welcome indeed…we have all here followed your work, which has led to so many promising results, with keen interest." It is the opinion of many of his colleagues that Foldy could easily have found a professorship at any of the larger and more prestigious American universities. He decided, however, to remain with his friends at Case and CWRU and, with his wife, Roma, to raise his family in his hometown.

In 1993, the American Institute of Physics and the American Physical Society celebrated the 100th anniversary of the *Physical Review*. As part of that event, a committee of the AIP and APS compiled a list of the most important papers appearing in that journal. About 1000 papers were included. Les Foldy took great pleasure in going through the complete list to identify those with a departmental connection. He wrote: "Of the 979 papers, I found that at least 21 of them had an author or authors who had a connection with CWRU in that they had been students, professors, or research associates at CWRU or one of its predecessor institutions." His list of these 21 papers appears in **Appendix E**. It includes two of his own papers: the famous FW paper and the paper on many-body scattering which he wrote with Bill Tobocman.

Foldy remained active in the CWRU physics department well beyond his retirement in 1990. In April of 2000, the department celebrated Foldy's long career of scholarship and teaching in a day-long commemoration of his 80th birthday. The six principal speakers traced the development of the quantum theory of particles and Foldy's contributions to it. (Gerardus 't Hooft, James D. Bjorken, Kenneth L. Kowalski, Frank Wilczek, Mark B. Wise, and Philip L. Taylor.) Les Foldy died in January of 2001.

Funding

Between around 1950 and 1971, the experimental and theoretical nuclear and particle physics programs at Case, and then CWRU, were supported by a blanket contract with the Atomic Energy Commission. This covered the work of Schrader *et al.* at the van de Graaff (Chapter 7) and Willard's "medium energy" group (Chapter 16), as well as all the theorists of this chapter. This arrangement ended in 1971 when Willard got DOE funding; and Thaler and Tobocman joined Ken Kowalski and Carl Shakin (Chapter 13) to establish a nuclear theory contract with the National Science Foundation. The NSF also funded Bob Brown (Chapter 13) in a separate particle theory contract. Some of the later theoretical work was funded by NASA.

Martin Klein

Chairman Robert Shankland had come to know Martin Jesse Klein when he was working on underwater sound detection during the war. On his return to Case, he would offer a position to the young theorist who was working for the government in Washington. Klein, who was five years younger than Foldy, would add a new and complementary dimension to the theory program at Case. While Foldy was, at that time, concerned mostly with nuclear and particle physics, Klein was interested in thermodynamics and the statistical mechanics of many-body systems. After finishing his doctoral work at MIT in 1948, Klein came to Case the following year. He was to play an important role in the department for almost two decades, including acting as co-chairman with WRU's Chandrasekhar at the time of the federation. His photo is shown in **Fig. 9-4**.

Fig. 9-4. Martin J. Klein

Statistical Mechanics

In the 1950's, Klein became interested in some of the condensed matter experimental work being done in the department. He published two papers on thin ferromagnetic films, written with grad students Robert Smith and Solomon Glass. They presented calculations, based on Bloch spin-wave theory, of the magnetization of such films as a function of temperature and film thickness, exactly the properties being measured by his new colleague Richard Hoffman (Chapter 12). "Thin Ferromagnetic Films" *Phys. Rev.* **81** 378 1951 and *Phys. Rev.* **109** 288 1958.

Most of Klein's work done at Case concerned basic principles of thermodynamics and statistical mechanics, often consisting of commentaries on and clarifications of earlier work by key players in these fields. The following selection of titles illustrates the wide scope of his work in the 1950's. "Classical Spin-Wave Theory" *Phys. Rev.* **80** 1111 1950. "The Ergodic Theorem in Quantum Statistical Mechanics" *Phys. Rev.* **87** 111 1952. "Principle of Minimum Entropy Production" *Phys. Rev.* **96** 250 1954. "Principle of Detailed Balance" *Phys. Rev.* **97** 1446 1955. "Generalization of the Ehrenfest Urn Model" *Phys. Rev.* **103** 17 1956. "Negative Absolute Temperatures" *Phys. Rev.* **104** 589 1956. "Grüneisen's Law and the Third Law of Thermodynamics" *Phil. Mag.* **3** 538

1958. "Thermal Expansion Coefficient of Solid He^3" *Phys. Rev. Lett.* **5** 363 1960. "The Laws of Thermodynamics" *Rendiconti Soc. Ital. Fis.* **10** 1 1960.

History

In the 1960's, Klein began to concentrate on the *history* of theoretical physics, an occupation he would pursue for the next four decades. He spent a year as a National Research Council Fellow in Dublin and later he was a Guggenheim Fellow at the Lorentz Institute in Leiden. While there, he wrote on Paul Ehrenfest's contributions to quantum statistics and edited Ehrenfest's collected papers. Returning to Case, he published a series of papers on the contributions of Planck and Einstein to quantum theory. He translated and wrote commentaries on letters about wave mechanics written by Einstein, Schrödinger, Planck and Lorentz. "Ehrenfest's Contributions to the Development of Quantum Statistics I" *Proc. Kon. Ned. Akad.* **62** 41 1959. "Max Planck and the Beginnings of the Quantum Theory" *Arch. Hist. of Exact Sci.* **1** 459 1962. "Einstein's First Paper on Quanta "*The Natural Philosopher* **2** 1963. "Einstein and the Wave-Particle Duality" *The Natural Philosopher* **3** 1964. "Einstein, Specific Heats, and the Early Quantum Theory" *Science* **148** 173 1965. "Thermodynamics in Einstein's Thought" *Science* **157** 509 1967.

After his short stint as acting chairman of the Case department, Klein moved to Yale University in 1967, joining the faculty as Professor of the History of Physics. He was appointed the Eugene Higgins Professor in 1974. Klein continues to write extensively on major physicists of the 19th and early 20th centuries. He has served as senior editor of four volumes of Einstein's collected papers.

Milford – nuclear models

Frederick W. Milford completed his bachelor's degree at Case in 1949, having published two papers with his advisor Les Foldy. These papers, mentioned above, were on very different topics: nuclear magnetic moments and energy loss by cosmic rays. The first of the two was referenced in each of the 1975 Nobel lectures by A. Bohr and B. Mottelson, not bad for an undergraduate research project.

Milford went on to MIT where he earned his doctorate in 1952; he returned to Case as an assistant professor. His work at Case continued his earlier research in nuclear theory. "The odd-nucleon-plus-liquid-drop-model of heavy odd nuclei" (*Phys. Rev.* **93** 1297 1954) and "Projection operator for the Rarita-Schwinger equation" (*Phys. Rev.* **98** 1488 1955) Milford coauthored a textbook with his colleague John Reitz: "Foundations of Electromagnetic Theory" (Benjamin Cummings 4th ed. 1993). He left the department in 1959 to take a position in industry. He eventually became the Director of the National Security Programs of Battelle, the international contract-research and technology think-tank in Columbus, where he directed 1200 staff members doing research for government and industrial sponsors.

Winterberg: nuclear dreams

In the fall of 1959, a young German nuclear theorist was invited to join the department. Thirty-year old Friedwart Winterberg had done his doctorate in natural science (Göttingen, 1955) on shell-model analyses of nuclei. He came to the United States under the "Defense Scientists Immigration Program" of the Department of Defense. This agency recruited foreign scientists "to assist in maintaining the United States in the foremost position in all phases of research and development."

Winterberg's interests were much wider, however, than nuclear theory. In Germany, he had published extensively on nuclear reactors and on nuclear propulsion of rockets, research areas clearly of interest to the US DOD. The titles of a few of his papers illustrate his potential value as a theoretical physicist, expert in strategic applications of physics: "Relativistic time dilation in an artificial satellite", "Non-linear behavior of reactors", "Nuclear fission and maximum power rockets", "Nuclear plasmas and magnetic combustion-chambers for rockets". At Case, Winterberg worked in magneto-hydrodynamics, publishing a paper in which he presented a design for a laboratory demonstration of self-sustaining hydrodynamic generators of magnetic fields. "Experimental test for the dynamo theory of earth and stellar magnetism", *Phys. Rev.* **131** 29 1963. After four years at Case, Winterberg moved on to the Desert Research Institute of the University of Nevada, where for forty years he would continue to generate ingenious ideas for the applications of nuclear fission and fusion. *Physics Today* recently reported that he was the 1991 recipient of the Oberth - von Braun Medal for his achievements in thermonuclear propulsion.

Tobocman: scattering problems

In 1960, Foldy and Klein were joined by two nuclear/particle theorists: William Tobocman and Roy Thaler. They were hired by Fred Reines who had just taken over as chairman. Nuclei and particles were indeed the hot topics in theoretical physics. Particle accelerators in the billion electron volt (GeV) range were producing beams of electrons, protons, mesons, and even light nuclei. And these beams were being used to probe all manner of targets, sometimes to learn about the internal structure of the target, sometimes to create new types of particles, sometimes to learn about the forces behind the interaction between the beam and the target particles. As was the case for Les Foldy, the mathematical techniques for analyzing scattering processes were the major interest of each of these young researchers.

Fig. 9-5. Bill Tobocman

William Tobocman came to Case in 1960. Born in Detroit in 1926, Tobocman had done his bachelors degree and his PhD (1953) at MIT. His doctoral work on the theory of deuteron stripping reactions was done under the direction of Francis Friedman. This was followed by a one-year post-doctoral position at Cornell, two years at the Institute for Advanced

Study at Princeton, a year as an NSF fellow in England at Birmingham, and three years on the faculty of Rice University. The photo of Tobocman in **Fig. 9-5** was taken around 1980.

While at Princeton, Tobocman derived a rigorous "sum-over-histories" formulation of quantum mechanics which provided an alternative approach to the formulation devised by Feynman. The sum-over-histories technique describes the passage of a system of interacting particles from a given initial state to a particular final state by summing over all possible intermediate paths. The Feynman approach was applicable to systems which can be described "classically". Tobocman's contribution successfully extended the approach to systems such as those involving spin ½ particles. "Transition Amplitudes as Sums Over Histories" *Nuovo Cim.* **3** 1213 1956.

Stripping reactions

During his postdoctoral years and the three years at Rice, Tobocman continued work on "stripping reactions". In these reactions, an incident deuteron breaks up upon collision with a nucleus. It leaves its neutron in the nucleus, while the remaining proton leaves the scene and can be detected. The angular distribution of the outgoing protons is sensitive to the angular momentum of the captured neutron, so that these measurements could provide a probe of the properties of the nucleus.

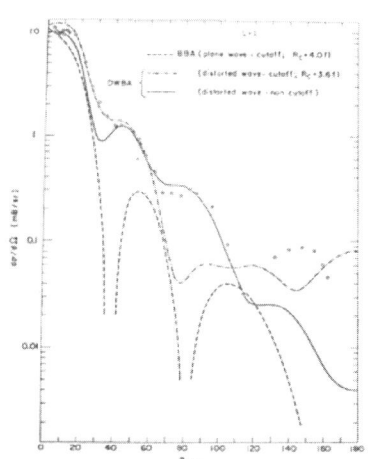

Fig. 9-6. Comparison of theory (top curve) with experiment (circles).

These calculations become more complex as one takes into account the forces between the nucleus and the incoming deuteron and outgoing proton. "Theory of the (d,p) Reaction" *Phys. Rev.* **94** 1655 1954 (at Cornell); "Numerical Evaluation of Deuteron Stripping Cross Sections and Polarizations" *Phys. Rev.* **115** 98 1959 (at Rice). Earlier calculations of this type of reaction were done "in the Born Approximation", *i.e.* the incoming and outgoing particles were described as plane waves, unaffected by Coulomb and nuclear forces. With the advent of "powerful" computers (in this case the IBM 650), Tobocman and his collaborators were able to refine the calculations, introducing suitable distortions into the single particle wave functions. "Distortion Effects in Deuteron Stripping Reactions with Low Q Values" *Phys. Rev.* **124** 1496 1961 (at Case). Tobocman was a pioneer in the use of electronic computers for nuclear physics calculations.

The results of a typical calculation are shown in **Fig. 9-6**. The theory is compared with experimental data for the deuteron stripping reaction on carbon-12. The solid curve shows how the theory can track the considerable structure in the data. "Distorted-wave Born approximation analysis of C^{12}(d,p)C^{13}" *Phys. Rev.* **136B** 1682 1964. The results of these and similar calculations were brought together in a monograph published in 1961 by Oxford University Press: "Theory of Direct Nuclear Reactions".

In the mid-1960's, Tobocman authored a series of papers on various aspects of the theory of nuclear reactions, including a half dozen with M. A. Nagarajan. This younger colleague will be introduced in Chapter 13. Tobocman spent the school year 63-64 as a Sloane Fellow at the Weizmann Institute in Israel.

Elementary particle interactions

Tobocman briefly moved up the energy scale to apply his techniques to reactions in the GeV range which were then being measured at the new particle accelerators. The observation that many collisions were "peripheral" or "glancing" indicated that "virtual" mesons hovering around a proton or other projectile were responsible for the interaction. Some of these experiments will be described in Chapter 16. A great deal of accelerator-based high energy data was well described by theories based on this idea, where the mass and even the spin of the "exchanged" particle would determine the outcome of the scattering. Tobocman examined several different interactions in the GeV range: p-p, π-p, pbar-p, etc. Two papers written with grad student David Giltinan on ρ production presented similar calculations. Here the incident π combines with the virtual π to form the ρ, and the target proton acts pretty much as a spectator to the event. "Distorted Wave Theory of the One-Meson-Exchange Reaction" *Phys. Rev.* **143** 1252 1966. "One-Meson Exchange Calculation of the $\pi^+ p \to \rho^+ p$ Reaction " *Phys. Rev.* **185** 1849 1969.

Three undergraduate physics majors were taken under Tobocman's wing, each one co-authoring with him a *Phys. Rev.* paper on nuclear reactions. So positive was the experience that all three, Myron Pauli, Richard Goldfinger, and Andrew Lewanski, remained at CWRU to complete their PhD's with him.

Many-body scattering

Tobocman moved on to calculations which treated the target nucleus as a collection of individual particles, in an approach described as "many body theory". An early paper written with Les Foldy was described above. It pointed out inadequacies in the accepted theoretical approach to this problem.. This work would serve as the starting point for a program which Tobocman would later pursue at Case for more than a decade. The calculations, as detailed in over 80 papers, would take into consideration more and more of what happens in nature: distortion of the particle waves, absorption effects, structure in the target nucleus, and the role of excited states. The culmination of this work was a comprehensive 89-page review (including 244 references), entitled "Calculable Methods for Many-body Scattering" *Rev. Mod. Phys.* **55** 155 1983.

Heavy ions

The introduction of heavy ion accelerators at several laboratories allowed the study of more complex collisions (nuclei on nuclei), and Tobocman and his colleagues and students would take an interest in their analysis. "Elastic Channel Contributions from Particle Transfer Between Heavy Ions" *Nucl. Phys.* **A202** 561 1972.

In some reactions, the incident projectile would leave two or more nucleons in the target. An example is ^{6}Li + ^{16}O → ^{4}He + ^{18}F, in which agreement with the data required that the theory should include both direct and exchange mechanisms: *i.e.* a deuteron jumps out of the lithium and into the oxygen to make the fluorine **or** an alpha jumps out of the oxygen and the oxygen absorbs the lithium to make the fluorine. "Analysis of exchange effects in the ^{16}O(^{6}Li,^{4}He)^{18}F and ^{16}O(^{3}He,^{1}H)^{18}F reactions" *Phys. Rev.* **C15** 686 1977.

A different kind of nuclear reaction is spontaneous radioactive *decay*, where sufficient energy is already sitting in the nucleus. The process will occur if and when the components of the nucleus jiggle themselves into a configuration where some sort of energy barrier is overcome – like shaking a bowl full of marbles until one collects enough energy to fly over the edge. "Calculation of the Lifetime of a Metastable System" *Phys. Rev.* **174** 1115 1968. "Comparison of methods for calculating decay lifetimes" *Phys. Rev.* **C17** 2205 1978. "Alternative treatment of exchange effects in the theory of radioactive decay" *Phys. Rev.* **C18** 1857 1978.

Scattering of ultrasound

By the 1980's, interest in scattering from nuclei had largely been replaced by high energy studies with multi-GeV probes incident upon simple targets, where the collisions are best described by quantum chromodynamics and more fundamental components: leptons, quarks and gluons. Many of the calculational techniques were similar to those used for lower energy nuclear work. Tobocman decided, however, rather than moving into "particle physics", he would apply his skills to a very different research area: the scattering of ultrasound waves.

Starting around 1985, he worked in the exciting field of "medical imaging". With the development of intense and well-controlled sources of ultrasound and of fast computers, the medical applications of ultrasonic imaging were proliferating. Tobocman found a home in this new field, successfully applying inverse scattering techniques. A *direct* scattering analysis determines the properties of the scattered waves from knowledge of the incident beam and the target profile. An *inverse* scattering analysis determines the properties of the target profile from knowledge of the scattered waves and the incident beam. His work on the application of inverse scattering theory to medical imaging with ultrasound has been published in two dozen articles in the *Journal of the Acoustical Society of America, Ultrasonics*, and similar dedicated journals. Methods of reconstruction developed by Tobocman and his students have been "found to yield high resolution images of small tissue structures that are free of speckle". Their method, in fact, has been awarded a U.S. patent.

Tobocman's experience in being able to move from traditional "pure" physics research into a very different area involving applications to technology has been shared by an ever growing number of theorists and experimentalists. Several other members of the CWRU physics department have succeeded in similar transitions, for example to magnetic resonance imaging and to biomagnetic diagnostic technology. We shall describe some of this work in later chapters.

Bill Tobocman retired from the department in 2002 after more than 40 years of teaching, service and research at CWRU. He was advisor to eleven doctoral students and a long series of post-doctoral associates.

Raphael (Roy) Thaler

Roy Thaler was born in Brooklyn in 1925, earned an AB at New York University in 1947 and his doctorate at Brown University in 1950. His dissertation was on a problem in atomic physics: calculation of the electron affinity of the sodium atom. He held a post-doctoral position at Yale in 1952-54 with Gregory Breit, studying relativistic corrections to magnetic moments of nuclei. He then produced a series of papers on nucleon-nucleon scattering in the 200 MeV range. He moved to a position at the Los Alamos Scientific Laboratory from 1955 to 1957 where he studied Coulomb excitation of nuclei. *(This involves calculating the probability that a target nucleus would be raised to a particular excited state when a charged particle passes nearby.)* A photo of Thaler, taken around 1980, is shown in **Fig. 9-7**.

Fig. 9-7. Roy Thaler.

During the following two years, Thaler alternated between MIT and LASL, continuing his study of "intermediate energy" scattering processes, before joining the Case department in 1960. During his tenure at Case, Thaler would spend many summers at the Los Alamos Laboratory, maintaining a connection with the research program there throughout his career. His habitual western-style string-tie declares his fondness for the lifestyles of New Mexico. Thaler would spend twenty years in residence at CWRU. He co-authored about fifty papers during that period. He worked with his colleague Carl Shakin during the three years Shakin was at Case, then with post-doc Alan Picklesimer and later with Peter Tandy who was at Kent State. Most of these collaborative works included comparisons of the theoretical calculations with experimental data coming from a variety of accelerator-based experiments.

Analysis of accelerator data

An early paper, published with Tobocman's graduate student Giltinen, was on the proton-proton interaction. They compared their model, one based on a potential with a hard-core surrounded by monotonic attraction, with experimental phase shifts at 310 MeV. The data came from pp-scattering experiments at the cyclotron of the Radiation Laboratory at UC Berkeley. "Nonlocal Nucleon-nucleon Interaction" *Phys. Rev.* **131** 805 1963.

Working with Case colleague John Rix, Thaler proposed techniques for unraveling the strong and electromagnetic contributions to the scattering process. From the abstract: "Specific prescriptions, involving only the use of observed scattering data, are derived for obtaining the connection between the idealized "strong" scattering amplitude and the observed full scattering amplitude." "Separation of Strong and Electromagnetic Effects in Charged-Particle Scattering" *Phys. Rev.* **152** 1357 1966. At about this time, Thaler co-authored a book with Leonard S. Rodberg. "Introduction to the Quantum Theory of Scattering", Academic Press, New York 1967.

The backward scattering of neutrons with energies up to 750 MeV from protons was the subject of a subsequent paper. The data came from experiments at the Penn-Princeton accelerator. Collisions in which the neutron emerges from the collision traveling (in the center of momentum frame) in a direction opposite that of the incident neutron are interpreted as "charge-exchange", in which the neutron and proton switch identities by exchanging a charged pion as they fly by one another. The "One-pion Contribution to Neutron-proton Charge-exchange Scattering" *Phys. Rev. Lett.* **25** 1065 1970.

In a paper written with colleagues Kowalski and Shakin, Thaler discussed the effects on the scattering process of the existence of bound states in the projectile-target system. Even though the total energy of the collision may be quite different from the energy of the bound state, the latter's existence will affect the interaction. This paper offered ways to take such effects into account. "Off-Shell Contributions of the Two-Particle Transition Matrix with Bound States" *Phys. Rev.* **C3** 1146 1971.

Scattering from nuclei

Thaler worked with a series of collaborators, including colleagues from within the Case department and post-doctoral assistants. Between 1971 and 1974, Thaler would publish 14 papers in APS journals with Carl Shakin (Chapter 13). These concerned various details of scattering processes, e.g. off-shell effects and center-of-mass motion in many-particle systems. For the next four years, Thaler and post-doc D. Ernst continued the program, looking at scattering of nucleons from nuclei and developing techniques for tracking the incident particle through the target nucleus, as it interacts with one nucleon after another. A paper with post-doc E. R. Siciliano proposes an interesting technique for describing this multi-step process as an expansion of terms, the first for collisions with a single nucleon, the second involving pairs of target nucleons, and so on. "Spectator expansion in multiple scattering theory" *Phys. Rev.* **C16** 1322 1977.

Further developments in this work involved the inclusion of distortions of the scattered waves, similar to the calculations by Tobocman as described above. Thaler and Siciliano were joined by Kowalski in looking at what happens to π mesons as they plough through nuclear material. "Composite-particle structure of pion-nucleus amplitudes" *Phys. Rev.* **C19** 1843 1979. *Phys. Rev.* **C22** 2321 1980. For the next four years, Thaler worked mainly with Siciliano, Pickelsimer, and Tandy. Quoting Peter Tandy: "In about mid-late 1981, Alan (Pickelsimer) joined with Roy and me in a major research project on elastic and inelastic N-nucleus scattering in response to some interesting high-precision data starting to come out of LAMPF at Los Alamos." These data included measurement of the analyzing power (left-right asymmetry) in the elastic scattering of polarized protons by a variety of nuclei. The very high statistics of the data allowed one to take the theory to a more detailed level. **Fig. 9-8** shows a sample of the data along with two sets of calculated theoretical predictions, the dashed curves representing a theory without "relativistic" corrections, the solid curve one with these corrections. The relativistic (or "Dirac Signature") corrections included the treatment of the incident proton as an emitter and re-absorber of virtual nucleon-antinucleon pairs in the field of the target nucleus. Even though the predicted effects are tiny, it can be argued that the Dirac model wins out (especially in the analyzing power data). "Characteristic Dirac Signature in Elastic Proton Scattering at Intermediate Energies" *Phys. Rev. Lett.* **52** 978 1984.

The Los Alamos accelerator continued to pump out high statistics data on scattering from nuclei, much of which attracted the attention of Thaler and his partners. Pickelsimer, Siciliano, Tandy, Case PhD Gary Chulick and Thaler all became members of the LAMPF scientific staff, some remaining for many years.

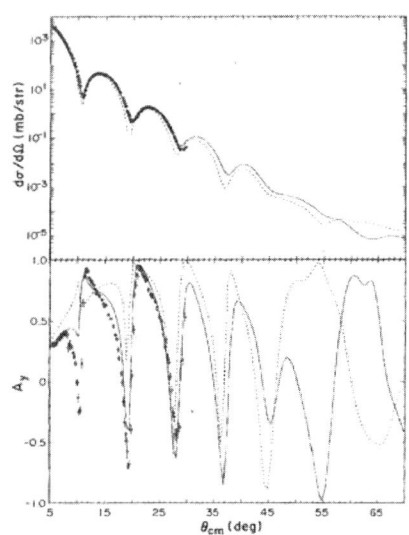

Fig. 9-8. Polarized proton scattering data compared with theory.

In the mid-80's, Roy Thaler took an extended leave of absence from CWRU, eventually resigning his position on its faculty. He continued to publish work on nucleon-nucleus scattering theory until the mid-1990's, while at LAMPF or Vanderbilt.

Chapter 10 WRU Experiment Takes Off

McCarthy,   **Beth,**   **Meeks,**   **Major,**   **Winter**
1937-56   1946-57   1948-55   1955-66   1951-54

Harry Mountcastle was the mainstay of the Western Reserve department for forty years, having been hired at age 32 by chairman Whitman in 1907. While Mountcastle had done some atomic spectroscopy early in his career, he was principally concerned with undergraduate teaching. He had, at most, the aid of one or two junior faculty. He succeeded Whitman as chair in 1919. He was assisted in the late 1930's by Cassius Curtis (Chapter 5). In 1937, Mountcastle was joined by John McCarthy who would remain in the department for two decades.

At the beginning of Chapter 7 we commented on the rapid growth in research funding which occurred during and following the second world war. The impact on the two physics departments was enormous, leading to the significant expansion of each. When the two institutions federated in 1967, largely as a result of pressure by the funding agencies, it would be the WRU physics faculty who would be most affected by the subsequent downsizing of the new CWRU department.

Fig. 10-1. John T. McCarthy.

McCarthy: the new electronics

John McCarthy (born 1912 in Canandigua, NY, BS Hobart College 1934) was hired in 1937 soon after completing his doctorate at Yale. **Fig. 10-1.** His dissertation was a study of α scattering by Ne and D atoms in a Wilson cloud chamber. From the observation of 600,000 tracks of α's from a thorium source, McCarthy found 25 examples of elastic scatters by neon nuclei and 30 by deuterons. From the measurement of the scattering angles and the lengths of the recoil tracks in the stereoscopic photographs, he was able to produce range vs. velocity curves for recoil neon ions and deuterons. (There is no mention of how much charge is on the neon ions; the plotted data look like they were all equally ionized, presumably singly.) These are shown in **Figs. 10-2 and 10-3**, where the ranges have been re-scaled to air at one atmosphere. This was cutting-edge work in nuclear physics for the time, making use of the radiation sources and the detector technology available in the 1930's. (*Phys. Rev.* **53** 30 1938) (Twenty years later, I was another Yale grad student, analyzing thousands of proton-proton collisions in an accelerator-based *bubble* chamber experiment.)

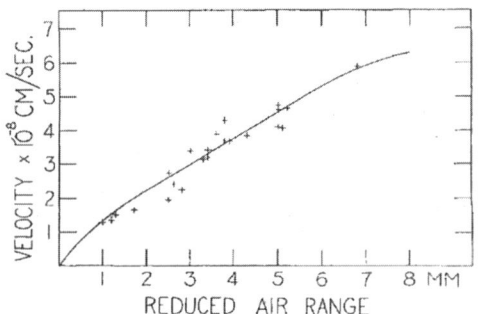

Fig. 10-2. Velocity vs. range for neon ions.

At Western Reserve, McCarthy and Chairman Mountcastle were responsible for most of the physics teaching duties in the period before and during the world war. In the mid-1940's, McCarthy became interested in electronics and the development of teaching-laboratory instrumentation. He published papers on improvements in vacuum-tube voltmeters and current stabilizers. His expertise with electronic circuits brought him into several collaborations, including one on electrolytes with Ernest Yaeger and Frank Hovorka of the chemistry department. McCarthy's main contribution was a circuit which produced ultrasonic waves.

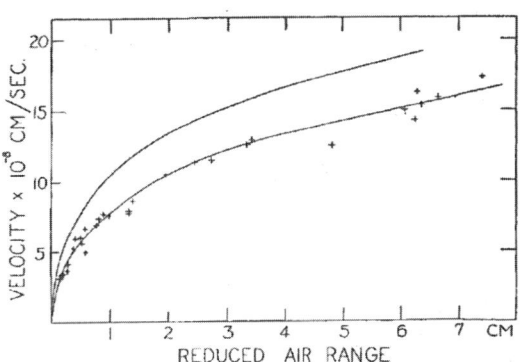

Fig. 10-3. Velocity vs. range for deuterons.

Chairman Mountcastle retired in 1945 and McCarthy would be the be the bridge to the new, post-war department, under the chairmanship of Richard Beth.

In a 1950 paper, McCarthy describes how he used the timing signals broadcast by the Bureau of Standards to calibrate a pendulum clock. (He mentions that the laboratory standard clock was out for repair.) The government had been broadcasting timing signals over their dedicated radio station, WWV, since 1920. McCarthy developed a vacuum-tube receiver and relay combination which would produce an audio signal when the pendulum and the WWV signal came into coincidence. He was essentially "beating" the two signals against one another. The pendulum turned out to have a period of about 1.0033 s, which means that it would get into phase with the WWV signal about once every five minutes. The observer would just measure the time between the "coincidence clicks" to determine the difference between the pendulum and WWV frequencies. The resulting measurement of the period was good to three parts per million. McCarthy concludes that using the WWV signals produced data "more consistent and reliable" than using the standard clock. That would hasten the demise of pendulum clocks as standards. *Amer. J. of Phys.* **18** 306 1950.

TABLE I. Mass differences for stable isobars which are most likely to exhibit double-beta disintegration.

A		ΔM_{calc} mMU	ΔM_{exp} mMU	Ref.
48	Ti–Ca	–4.50	–4.66	a
76	Se–Ge	–1.86	–2.5	b
78	Kr–Se	2.80	3.19	b
82	Kr–Se	–3.04	–2.7, –3.5	b, c
96	Mo–Zr	–4.12	–3.6	b
96	Ru–Mo	3.12	3.0	b
100	Ru–Mo	–2.48	...	
106	Cd–Pd	3.16	3.0	d
116	Sn–Cd	–2.96	–2.7	d
124	Te–Sn	–2.24	–2.1	d
124	Xe–Te	3.14	3.0	d
130	Xe–Te	–2.60	–3.5, –2.9	d, e
130	Ba–Xe	2.66	2.74	f
136	Ba–Xe	–2.76	...	
136	Ce–Ba	2.26	...	
150	Sm–Nd	...	–4.3	g

McCarthy found the time in the mid-1950's to do some nuclear physics research. This work was an effort to identify those nuclei which might possibly decay by

Fig. 10-4. Searching for candidates for double beta decay.

the simultaneous emission of two electrons (called double beta decay). He collected all the available information on the masses of pairs of stable nuclei having the same atomic mass number, A, but with atomic charge numbers, Z, differing by two units. In his Table (**Fig. 10-4**), he lists sixteen such isobaric pairs, along with the experimental mass difference (in milli-atomic mass units) and the theoretical mass difference based on the Wigner mass formula. In most cases, the experimental and theoretical mass differences were quite close. Ten of the states could decay by emission of two electrons (ΔM negative) and six by emission of two positrons (ΔM positive). (*Phys. Rev.* **95** 447 1954) Later in this chapter, we shall describe experimental searches performed in 1955 by another young experimenter, Rolf Winter. (In Chapter 8 we described an experimental search for double beta-decay done a decade later by Tom Jenkins in the low-background environment of a salt-mine.)

McCarthy left WRU in 1956 to take a position at the University of Cincinnati where he would spend the rest of his teaching career.

Beth: the angular momentum of light

Fig. 10-5.
Richard A. Beth.

In 1946, 38-year-old **Richard A. Beth** was appointed the seventh Perkins Professor and chair of the WRU physics department. (Seventy-year-old chairman Mountcastle had stepped down the previous year, and Professor Frank Hovorka of the chemistry department was acting chair of physics.) Beth was born in New York City, did his BS at Worcester Polytechnic Institute in 1929 and a Doctorate of Natural Philosophy in Frankfurt in 1932. He taught at WPI for eight years while working in research at Princeton. During the war he was head of a group studying "terminal ballistics and explosive effects" for the National Defense Research Commission. His photo is shown in **Fig. 10-5**.

At Princeton Beth designed and executed an important and historical experiment. While this work was done before Beth came to WRU, we shall describe it briefly. No doubt, it had a lot to do with his being offered the job. The idea was to send a beam of polarized light through a doubly refracting quartz plate suspended from a fine quartz fiber. Since the plane of polarization is rotated as the light passes through the quartz, one expects to observe a torque on the plate. This can be measured by observing the twisting of the suspension. In the diagram, **Fig. 10-6**, one can see the illuminating filament at the bottom, the parallelogram shaped Nicol prism which polarized the light, the suspended quartz disk (M), and a fixed disk (T). This second disk was silvered on its top to reflect the beam back through M, after shifting the phase of the light in such a way that the reflected light would re-enforce the torque on M. The measured torques were in the order of 10^{-9} dyne-cm. They were compared with theoretical values calculated from the measured intensity of the light, the rate of energy deposited in the disk, and the amount of rotation of the plane of polarization. The measurements were made for different light intensities and for different directions of polarization. In all cases the observed torque agreed with the calculated value. This 1936 experiment was the first direct observation of the angular momentum associated with a beam of polarized light. (*Phys. Rev.* 48 471 1935, and *Phys. Rev.* 50 115 1936) "Mechanical Detection and Measurement of the Angular Momentum of Light".

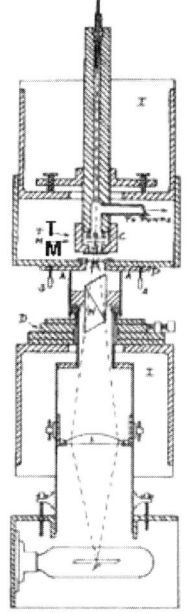

At WRU, Beth took an interest in a very different area: the mathematical analysis of stress and deformation in structural elements, e.g. the bending of beams under loading. He, along with two collaborators, published papers on the subject in the Journal of Applied Physics and the Journal of Applied Mechanics. He soon returned to the question of the mechanical effects of electromagnetic waves In 1952, he and Wilkison Meeks submitted a proposal to the Office of Naval Research to improve upon the 1936 experiment.

Beth stepped down as chairman in 1955 when John Major arrived to take over that responsibility. Beth subsequently took an extended leave to go to Brookhaven Laboratory. He and assistant professor Wilkison Meeks collaborated on the investigation of the focusing action of wave guides to be used in the alternating gradient proton accelerator. The AGS would become the centerpiece of the particle physics program at BNL. By 1957, Beth had decided to resign his professorship and to take a position at Brookhaven where he became part of the AGS design group.

Fig. 10-6. Observation of the angular momentum of a beam of light.

Meeks: the teaching labs

Wilkison Winfield Meeks (born in 1915) was hired by Beth as an assistant professor in 1948. **(Fig. 10-7)** He had spent the war years at the Naval Ordnance Laboratory. He had then completed his PhD at Northwestern University in 1947. His dissertation was on the properties of the nucleus of columbium, ^{93}Cb. "Hyperfine Structure and Nuclear Moments of Columbium" *Phys. Rev.* **72** 451 1947.

Meeks was to be a member of the WRU faculty for eight years, responsible for a large portion of the teaching duties. Chairman Beth wrote in Meeks' evaluation that he had taught expertly in nine different laboratory courses, including setting up three new labs. He was appointed University Marshall (antecedent to physicist Keith Robinson's appointment forty years later, cf. Chap. 16). During his tenure he worked with Beth on the design and construction of a device to measure the torque in a rotat-

Fig. 10-7. Wilkison W. Meeks.

ing shaft. The idea is based on the fact that the magnetic permeability of iron changes slightly if the sample is subjected to tension or compression. The application of torque to a shaft produces both tension and compression in the shaft which increase linearly with radius. In their invention, Meeks and Beth placed driving and pickup coils near the torqued shaft, so that the B field caused by a 500 Hz current in the driver is picked up by a galvanometer circuit on the opposite side of the shaft. The response on the galvanometer (deflection in mm in **Fig. 10-8**) was remarkably linear as a function of torque (shown in inch-lb). They applied for a patent for this invention in 1952. "Magnetic Measurement of Torque in a Rotating Shaft" *Rev. Sci. Instr.* **25** 603 1954.

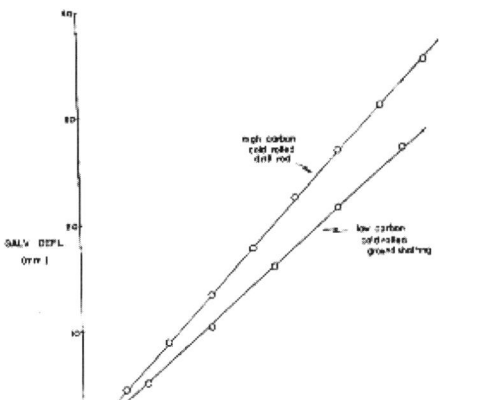

Fig. 10-8. Magnetic measurement of the torque on a shaft.

Meeks became interested in speech synthesis and presented papers on this subject at meetings of the Acoustical Society of America. He resigned his position at WRU in 1955 to go to the B.F. Goodrich Research Labs.

Winter: double beta decay

Rolf Gerhard Winter joined the Western Reserve department as instructor in 1951, just after receiving his doctorate from Carnegie Tech. **Fig. 10-9**. The 23-year old, Düsseldorf-born Winter had done his doctoral research on double beta decay. His thesis is a classical combination of theory and experiment. He had worked at Carnegie with E. Creutz and Lincoln Wolfenstein. "A Search for Double Beta Decay in Palladium" *Phys. Rev.* **85** 687 1953.

Winter continued this work at WRU and was promoted to assistant professor in 1952. As we described in Chapter 8, the theoretical interest in double beta decay concerned the "neutrinoless decay" hypothesis: if the neutrino is its own antiparticle, then the two neutrinos emitted along with the two electrons can devour one another. In a typical one of Winter's "runs", 42 grams of thin molybdenum foil were placed in a 24-cm diameter Wilson cloud chamber which was in a magnetic field of 790 Gauss. The chamber was expanded every 30 seconds, or so, until about 12 thousand stereo photographs were taken. The pictures were scanned for events in which either two electrons or two positrons were seen to come from the same point in the metal foil. Of the few dozen candidates, only one or two events had the total energy expected in the sought-after decay. Knowing the total number of atoms in the foil and the total amount of time during which the chamber was sensitive, Winter was able to place lower limits on the double-beta lifetimes. These were typically

Fig. 10-9. Rolf G. Winter.

in the 10^{17} years range. Results such as these gave support to the conclusion that the antineutrino is different from the neutrino. "Search for Double Beta Decay in Cadmium and Molybdenum" *Phys. Rev.* **99** 88 1955.

While at WRU, ROTC 2nd Lt. Winter was called to active duty in the US Army and many letters between the Defense Department and chairman Beth and even WRU president Millis seem finally to have kept young Rolf out of Korea. After only three years at WRU, however, Winter decided to accept a position at Pennsylvania State University where he remained for about 15 years before going on to spend the rest of his career at the College of William and Mary.

Fig. 10-10. McCarthy, Meeks, Winter and Beth

A photo from the 1954 WRU yearbook shows the four-man WRU physics faculty, with some of their electronics equipment. **Fig. 10-10.**

John Major: a new young chairman

Fig. 10-11. John Keene Major.

John Keene Major came to WRU as an associate professor and chair in 1955. (**Fig. 10-11**) He was only 31 years old, but with the impending departure of essentially all the WRU physics faculty (Beth, McCarthy, Winter and Meeks), new talent was urgently needed. Major had completed his BS at Yale in 1943 and, after a two-year stint in the sonar analysis program at Columbia, he had earned his doctorate under F. Joliot and I. Joliot-Curie at Collège de France in 1951. *(It is possible that he interacted with Shankland or Foldy who were in the underwater acoustics program at Columbia during the same period.)* After a Fulbright in Paris, Major was awarded an NSF Scholarship which took him to Munich to work on Mössbauer spectroscopy. He returned to Yale for a short period to work on a comprehensive compilation of nuclear electron-capture data. (*Rev. Mod. Phys.* **26** 321 1954)

At WRU, Major was able to continue a modest research program on Mössbauer spectroscopy. This technique will be described in Chapter 12. "Recoil-free resonant and non-resonant scattering from Fe^{57}" *Nucl. Phys.* **33** 323 1962. His main activity, however, would be the tripling of the size of the department.

Rapid Expansion

After two years on the WRU faculty, Major was appointed as the eighth Perkins Professor. He was the principal player in the creation of a research-oriented physics department at WRU. With the support of the WRU administration and president John S. Millis, and generous funding from the government in the "Sputnik era", Major transformed a four man department which was principally occupied with the teaching of hundreds of pre-med students into a department of a dozen faculty researchers. The expanded department was housed in the large new Millis Science Center, which they shared with the WRU chemistry and biology departments. (Whitman's 1895 building, described in Chapter 5, was torn down around 1969.) The WRU physics PhD program was initiated during this period and the first doctoral degree was granted in 1962. (E. Brooks Shera wrote his dissertation on experimental nuclear physics under the direction of Berol Robinson, c.f. Chapter 14. The first physics PhD at Case had been granted in 1949 to Earle Gregg, Shankland's student.)

The Western Reserve department was for the first time becoming a worthy rival of the Case department. By 1963, Major's twelve man team and Reines' nineteen man team were beginning gradually to interact. (The dingy restaurant in the basement of Eldred Hall provided a most convenient locale, a few steps from each department.) They had dissimilar missions: one, part of a liberal arts and sciences university, and the other, part of a school of mainly engineering technology. But they had become similar in size and research activity. Each institution submitted a proposal to the National Science Foundation for a multi-million dollar "Science Development Program" grant. One component of the WRU proposal was a Condensed State Center while Case Institute proposed a Center for the Study of Materials. We shall see later how the NSF funding played a role in the eventual union of the two departments and the federation of the two institutions.

It was during John Major's tenure as chairman that all of the seven theorists whose work will be described in Chapter 11 and ten of the experimentalists we shall meet in Chapters 14 and 15 were recruited to the WRU department.

John Major decided to leave Western Reserve in 1966 to take a position with the National Science Foundation in Washington. He was on the executive council of the Federation of American Scientists (a progressive liberal organization which is still active in monitoring the role of science in society). He moved on to administrative positions at the University of Cincinnati, New York University, Northeastern Illinois University. Ultimately, he decided to combine his love for music with his talents in electronics and moved to the world of FM broadcasting. He founded classical music station KCMA at Tulsa, Oklahoma. He died in 2003.

Chapter 11   Theory at Western Reserve

Tauber,   **Kisslinger,**   **Machlup,**   **Weinberg,**   **Zilsel,**   **Goswami,**   **Chew**
1954-68   1956-69   1956-00   1959-69   1960-70   1963-69   1964-67

Between 1954 and 1963, the Western Reserve department would emulate Case's rapid expansion. It added seven theorists to its ranks. Only one, however, Stefan Machlup, would remain for more than three years beyond the federation with Case in 1967.

Tauber: nuclei and gravity

Gerald Erich Tauber was hired in 1954. He was the first theorist in the WRU department. (Foldy, the first theorist at Case, had joined that department six years previously.) Tauber was born in Vienna in 1922. He escaped to England and subsequently to Canada during the war. He completed his BA at Toronto and his PhD at Minnesota in 1951. **Fig. 11-1**.

Tauber did theoretical nuclear physics while at Western Reserve as represented by the following two papers. "Energy Levels of Pb^{208}" presented calculations of the energy levels in this "doubly magic" nucleus. *(This means that both the neutrons and the protons form closed shells within the nucleus.)* The second work, "Self-Consistent Treatment of the Independent-Particle Central-Field Nuclear Model", involved calculations of the nuclear energy levels in several isotopes of oxygen. *Phys. Rev.* **99** 176 1955; *Phys. Rev.* **105** 1772 1957.

Fig. 11-1. Newspaper photo of Eric Tauber.

A later paper, written with colleague Joseph Weinberg (whom we shall meet presently) concerned general relativity and the lower limits on the size of white dwarf stars: "Internal State of a Gravitating Gas" (*Phys. Rev.* **122** 1342 1961). This 23-page paper predicted the gravitational collapse of stars smaller than about one-third the earth's diameter. The authors may have struggled to get the paper accepted, as it was published thirty months after it was first submitted. However, the work, retitled "Gravitational Stability of Large Masses", won the 1963 $1000 prize from the Gravitational Research Foundation. The citation stated that the work was "expected to lead to a new experimental test of the theory of General Relativity". Tauber was the recipient of a two-year grant from the Army Research Office to study gravitational radiation. He took over as chair of the department when John Major left in 1964, but soon took an extended leave of absence to work as a visiting professor at the Technion in Haifa. By 1968 he had resigned his position at WRU and taken a permanent faculty position in Israel.

John Major hired three more theorists in quick succession: Leonard Kisslinger and Stefan Machlup in 1956 and Joseph Weinberg in 1959. Their principal areas were particle physics, statistical physics, and relativity respectively.

Kisslinger: nuclei and mesons

Fig. 11-2. Leonard Kisslinger.

Leonard Kisslinger was born in St. Louis in 1930. He did his BS at St. Louis University and then matriculated as a graduate student at the University of Indiana. While there he published a paper on the theory of the scattering of mesons by light nuclei. His calculated differential cross-section for elastic scattering from ^{12}C was based on an optical model potential which assumed a Gaussian distribution of nuclear material. This sort of calculation was similar to those done by the Case theorists as described in Chapter 7. Kisslinger's results compared well with experimental data from a Columbia experiment with 62 MeV pions, including agreement with the observed substantial backward scattering. "Scattering of Mesons by Light Nuclei" (*Phys. Rev.* **98** 761 1955). The potential he proposed was further explored by other authors, becoming known as the "Kisslinger potential". He completed his doctorate in 1956 and came directly to the WRU department.. His dissertation was a calculation of the spin-orbit interaction in nuclei and was published just after he arrived at WRU. "Spin-orbit Interaction in Nuclei" (*Phys. Rev.* **104** 1077 1956). **Fig. 11-2** shows Leonard in the mid-1960's.

One area of particular interest to Kisslinger and his collaborator R. A. Sorensen (at Carnegie Institute of Technology) was the study of collective motions of nucleons induced by bombarding nuclei with energetic protons, alphas or other particles. This work addressed the large amount of experimental data in the 50 MeV range which were being accumulated at accelerator laboratories. Kisslinger and Sorensen had met when they were both on sabbatical at the Nils Bohr Institute in Denmark. Their first collaborative effort was published there. *Kgl. Danske Videnskab. Selskab. Mat-Fys. Medd.* **32** No. 9, 1 1960. Subsequently a full four years of work was reported in a comprehensive 62-page paper, "Spherical Nuclei with Simple Residual Forces" *Rev. Mod. Phys.* **35** 853 1963. This paper was, at one time, one of the ten-most-cited papers in that journal. As an example of the calculations described in this paper, we show one of their figures in **Fig. 11-3**. The measured energies (open

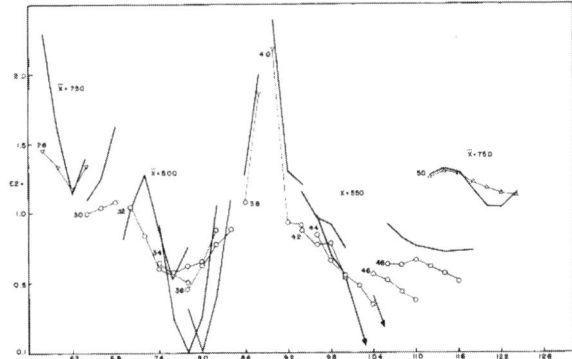

Fig. 11-3. Comparison of measured and calculated energies for nuclear excited states.

circles) and the calculated energies (dark lines) for a selected excited nuclear state are shown for nuclei with even Z from 28 to 50, for various isotopes.

Later work led to ever more detailed models for nuclear excitations over a wide range of nuclear masses. "Particle Model of Scattering from Collective States of Nuclei *Phys. Rev.* **129** 1316 1963. "Static Quadrupole Moment of Vibrational, Even Nuclei and the Coupling Scheme for Odd Nuclei" *Phys. Rev. Lett.* **19** 1239 1967.

Toward the end of the 1960's Kisslinger, along with many others in the field, was moving away from nuclei and toward "high-energy" particle physics. A representative paper from this period examined the scattering of K mesons by nuclei. Here the role of the Λ, a baryon with strangeness -1 which is formed within the nucleus, was explored. Cross sections were calculated for reactions of the type $K^- + (Z,N) \rightarrow \pi^- + (Z, N-1, \Lambda)$ for a wide range of nuclear masses. In this type of reaction, the negative strangeness of the kaon is transferred to one of the neutrons, making it into a lambda hyperon. "Tests of unitary symmetry in nuclei by meson-nucleus reactions" *Phys. Rev.* **157** 1358 1967.

Kisslinger became a leading spokesman for the study of hypernuclei (i.e. nuclei containing a strange baryon, usually a Λ or a Σ^+). With the substitution of a neutron or a proton by a hyperon, the structure of the nucleus changes drastically. This happens because the hyperon, not being constrained by the exclusion principle, resides in the lowest energy level, whereas the nucleon it replaced would usually have been in an upper level. Hypernuclei can provide an alternate path to the unraveling of nuclear structure.

In a related study, the effects of baryonic excited states were included in an analysis of proton-deuteron scattering at 1 GeV. "High Energy Backward Elastic Proton-deuteron Scattering and Baryon Resonances" *Phys. Rev.* **180** 1483 1969. This time, Kisslinger considered the role of the N(1688) baryon resonance (an excited state of the proton) as a possible intermediary in the interaction.

Kisslinger was advisor to seven Western Reserve doctoral students. Their research areas ranged from vibrational states in spherical nuclei to high energy proton-deuteron scattering. During his thirteen years on the Western Reserve faculty, Kisslinger typically spent summers at other institutions where he could interact with other nuclear theorists. Among these were Oak Ridge, Los Alamos, University of Colorado, Brookhaven Lab, and the Radiation Lab at UC Berkeley. In addition, he benefited from extended leaves at the Bohr Institute in Copenhagen, the Weizmann Institute in Israel, and at MIT. After spending two years at MIT on leave of absence from the newly federated CWRU department, he resigned in 1969 to join the physics department at Carnegie-Mellon, where he has been extremely productive in particle theory and cosmology for the past three decades.

Machlup – irreversibility and negative temperatures

Stefan Machlup joined the Western Reserve department in 1955. He was born in Vienna in 1927 and was brought as a child to America where his father, Fritz Machlup, a world renowned economist, would take a professorship at the University of Buffalo. Stefan did his baccalaureate at Swarthmore, spent a year in the navy, and one at the University of Paris before beginning his doctoral studies at Yale. His dissertation "Fluctuations and Irreversible Behavior in Thermodynamic Systems" led to two publications with his research director, Lars Onsager, the physical chemist who would subsequently win the 1968 Nobel Prize in chemistry. *Phys. Rev.* **91** 1505 and 1512 1952. This work was an elaboration of a theory of irreversible processes developed two decades earlier by Onsager. As a Yale graduate student, Machlup followed his professor to Cambridge when Onsager took a year-long sabbatical there. Subsequently, Machlup held post-doctoral positions at Bell Laboratories, at the University of Illinois Urbana, and at the University of Amsterdam, before accepting an invitation by John Major to join the WRU department. **Fig. 11-4** is a photo of Machlup from around 1980.

Fig. 11-4. Stefan Machlup.

The thermodynamic behavior of systems of large numbers of particles subjected to a variety of driving forces would be central to Machlup's research from the 1950's through to the 1990's. The abstract of the first 1952 paper describes the problem: "The probability of a given succession of (nonequilibrium) states of a spontaneously fluctuating thermodynamic system is calculated, on the assumption that the macroscopic variables defining a state are Gaussian random variables whose average behavior is given by the laws governing irreversible processes." The chemical-physics literature refers to the "Onsager-Machlup-Laplace approximation".

The theory applies to a very wide variety of constituents and forces: nuclei with spins aligned by magnetic fields, excited atoms in a laser, molecules undergoing chemical reactions, vortices in a liquid, ions transported through biological membranes. Machlup would explore the applicability of some of these ideas to biology: "Biological clocks share with excitable membrane (nerve, muscle) the requirement that the chemical systems that underlie them have unstable steady states and hence are capable of limit-cycle oscillations.... The oscillators responsible for biological clocks are surely not mechanical mass-and-spring systems, nor are they electrical inductance-capacitance combinations. They are chemical oscillators." "Oscillatory Chemical Reactions: the Tomita-Kitahara Model" *BioSystems* **8** 241 1977.

In a 1975 paper, Machlup discusses the common features of systems which may be described as having negative temperatures. "Negative temperatures and negative dissipation" *Amer. J. Phys.* **43** 991 1975. He writes: "If we think of absolute temperature as a measure of kinetic energy per (classical) degree of freedom, then a negative absolute temperature seems absurd. If, however, we use the (quantum-mechanical) idea of the population of energy levels and measure this population with a Boltzmann factor $\exp(-E/kT)$, then a negative T makes sense: It means higher energy levels are more populated than lower ones." The paper concludes: "This article has attempted to make more intuitive the connection between negative temperature and negative resistance, and to suggest that a large class of nonlinear systems involves negative-temperature subsystems." One such system, he suggests, might be current vortices in type II superconductors.

In a separate project, Machlup and collaborators investigated how the rules of statistical fluctuations could be applied to the frequency distribution of random noise in such structures as semi-conductor devices.

In the 1980's, Machlup would join colleague T. Hoshiko of the CWRU School of Medicine's Department of Physiology and Biophysics in a research collaboration which wedded the statistical mechanics approach to the analysis of a biological systems. One such paper reports a study of ionic transport in frog skin cells. (*Biochimica et Biophysica Acta* **942** 186 1988)

In a somewhat related area, Machlup investigated the impact of low-frequency electromagnetic fields on biological systems. Reports by researchers that various physical disorders in humans and animals appear to result from exposure to low-field electromagnetic radiations from devices ranging from kitchen appliances to cell phones to power transmission lines have encouraged extensive measurements and analysis.

Machlup and his collaborator, Carl Blackman of the Environmental Protection Agency facility at Research Triangle in North Carolina, have conducted studies in this area. They approach the problem at a very elemental level. They tracked the growth of rat liver cells as a function of the intensity, frequency and angle of application of quite low magnetic fields. The cells were arranged in a flattened mono-layer on a culture dish, and the communication between neighboring cells was measured by observing the transfer from cell to cell of a dye. Machlup had proposed a model for this process in 2000, and the results presented in 2003, specifically the dependence on the angle of the applied fields, were in agreement with his predictions

Stefan Machlup taught at CWRU for four decades. He published an introductory text, "Physics" (Wiley, 1988), which was based on his long class-room experience and which emphasized biological examples of special interest to health-science students. He was a frequent contributor to conferences of the AAPT. Complementary to Machlup's long career in teaching and research has been his lifelong passion for music. He is an accomplished cellist who has enjoyed performing with chamber groups since his college days. He assumed emeritus status in 2000 and continues his work on the biological effects of magnetic fields.

Weinberg: gravity and MRI

Joseph Woodrow Weinberg was born in New York City in 1917, completed his BS at CCNY in 1936 and his PhD with J. Robert Oppenheimer at UC Berkeley in 1943. He worked at the UC Radiation Lab until 1947, when he was appointed associate professor at the University of Minnesota. During this period, Weinberg began working with graduate student Gerald Tauber (see above), on the gravitational stability of white dwarf stars.

In 1949, the young Weinberg fell victim to Senator Joseph McCarthy and the House Un-American Activities Committee. He was accused of being the mysterious "Scientist X", who was purported to have given nuclear secrets to the Soviets while at the Radiation Lab. Though he denied being a spy, he was dismissed by the University of Minnesota Board of Regents in 1951. Weinberg was completely exonerated two years later after a long series of humiliating hearings and trials, but he was *not* reinstated at UM. Subsequently, Weinberg worked for the American Institute of Physics, then as a research engineer for an optical manufacturer, and later for the Pioneer Scientific Company before resuming his academic career by accepting a position at WRU in 1959. His appointment at Reserve was made possible through the efforts of President John S. Millis who was himself a PhD physicist. Weinberg's photo is shown in **Fig. 11-5**.

Fig. 11-5. Joseph Weinberg.

Soon after taking up his post at WRU, Weinberg and Tauber published the paper on the gravity of collapsing stars which was described above. Of Weinberg's four doctoral students, one wrote on relativity, two on particle theory, and the fourth, Clyde Bratton, on "Nuclear Magnetic Resonance Studies of Living Muscle". This work, unusual in 1964, was done in collaboration with Amos L. Hopkins of the WRU Department of Anatomy. Samples of "living" (though excised) frog muscles were studied. Magnetic resonance techniques were used to determine the disposition of water molecules in the tissues as the muscles were electrically stimulated to contract and relax. (*Science* **147** 738 1965) Thirty years later, magnetic resonance imaging would become a major area of research in the CWRU department. In 1970, Weinberg left CWRU to join the faculty at Syracuse University.

Paul Zilsel: superfluidity theory

In 1958, John Major invited **Paul R. Zilsel** to join the WRU department as a visiting lecturer. Thirty-five year-old Zilsel had received his PhD ten years earlier, and already had a reputation as a significant contributor to the theory of superfluid helium. Zilsel was born in Vienna in 1923 and had come to the United States with his family in 1939, a refugee from Nazism. He did his undergraduate degree at the College of Charleston (South Carolina) and a master's at Wisconsin. He accompanied his professor, Gregory Breit, when Breit moved from Wisconsin to Yale. Zilsel published three papers at Yale: two on the scattering of slow neutrons by bound protons (as in water molecules), and one on proposed corrections to the potential for low energy proton-proton scattering. He completed his doctorate in 1948.

As a post-doc with Fritz London at Duke, Zilsel would change fields from particle theory to the theory of helium at low temperatures. He and London wrote an important paper on heat flow in superfluid liquid helium in which they compared experimental measurements with the predictions of the widely accepted two-fluid model. "Heat transfer in liquid helium II by internal convection" *Phys. Rev.* **74** 1148 1948. In 1950, Zilsel took a faculty position at the University of Connecticut where he continued this work, publishing several papers on the two-fluid model. Zilsel then spent two years at the Israel Institute of Technology and two at McMaster before coming to WRU. His photo is shown in **Fig. 11-6.**

Fig. 11-6. Paul Zilsel.

Zilsel's most important contribution to the theory of superfluidity followed from work done at WRU with graduate student, Richard Whitlock. The question of how helium atoms interact at very low temperatures and how they collapse into a single quantum mechanical ground state had been studied by many of the world's leading theorists (e.g. Lee, Yang, Dyson, Bogoliubov, Bloch). Zilsel and Whitlock proposed a new approach for the hard-core plus weakly attractive interaction. It is based on a mathematical analogy with the two-valued quantum number for spin. The authors end their paper with the hopeful comment: "….the fact that the model….is able to describe rather well at least *some* features of the λ transition in liquid helium, give(s) rise to the hope that refinement of the approach….may lead to a feasible method of treating the liquid He^4 problem." "Pseudospin model for hard-core bosons with attractive interaction. Zero temperature" *Phys. Rev.* **131** 2409 1963. Chairman Major sought the opinion of CIT's Foldy on this paper when Zilsel was up for promotion to professor. From Foldy's letter: "I believe that this is a paper of which anyone in this field would be happy and proud to be the author."

In October, 1965, Zilsel was awarded a Science Faculty Fellowship from the NSF and he was granted a 15-month leave of absence to allow him to work at Stanford and at Princeton. He extended his work on the "pseudospin model", this time using it to calculate such features as the liquid-vapor phase curve, the λ temperature, the degree of superfluidity, and the low energy excitations. "Pseudospin model of liquid He^4" *Phys. Rev. Lett.* **15** 476 1965. Returning to WRU in 1966, Zilsel found his department very much caught up in working out the federation with the Case department. He took on a new PhD student, Ranendra Roy, with whom he advanced the work on superfluidity. In a variation on the Bose condensation topic, Zilsel and post-doc Michael Schick investigated the problem of a superfluid in an annular container. In analogy to work done by Felix Bloch on persistent electrical currents in superconducting rings, they considered what will happen to the angular momentum of the superfluid when a torque is applied to the container. They predicted that the torque will produce an angular acceleration, but that there would be regularly spaced jumps in the acceleration. This occurs because the total moment of inertia of the fluid and container has a non-classical piece which steps upward as the angular velocity is increased. The extra term results from the creation of quantum vortices in the superfluid. "New manifestations of the Josephson effect in helium" *J. Low Temp. Phys.* **1** 385 1969. Zilsel's final paper at CWRU was written with Mike Schick. "Order Parameter, mean-field theory, and the ideal Bose gas" *Phys. Rev.* **188** 522 1969.

In the summer of 1968 there were serious anti-war riots in Cleveland, as in other American cities. During the following year a CWRU professor of political science, Louis Masotti, would become head of the CWRU Civil Violence Research Center. He was responsible for the preparation of a report on the disturbances for a federal commission on violence. In May of 1969, a participant in the riots, Ahmed Evans, was sentenced to death for murder. The supporters of Evans called for the immediate release of the Masotti report and disrupted a seminar which Professor Masotti was conducting on campus. Among the more vociferous protesters was Paul Zilsel. The affair escalated over the following days, including occupation of various university offices. A few weeks later a related demonstration which took place at the Cleveland Criminal Court Building culminated in the arrest of several protesters, including Zilsel. Although supported in court by five members of the CWRU physics department, he was fined, sentenced to ten days in jail, and required to seek psychiatric help. In 1971, Zilsel spent several months at the University of Washington in Seattle, working with his friend and collaborator, Mike Schick. He left the CWRU department in 1974, resettling in Seattle, where, as one of the founders of the Left Bank Books Collective, he would continue his lifelong campaign for peace and social justice.

Goswami - nuclei

In 1963, the department added a second nuclear theorist. **Amit Goswami** arrived with a fresh doctorate from Calcutta. He and Kisslinger soon collaborated on the calculation of nuclear energy levels. Their first paper looked at the effect of "isospin pairing". The quantum number, isospin, was invented to join the neutron and proton in a family of two. Because they each react similarly to the force which holds nuclei together, it is reasonable to lump them together and invent a new quantum number for them – isospin. A family of two would have $T = \frac{1}{2}$, because that has two substates: $T_z = +1/2$ or $-1/2$. Therefore, any *pair* of nucleons can have a net $T=0$ or $T=1$. It turned out that the energies of nuclear excited states depend on how they pair up. "Particle correlation arising from isopin pairing in light nuclei" *Phys. Rev.* **140** B26 1965.

Goswami continued similar fundamental studies of nuclear states until his departure from the newly merged department in 1969. He took a faculty position at the University of Oregon where he has taught physics for over thirty years. His interests would eventually turn from nuclear theory to studies not found among traditional physics topics. He has become well known as a writer and lecturer on the intersection of science with the study of human consciousness.

Chew - particles

One more particle theorist would spend a rather short time in the WRU department: Chinese-born **Herman W. Chew**. He had done his doctorate in 1961 at the University of Chicago under R. H. Dalitz and spent one year each at the University of Pennsylvania, Columbia, and Princeton before his appointment as assistant professor at WRU in 1964. His doctoral work was on the effects of $\pi\pi$ scattering in the rare radiative kaon decay: K $\rightarrow$ $\pi\pi\gamma$. "Structure of radiative decay amplitudes" *Phys. Rev.* **123** 377 1961. Chew contributed to a wide variety of problems in particle theory. His first paper at WRU was largely mathematical, "Analytic expression for a class of Green's functions", *Phys. Lett.* **23** 85 1966. A second work examined the electromagnetic mass difference in the Ξ strangeness-minus-two baryon doublet (another family of two), building upon published works on the neutron-proton mass difference. *(This work predated the development of quantum chromodynamics in which particle properties follow from their quark content.)* "Ξ^- - Ξ^0 mass difference" *Phys. Rev.* **150** 1249 1966. Chew spent the 66-67 academic year on leave at Clarkson College of Technology, and deciding to remain there, resigned from WRU in August 1967.

Chapter 12   Case Experimenters: Round 2

Hoffman, Benade, Gordon,   Eck,   Schuele, Chottiner
1952-92   1955-87 1955-94 1957-2002 1963-2005 1980-

The postwar funding of university science, especially for experimental physics, grew at a rapid pace through the 1950's and the Case department grew accordingly. In parallel with the new hires in experimental cosmic ray and particle physics (Chapter 8) was the addition of several young condensed matter experimentalists who would continue the work of Smith and Crittenden. The first of these was **Richard W. Hoffman**, Crittenden's first graduate student, who joined the faculty in 1952 at age 25. **William L. Gordon** came in 1955 at age 28, **Thomas G. Eck** in 1957 at age 28, and **Donald E. Schuele** in 1963 at age 29. Each of these four men would remain on the Case faculty for forty years or more. In this chapter, we shall describe their research on the properties of condensed matter. At the end of the chapter, we shall include a section on **Gary S. Chottiner** who joined the department two decades later, in 1980. His work on thin film and surface physics is in some respects a continuation of Hoffman's work. But first, let's describe the work of **Arthur H. Benade**, who arrived at CIT in 1955, and who would continue the department's tradition in acoustics research.

Benade: the sounds of music

Benade was born in India in 1925 of American missionary parents. He completed college in India and earned his PhD at Washington University St. Louis in 1952. He joined the Case department in 1955, ostensibly to build upon the experimental nuclear program of Gregg and Shrader. We met Benade briefly in Chapters 7 and 8, once as a graduate student at Washington University and then as an assistant professor at Case doing neutron work at the Case betatron. A second major interest had long been the physics of musical instruments. Benade's widow, Virginia Belveal, recently told me that Marshall Crouch phoned Benade in St. Louis to tell him that Case was looking for new physics faculty. Benade hesitated, but when Crouch told him that the attic was filled with Miller's acoustics equipment, he couldn't say no.

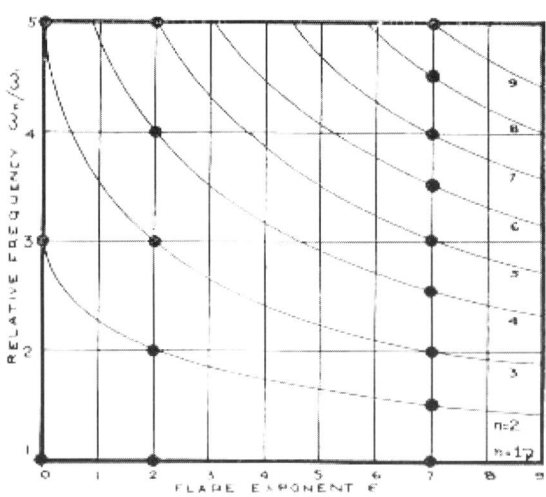

Fig. 12-1. Normal mode frequency analysis for woodwinds.

A paper by Benade published in 1959 (*J. Acoust. Soc. Am.* **31** 137 1959) was titled "On Woodwind Instrument Bores". It concerned the ideal shape of the instrument, and was not a put-down of a whole class of musicians. Benade writes a wave equation for the pressure variation within a flaring horn. One of the parameters of the equation is ε. The cross-sectional area of the horn increases along the length of the horn as distance raised to the ε power. The solution to the wave equation includes Bessel functions for the order $-1/2$ $(1-ε)$, so that only certain integral values of ε correspond to well-behaved solutions, i.e. ones which satisfy the boundary conditions. **Fig. 12-1** shows the ratios of the frequencies of the normal modes to the fundamental as a function of ε. Only for ε = 0, 2, or 7 do these ratios fall at the desired integral values. Such a horn is called a Bessel horn. (Parameters raised to the 7th power don't appear very often in physics!) Benade goes further to look at various combinations of cylinders and cones, the effects of side holes, the shape of the mouthpiece, and the properties of the reed which drives the whole system. He reports on laboratory measurements made to test the theory. He concludes that Bessel horns "constitute the only family of bores in which the normal mode frequency ratios are independent of the length of the horn." The footnotes in this paper refer to publications dating from 1860 to 1940, so this type of analysis had had a long history.

Benade, Shankland and Severance Hall

Just across Euclid Avenue from Case Institute is Severance Hall, the home of the world-famous Cleveland Orchestra. This very handsome concert hall with about 2000 seats had been completed in 1931. Dayton Miller played a major role in its acoustical design. (*J. Acoust. Soc. Am.* **3** 312 1932) However, according to Shankland, several changes made to Miller's design were made which proved deleterious to the acoustics of the hall: a large Skinner organ was added, luxurious sound-absorbing upholstery and carpeting were added, and, for Cleveland's opera lovers, a proscenium arch and high flies were added. As a result, much of the sound never got out into the hall. "Acoustics of Severance Hall" *J. Acoust. Soc. Am.* **31** 866 1959. As we mentioned in Chapter 6, Shankland had measured reverberation times in many halls around the world in an effort to quantify "good acoustics". He discussed with George Szell, "The Cleveland's" musical director, how the reverberation time at Severance might be increased. In 1993, Jack A. Kremers of Kent State University reported in the electronic journal, *Architronic*, on the saga of Severance acoustics. He writes that the well-traveled Szell and his musicians were very much aware of the hall's short reverberation times, relative to those of comparable halls. (www.architronics.com; vol 2 no 1.05 May 1993)

In the summer of 1958, the interior of the hall was redesigned to make its sound "warmer". A new wooden shell was placed around the stage. It consisted of a series of concave surfaces, backed by a heavy layer of sand fill. This greatly improved the reflection of sound into the hall. The combination of this reflected sound with the direct sound resulted in longer reverberation times. Shankland and his new colleague, Arthur Benade, were invited to make a series of measurements in the hall, before and after the 1958 renovations.

Reverberation time measurements, as a function of frequency, were made for gun shots, for organ sounds, for full orchestral chords, in both the empty hall and the fully occupied hall. Some measurements were made simply "by ear and stopwatch" and later ones by more quantitative analysis of tape recordings. **Fig. 12-2** shows some of the results: the reverberation time in seconds vs. sound frequency in Hertz, the lower curve made before the modifications to the hall, the upper curves made afterward. For example, at 1600 Hz, the time increased from about 1.2 to 1.8 seconds, a very significant change in the perceived sound. "Reverberation Time Characteristics of Severance Hall", *J. Acous. Soc. Amer.* **32** 371 1960. Unfortunately, the newly installed wood panels around the stage (the Szell shell) covered up the pipes of the great Skinner organ.

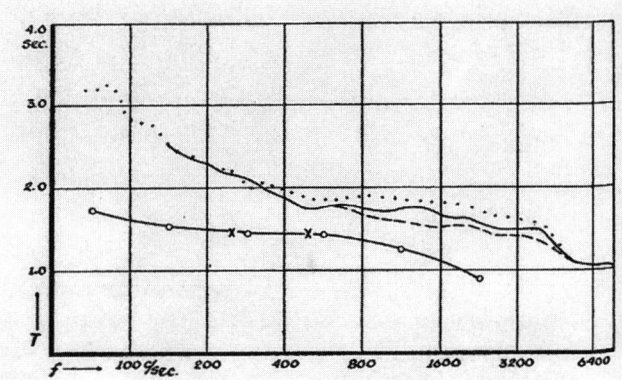

Fig. 12-2. Reverberation times vs frequency for Severance Hall before and after 1958 renovation.

Kremers' essay described the vastly improved measurements made in the 1990's in which modern electronic devices were used to establish reproducible frequency dependent decay times for specific decibel level reductions. These data were used in planning a major renovation which began in the late 1990's.

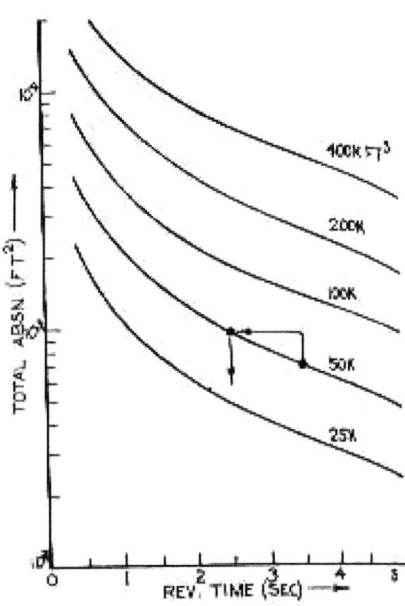

Fig. 12-3. Dependence of reverberation time on total absorption parameter for different frequencies.

When Severance reopened in 2001, the wood panels had been removed from the back of the stage and the voice of the organ was once again allowed to fill the hall. A new shell around the stage and other changes in the hall resulted in even further increases in the reverberation time. More importantly, the new design vastly improved the ability of the musicians to hear one another. Later in this chapter, we shall find that Benade investigated the effect of organs on concert hall acoustics.

Shankland and Benade also presented the results of *calculations* of the reverberation times, based on the areas, volumes, and absorption coefficients of all the materials in a hall: walls, ceilings, seats, and people. Fig. 12-3 shows the relation between reverberation time and total absorption for various sized halls. The agreement with measurements was very good. Another element in the study was the difference between the empty hall and the full one; this was important because most recordings were made in the empty hall, and the missing contribution of 2000 somnolent sound absorbers might be significant. It wasn't.

Equations for musical instruments

In recognition of Benade's prominence in the field of the acoustics of musical instruments, *Scientific American* published his article, "The Physics of Woodwinds" in October 1960. A telling paragraph in that article, one which would greatly have pleased Dayton Miller, gives us some idea of what Benade had in mind. After an introduction describing work done from the time of Pythagoras to that of Lord Rayleigh, he writes: "Since the mid-1920's, however, the engrossing new questions of quantum physics have diverted the energy of both theoreticians and experimenters from the more traditional lines of study and so brought active musical research largely to an end. Still, the stage is set for a revival of musical physics. The techniques of measurement and calculation that have developed during the last 40 years in other areas of physics may now make it possible to solve problems in music that have withstood the best efforts of the past." (Recall that Benade had started out doing (quantum) nuclear physics.) The same material was presented in more technical detail in the article, "On the Mathematical Theory of Woodwind Fingerholes" *J. Acoust. Soc. Am.* **32** 1591 1960.

Benade used impedance methods to tease out the effects of closed and open holes on the "effective length" of the woodwind's main cylinder or cone. I was surprised to learn from his discussion of "radiation from open holes" that most of the "music" comes from the open holes rather than from the "bell" at the end of the instrument. Benade calculates the dependence of the radiated intensity as a function of angle relative to the principal axis, noting the similarity to the single-slit Fraunhofer optical diffraction pattern. While the calculations are mathematically complex (for me anyway), it is clear that it was unnecessary to borrow any techniques from 20th century non-classical physics.

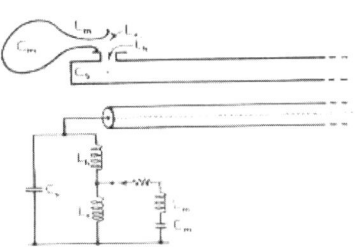

Fig. 12-4. LC circuit analogy to a "flute head joint".

Five year later, Benade reported on an investigation of the flute. *Acoust. Soc. Am.* **37** 679 1965. Once again he studied changes of the effective length of the cylinder which stem from the geometry of the "flute head joint". This includes the mouthpiece, the embouchure hole, the lip placement, and the mouth cavity, as well as the joining of that end of the flute with the main body. The interesting little drawing in **Fig. 12-4** shows the impedance analogy between this acoustic system and an electrical LC circuit. Benade worked with J.W. French of New York University, measuring the characteristics of many types of flutes, old and new. Included among these was a selection from the Dayton C. Miller collection at the Library of Congress (see Chapter 4). A series of figures shows the results of both measurements and calculations. The paper ends with comments and suggestions for both the flute-maker and the flautist on the best techniques for construction and playing. *J. Acoust. Soc. Am.* **37** 679 1965. Benade explored the general problem of the "propagation of sound waves in a cylindrical conduit" in later papers. In these, he presented tables of such parameters as admittance, impedance, and reactance as functions of length, radius, frequency, and temperature. *J. Acoust. Soc. Am.* **44** 616 1968).

When I got to the next Benade paper in the Shankland collection, I thought, "oh, just another instrument", but it turns out that his analysis of the *absorption* of sound by a set of organ pipes (not the *production* of sound by organ pipes) impacts directly on Shankland's proposal to wall in the Severance Hall organ along with the Amontillado. Benade starts with the frequency dependence of the absorption of ambient sound by a single open-ended organ pipe of "standard" length and diameter. *That the "absorption" is given in units of area, ft^2, was somewhat puzzling at first, but as a particle physicist for whom reaction rates are given in barns, I should have known better.* The most enlightening figure in the paper is a plot of the absorption (as calculated from the characteristics of the given set of organ pipes) versus the reverberation time for various room volumes. For example, the installation of a set of organ pipes with total absorption of 300 ft^2 in a 50,000 cubic foot room will reduce the reverberation time from 3.5 s to 2.5 s. Although Shankland did not mention such a calculation in his Severance Hall paper, he must have known about it. Benade pointed out that the effect could be especially disastrous in a small practice hall where a set of organ pipes would eat up all the sound. *J. Acoust. Soc. Am.* **38** 780 1965.

In 1973, a second Benade article was featured on the cover of *Scientific American.* (*Sci. Amer.* July 1973) This time the subject was the "Physics of Brasses". Benade presented a summary, at the level appropriate for the reader of that magazine, of earlier measurements and analyses made by the author and a colleague from the Swedish Royal Institute of Technology, Erik V. Jansson. The measurements were of the frequency responses of the mouthpiece, the cylinder and the bell of a horn along with the transmission of the sound into the surroundings. The analyses were made at a fairly ad-

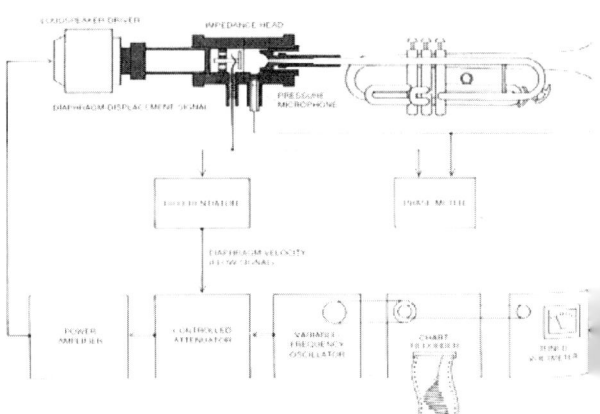

Fig. 12-5. Block diagram of electronic setup to analyze the acoustics of a horn.

vanced level, but in this paper, an effort was made to explain things on a level useful to the reader, and especially to the "practitioner". A block diagram of the experimental setup appears in **Fig. 12-5.** Several suggestions are made of little experiments which the player can perform to better appreciate what his instrument is capable of doing.

This article was followed in 1974 by two 20-page-long detailed technical contributions to the Stuttgart-based journal *Acustica*, both papers written in collaboration with Jansson. The summary begins: "The relation between axially symmetric plane waves in a cylindrical duct, and spherical waves in a conical horn, is reviewed initially as a basis for a study of waves in a horn of rapidly varying taper." The authors look at s-waves, i.e. those with no nodes between the horn axis and the walls, and at p-waves, those with one node. They are mostly concerned with the behavior of these waves as the horn transitions from cylinder to bell shape. Extensive measurements were made and compared with computer generated solutions. They conclude: "We find that the plane wave and spherical wave representations of horn acoustics both give excellent results in the calculation of resonance frequencies." "On Plane and Spherical Waves in Horns with Non-Uniform Flare. *Acustica* **31** 79 and 185 1974.

In 1960, Benade published *Horns, Strings and Harmony* (271 pp, Anchor Press 1960, reissued by Dover 1992). He later authored a widely-used text on musical acoustics: *Fundamentals of Musical Acoustics* (595 pp, Oxford University Press 1976, Dover 1990). He discussed in detail the properties of instruments, the sounds they produce, and the role of the room in which they are heard. Benade was the vice-president of the Acoustical Society of America in 1974-5. As an established world expert on the subject, Benade was invited in 1977 to speak at an international acoustics conference in Madrid. He examined the interactions among "oscillatory physicists", instrument craftsmen, concert hall architects, performers, and psychoacousticians. "Musical Acoustics Today: A Scientific Crossroads" paper at the 9th International Congress on Acoustics, Madrid 1977.

Fig. 12-6. Arthur H. Benade.

A photo of Arthur Benade taken around 1980 appears in **Fig. 12-6.** In his 32 years at Case, Benade built upon the tradition of acoustics research as practiced by Miller and Shankland. He advised seven doctoral students, his first in nuclear physics, and then six in the acoustics of musical instruments or concert halls. He was suddenly taken ill and passed away in 1987 at age 62. His papers and other materials, as organized by his widow, Virginia Benade Belveal, are available on the website of the Center for Computer Research in Music and Acoustics: http://ccrma-www.stanford.edu/marl/Benade/BenadeBio.html

Hoffman surfaces: magnetic thin films

Fig. 12-7. Richard Hoffman.

Richard Wagner Hoffman was born in Cleveland in 1927. He earned his BS at Case and in 1952 he completed his doctorate there with Professor Crittenden. His dissertation was titled: "Study of Ferromagnetism by Means of Thin Films". **Fig. 12-7** is a photo taken around 1980. We met Hoffman briefly in Chapter 7 where we described some of the early work in producing uniform and pure thin metallic films on substrates and the measurement of their electrical, magnetic and mechanical properties. We pick up now on the continuation of this work by Hoffman and his legions of graduate students.

An early Hoffman paper describes the measurement of the temperature dependence of resistivity, ρ, in thin films of Ni and Pd as a function of film thickness. An abrupt drop in $d\rho/dT$

as the thickness is reduced to about 200 Å was discovered, signaling a break-up in the uniformity of the film. "Variation of Temperature Derivative of Resistance" *Jour. Appl. Phys.* **29** 1512 1958.

Work which had more immediate application to contemporary technology concerned the laying down of thin films of iron for application to products like magnetic tape for the recording of data. Imperative for reproducible reading and writing of magnetic tapes is the spatial uniformity of magnetic properties along the tape, and this depended critically on the techniques used for creation of the film. Hoffman discovered, for example, that what he called the magnetic anisotropy depends on the angle between the stream of incident iron atoms coming from the hot filament and the plane of the substrate, a point of concern for manufacturers of magnetic tape. *(The magnetic anisotropy is proportional to the work done in reversing the magnetization and is determined from the area of the hysteresis loops in the M vs H plots.)* He concludes that the iron atoms collect in "fibers" which grow out from the substrate and that these tilt in the direction of deposition. "Dependence of Geometric Magnetic Anisotropy in Thin Iron Films". *Phys. Rev.* **113** 1039 1959. The magnetic anisotropy vanishes for deposition normal to the surface, and increases as the angle of deposition increases. Hoffman continued this study using not only the hysteresis measurements, but also electron diffraction and x-ray diffraction. These tools would become permanent features of the thin-film and surface physics laboratories. At this point, even with these new approaches, the role of "fibers" or "needles" eluded satisfactory understanding. "Fiber Texture and Magnetic Anisotropy in Evaporated Iron Films" *Jour. Appl. Phys.* **33** 949 1962.

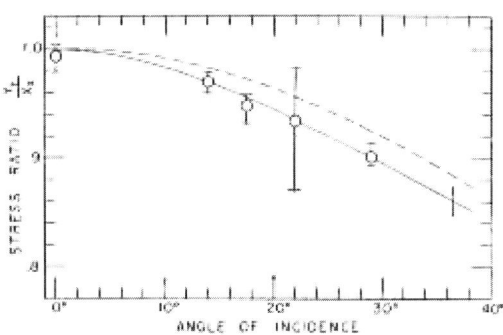

Fig. 12-8. Stress in thin magnetic films vs. the angle of the incident stream of iron atoms.

Still searching for an explanation of what is going on in a film which was deposited "at an angle", Hoffman turned to measurements of the *stress* which builds up in the film as it grows. He was able to quantify this stress by measuring the buckling of the circular, thin glass substrates. In a manner similar to that used to measure curvature of lenses or spherical mirrors, he looked at the spacing of the Newton ring interference patterns. Typically, the Newton rings were oval in shape because the stress in the film was different in different directions, relative to the direction of deposition (call it Y). **Fig. 12-8** shows a typical behavior. Plotted is the ratio of the stress in the film along the projection of Y on the film to the stress along the vector in the film perpendicular to Y. The ratio of these two orthogonal stresses is shown as a function of the angle of incidence, up to 36.5° from the normal, for a 650 Å thick film. The effect of subsequent annealing of the film by heating was also studied; this generally lessens the stress. What struck me about this paper, presented at an international vacuum congress was the interest shown in the question period by representatives from IBM, RCA and Redstone Arsenal. (Transactions of 8th Vacuum Symposium, Pergamon Press 1962.)

At a conference on thin films in Belgium, Hoffman presented data on the magnetization of thin *nickel* films. Here, the *torque* on a sample placed in a magnetic field was measured directly, as a function of film thickness and temperature. The data were compared with those of experiments in other laboratories, but no clear conclusion was reached on the underlying theory: "more detailed knowledge of the structure of the sample is needed before the understanding of the magnetization in the films is complete." One loose end seemed to be incomplete knowledge of the contaminants in the films and what efforts would be required to minimize them by creating the films at ever lower pressures.

We shall return later in this chapter to Hoffman's work on thin films and a series of new techniques which he used to learn about them. But first, we shall look at some parallel experimental work on the properties of matter in bulk.

Gordon: Fermi surfaces, polymers, liquid crystals

William L. Gordon (born 1927) did his BS at Ohio's Muskingum College and completed his PhD in 1954 at Ohio State University. At OSU he worked on the scattering of x-rays by liquid helium, above and below the temperature at which it becomes superfluid. He joined the Case faculty as an instructor in 1955 and began work on electrons in metals. He, and several other members of the department, both experimentalists and theorists, would continue the study of electrons in metals for many years. A photo of Bill Gordon, taken around 1980, appears in **Fig. 12-9**.

Fig. 12-9. William Gordon.

Electrons in metals

An aside on electron energies and the **Fermi Surface**. *If a particle is moving in a region where the potential energy is zero (e.g. a "free particle"), then its total energy is all kinetic and is equal to $p^2/2m$, where p is the momentum. The three components of the momentum can have any values as long as $p_x^2 + p_y^2 + p_z^2$ equals p^2. Therefore, if one draws a three dimensional plot of the three momentum components, the particle's momentum will be represented as a point somewhere on the surface of a sphere of radius p.*

When an electron is attached to an atom, it can take on only certain energies corresponding to solutions of the Schrödinger wave equation which incorporates the electric potential in which the electron moves. If the electron is put into a crystal, it again can have only certain energy values, this time determined by the periodic potential associated with the crystal lattice.

Because electrons obey the Pauli Principle, only two electrons (spin up and spin down) can be assigned to a given energy level. Therefore, in an atom or in a crystal, the electrons are distributed among many energy levels, up to some maximum energy. The Fermi Surface is the surface in momentum space which corresponds to this maximum energy.

Because the electron's kinetic energy changes as it moves through the crystal lattice potential, its total momentum must also change, and what was a spherical surface in momentum space for the free electron becomes a much more complicated surface. The bulges and valleys and even doughnuts and cigars of this surface reflect the lattice structure of the crystal. Condensed matter physicists find it convenient to relate the electric, magnetic and mechanical properties of the crystal to the shape of this "Fermi" surface. It turns out that at particular radii (i.e. momenta) specific ranges of energy are excluded altogether, giving rise to the **band structure** *in the energy. The allowed energy levels exist only in certain bands, separated by energy gaps where no electrons reside. The population of electrons in each band and the size of the gaps between neighboring bands determine the electrical properties of the crystal, e.g. insulator, semi-conductor, conductor. The experimental determination of the Fermi surface and the band-structure for a crystal is essential for understanding the material. In Chapter 17 we shall discuss theoretical work in this area.*

Torque measurements and the Fermi surface

In 1961, Gordon published a paper with his student Alfred S. Joseph, and two of his Case colleagues: theorist John Reitz and the newly arrived Thomas Eck. This work concerned the band-structures of zinc and cadmium. Clues to this structure come from the measurement of the torque on a sample of the material when it is placed at low temperatures in a strong external magnetic field. It had been discovered in 1930 that the magnetization of a material undergoes periodic variations as the applied magnetic field is changed. This is called the de Haas–van Alphen effect. In 1952, Lars Onsager at Yale had shown that the observed periodicity depends on the reciprocal of the applied field and on the orientation of the crystal relative to the field. By measuring the periodic variations of the magnetization as the strength of the field is changed, one can deduce details of the Fermi surface in the plane normal to the field. The periodicity in 1/H was shown by Onsager to be inversely proportional to the "*extremal cross-sectional area*" enclosed by the Fermi surface. Gordon's paper: "Evidence for Spin-orbit Splitting in the Band Structure of Zinc and Cadmium" *Phys. Rev. Lett.* **7** 334 1961.

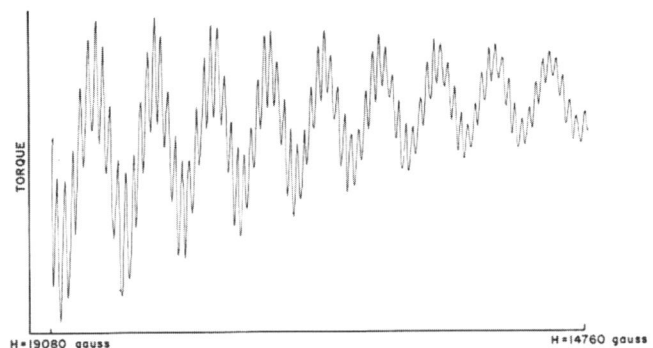

Fig. 12-10. Measured torque on a metal sample as a function of the reciprocal of the applied magnetic field.

In the Case experiment, a very pure sample of the metal was suspended by a fiber in a cryostat at liquid helium temperature. Sample purity was required so that the electrons could move freely through the crystal. The torque on the fiber necessary to balance the torque due to the applied field was measured as the field was raised as high as 23 kilogauss. This was done with the applied field, H, directed along the several axes of symmetry of the crystal. An example from Joseph's dissertation shows the rather complicated periodic variation of the measured torque as a function of 1/H. **Fig. 12-10.** A detailed description of the experiment and its analysis appeared a few months later. The departmental jack-of-all-trades, builder-of-instruments, engineer-machinist A. Hrushka, is thanked. "The low field DHVA effect in zinc" *Phys. Rev.* **126** 489 1962.

Torque measurements of the de Haas–van Alphen effect were conducted for several years, with significant improvements in the experimental apparatus. The work provided doctoral research for grad student Curtiss O. Larson "Low-field dHvA study of the Fermi surface of aluminum" *Phys. Rev.* **156** 703 1967, and for Paul M. Everett, "Fermi surface of beryllium and its pressure dependence" *Phys. Rev.* **180** 669 1969. An extensive study of dilute aluminum based *alloys* (less than one percent of Zn, Si, Ge, Mg or Ag) allowed comparison with theory. This work was done with post-doc John P.G. Shepherd. These solutes present an interesting variety of effects on the electron population in that the number of valence electrons ranges from one for silver up to four for silicon. Good agreement was found with the "rigid-band theory" which predicted very small changes in the Fermi surface relative to that for pure aluminum. "Study of the dHvA effect in dilute aluminum based alloys" *Phys. Rev.* **169** 541 1968.

Gordon was joined by post-doc John Tripp in studying changes in the Fermi surface of magnesium caused by the inclusion of about one percent of lithium or indium in the crystal. These two impurity atoms raise and lower, respectively, the local density of charge. The authors compared calculated and measured shifts in the resonant frequencies associated with several portions of the Fermi surface. "Effect of Alloying on the Fermi Surface of Magnesium" *Phys. Lett.* **54A** 463 1975. Two of Gordon's grad students were involved in this work: Larry J. Hornbeck and Wai K. Fung.

John Tripp would be connected with experimental condensed matter research in the CWRU department for many years. His work on superconductivity with David Farrell will be described in Chapter 15. Doctoral student Hornbeck subsequently took a position at Texas Instruments, where, two decades later, he would be the inventor of the digital micromirror device. With tens of thousands of movable mirrors on a tiny chip, the DMD is at the center of a new generation of video display technology.

Resistance measurements and the Fermi surface

In parallel with the torque measurements, Gordon, Eck and grad student R. W. Stark, determined aspects of electronic structure by looking at the effect of an applied magnetic field on the flow of current through the material. In this study, a constant current of 0.2 A flows along a half-inch long slab of a single crystal of Mg. A magnetic field is applied to the sample at right angles to the current in the sample. Then the potential difference between various pairs of electrodes is recorded as the magnetic field is rotated in the plane perpendicular to the current. **Fig. 12-11** shows a sketch of the sample and the position of the electrodes. The ratio of the potential difference between electrodes 1 and 2 (typically 10^{-5} V) and the current J is called the transverse magnetoresistance and it is found to change as the direction of the applied magnetic field is changed (**Fig. 12-12**). (As the magnetic field sweeps through the various symmetry axes of the crystal, the electron paths through the lattice are perturbed, resulting in this complicated behavior of the magnetoresistance.) The data can then be interpreted to give information about the electron energy levels in the material. "Magnetoresistance Investigation of the Fermi Surface of Magnesium" *Phys. Rev. Lett.* **8** 360 1962 and "Galvanomagnetic Investigation of the Fermi Surface of Magnesium" *Phys. Rev.* **133** A443 1964.

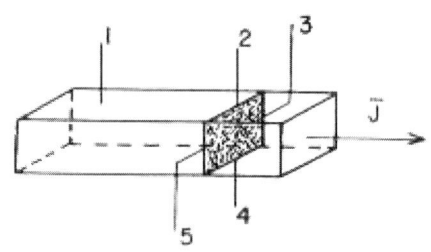

Fig. 12-11. Schematic of sample and probes used in magnetoresistance measurements.

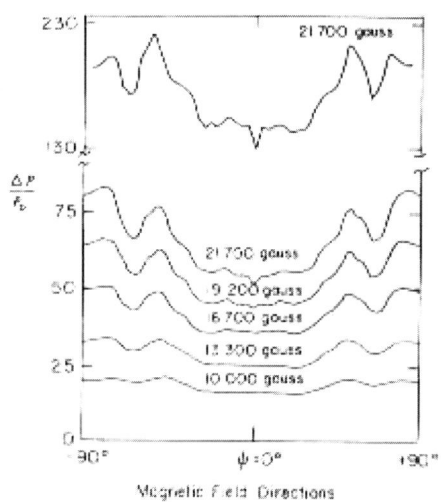

Fig. 12-12. Dependence of magnetoresistance on direction of applied magnetic field.

With grad student David Wagner, Gordon published a related study of the "intermetallic" compound, $MgCu_2$, once again using torque and pulsed magnetic field measurements to map out deHaas-van Alphen areas. "A Fermi surface study of $MgCu_2$" *J. Low. Temp. Phys.* **27** 37 1977.

Bill Gordon describes the motivation for Fermi surface measurements as follows: these "determinations are important to the understanding of transport properties of metals and semiconductors as well as their equilibrium and optical properties. An experimentally measured Fermi surface provides a target for band structure calculations and can be used to furnish data for fitting parameters in a phenomenological crystal potential which can be used to calculate other properties." In Chapter 17, we shall describe related theoretical work, including band structure calculations, by Segall and Lambrecht.

Probing polymer structure

Toward the end of the 1970's, Gordon moved away from the experimental study of Fermi surfaces and band structure in crystal materials. He joined Jerry Lando and Jack Koenig of CWRU's macromolecular science department in a study of the properties of polymers. Bill's colleague, Phil Taylor, had been exploring polymers on the theoretical side for several years (Chapter 17), and polymer physics would play a major role in the department's research through the 1980's.

The first experiments used infrared absorption spectra and x-ray scattering measurements to probe the structure of polyvinylidene fluoride, nicknamed PVF_2. The monomer, or building block, of this chain is $[-CH_2-CF_2-]$ so that one has a string of carbons with alternating pairs of hydrogens or fluorines hanging from either side. Grad student Michael Bachman's thesis describes four configurations (phases) of this polymer which have been identified, the simplest being a zig-zag chain with the H's and F's sticking out in alternating directions like leaves on a vine. The other three phases are more complicated arrangements of the same monomers. The different phases were produced in the form of thin films by subjecting the "melt" to controlled temperature and pressure changes along with applied electric fields. Their detailed structures can be deduced by interpreting infrared and x-ray absorption spectra. Certain frequencies are absorbed as the photons interact with phonon excitations in the polymer lattice. "An infrared study of phase III poly(vinylidene fluoride)" *J. Appl. Phys.* **50** 6106 1979. Similar work, including dielectric studies, with this and other polymers was continued through the 1980's.

Starting in the early 1990's, Gordon began work with colleague Don Schuele on the study of the dielectric properties of certain polymer stabilized liquid crystals. We'll describe these experiments later in this chapter.

In his nearly four decades of research at CIT and then CWRU, Bill Gordon was advisor to seventeen PhD candidates. About two thirds of these students studied the behavior of electrons in metals as determined by de Haas–van Alphen measurements; the other third studied the structure of polymers as deduced from x-ray absorption measurements.

Gordon was chairman of the department from 1979 until his retirement in 1994. Five faculty members were added during his term: four in condensed matter and one in particle theory. For the past ten years, as professor emeritus, he has applied his talents and knowledge of liquid crystal technology to the creation of a unique teaching aid. In connection with the activities of the Advanced Liquid Crystal Optical Materials Science and Technology Center (ALCOM), Gordon and a parade of talented undergrads have produced a CD-ROM and website (http://plc.case.edu) with text, image, audio and video presentations on the science of liquid crystals and their applications. In addition, Gordon has continued to support secondary-school physics teachers in the department's outreach program, long championed by the late John McGervey (Chapter 14).

Eck: crystals and atoms

Thomas G. Eck joined the Case department in 1957. He had done his PhD research at Columbia with Polycarp Kusch. Recall that Nobelist Kusch earned his BS at Case, and no doubt had something to do with Eck's choice of a position. Eck's dissertation concerned measurements of the hyperfine structure of selected excited states in two isotopes of the indium atom. These small shifts in the electron energies result from the delicate interaction between the orbiting electron and the magnetism of the nucleus. By comparing the effect in two isotopes, ^{115}In and ^{113}In, one can extract a measure of the nuclear radius and the contribution to the magnetic moment of the two additional neutrons. *Phys. Rev.* **106** 954 1957.

Electrons in crystals

At Case, Eck joined Bill Gordon in the study of electrons in metals, starting with two of the experiments described above: the one on the band structure of zinc and cadmium and the other on the Fermi surface of magnesium. Subsequent work on these three metals provided doctoral research for three of Eck's students, P. Sampath, M.P. Shaw, and D.A. Zych. In these studies, the technique called "cyclotron resonance" was used to further elucidate the details of the Fermi surface. In this type of measurement, the carefully cut crystals are held at low temperatures, typically a few kelvins. A magnetic field of several kilogauss is applied to the sample. The direction of this field is chosen to lie in the plane of one of the crystal's principal surfaces, and it can be rotated about the various principle axes. The electrons in the metal undergo circular (cyclotron) motion under the influence of the magnetic field. A high-frequency electric field is then applied to the sample. When the frequency matches that of the cyclotron motion of the electrons, energy is absorbed by the sample. In this experiment the radio-frequency was set at 24 gigahertz and the magnetic field was swept from zero up to 5 kilogauss. **Fig. 12-13** shows the variation of the cyclotron motion as one changes the direction of the applied magnetic field, thus causing the looping electrons to sample different features of the crystal lattice. (Actually shown is M*/M, the ratio of the "effective mass" of the electrons in the crystal to that of a free electron. This ratio is the same as the ratio between the observed cyclotron frequency and that for electrons looping in free space.) **Fig. 12-14** shows the behavior of dR/dH as one changes the strength of the applied magnetic field. (R is a measure of the absorption of the rf power.) This rather complicated structure can be related to the wave-functions of

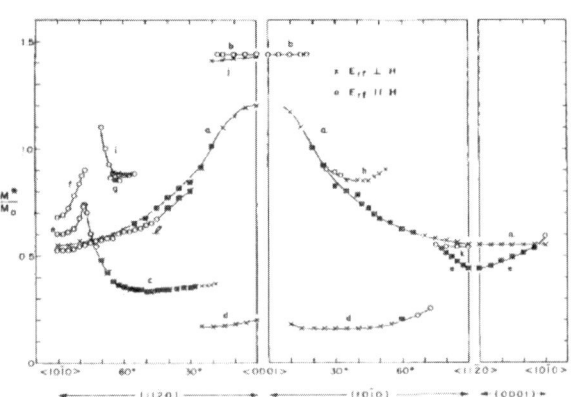

Fig. 12-13. Dependence of the cyclotron resonant frequency on the direction of the applied magnetic field.

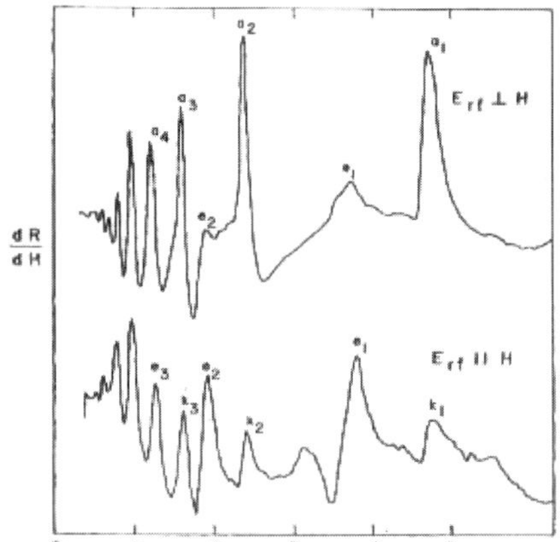

Fig. 12-14. Measure of rf power absorption as the magnetic field is varied.

the electrons in the crystal. *Low Temp. Phys.* **9** 759 and 761 1965 and "Cyclotron resonance in zinc" *Phys. Rev.* **142** 399 1966).

Electrons in atoms

Eck turned next to a series of measurements in **atomic physics**. The electrons in an atom reside in well defined energy states. When a magnetic field is applied to the atom, the energies of these states are shifted (the Zeeman effect). The amount of shift depends on the orientation of the spin and orbital angular momentum of the state. It is possible for two states to have the exact same energy for certain special values of the applied magnetic field. This is called "fine structure crossing". (See a simple example of the crossings of a $P_{3/2}$ level with two $P_{1/2}$ levels in **Fig. 12-15**.) When an atom in such an environment is subjected to electromagnetic radiation whose photons have exactly the same energy as the common energy of excitation, the two excited states interfere with one another; resulting in abrupt changes in the angular distribution of the outgoing photons. Eck and theorist Les Foldy determined that other, more subtle, interactions within the atom could break the mixing: an effect they called "anticrossing". "Observation of 'Anticrossings' in Optical Resonance Fluorescence" *Phys. Rev. Lett.* **10** 239 1963 and *Phys. Rev.* **153** 91 and 103 1967.

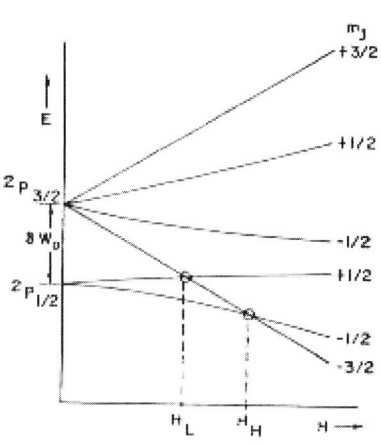

Fig. 12-15. Example of fine structure crossing.

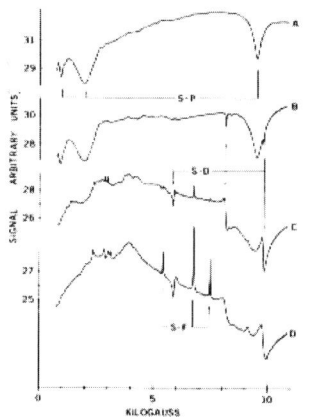

Fig. 12-16. Observation of atomic crossing signals in ionized helium.

Using a somewhat different approach, Eck looked at crossing among the substates of the n=4 level of ionized helium (an atom with only one electron). The sample gas is bombarded by a beam of electrons and the n=4 to n=3 radiation at 4686 Å is detected by a photomultiplier tube. Crossing signals are clear (the sharp structures indicated by the arrows) in the plot of the observed radiation vs. applied magnetic field. **Fig. 12-16.** The plots in the figure show the effect of applying a *static* electric field transverse to the magnetic field. (The four traces correspond to electric fields increasing from top to bottom.) *Phys. Rev. Lett.* **22** 319 1969 and "Level-Crossing Signals in Stepwise Fluorescence" *Phys. Rev.* **A2** 2179 1970.

In keeping with his interest in the interference between the excitation of various atomic substates, Eck later wrote a paper in which he developed a theoretical explanation for the experimental observation of beats in the intensity of radiation emitted by light atoms in beam-foil experiments. In this type of experiment, a beam of ions is incident on a thin foil where some of them are excited by grazing interactions with the foil atoms. The excited ions travel on and eventually decay, emitting a photon. The lifetime of each excited state can then be determined by measuring the number of decays as a function of the distance downstream from the foil at which the photon is emitted.

Eck describes his model as one "in which the coherence between different m_L states in a tilted-foil, beam-foil experiment is induced by the electrostatic interaction between each beam particle and the foil as a whole." "Coherent Excitation of S and P States of the n=2 Term of Atomic Hydrogen *Phys. Rev. Lett.* **31** 270 1973. "Coherent Production of Different m_L States in a Beam-Foil Experiment" *Phys. Rev. Lett.* **33** 1055 1974.

Returning to the study of the Fermi surface of metals, Eck measured the cyclotron resonance frequencies for the four principal orientations of the magnesium crystal. The sample, held at 2 K, was probed by a 22.9 GHz rf field and the applied magnetic field was swept from zero to as high as 12 kilogauss. The resulting spectra were compared with theoretical calculations of the Fermi surface and found to be in excellent agreement. "Cyclotron Resonance in Magnesium" *Phys. Rev.* **B1** 4639 1970. In a follow-up letter, Eck writes, "The purpose of this Letter is to present data for cyclotron-resonance signals in magnesium observed with H perpendicular to the specimen surface. These data shed considerable light on the nature of such signals and strongly suggest that they must arise from charge carriers on trajectories that lie entirely within the microwave skin depth at the surface of the specimen." "Cyclotron Resonance in Magnesium in the Field-Normal Geometry" *Phys. Rev. Lett.* **28** 440 1972.

Fig. 12-17. Thomas Eck.

Most of Eck's research was done between 1957 and 1975. Of his twelve PhD students, six would work on Fermi surfaces as determined by cyclotron resonance experiments, and six on atomic hyperfine structure and level crossings. A 1980 photo of Tom Eck is shown in **Fig. 12-17**. Eck was an extremely popular teacher, especially in the introductory courses. For 45 years, until his retirement in 2002, Eck filled the big lecture hall with his big voice and hearty laugh, teaching mechanics and e&m to thousands of science and engineering students.

Schuele: crystal properties

In Chapter 7 we looked at the experimental research of Chuck Smith. This consisted principally of the determination of mechanical properties of solids, first by x-ray diffraction and then by the measurement of the velocity of ultra-sound through the materials. Among Smith's eleven doctoral students was **Donald E. Schuele** who completed his doctorate in 1963. His thesis title was "The Thermal Expansion at Low Temperatures of Rubidium Iodide". He joined the Case faculty the following year, continuing the ultrasound measurements. Schuele is a native Clevelander who did his BS at nearby John Carroll University.

Thermal and mechanical

It is well known that solids expand when heated. The question is why and how much. What is the interplay between the electrons and the lattice of atoms and what changes take place when thermal energy is introduced? Schuele and Smith investigated the thermal expansion of several salts: RbI, KCl, and NaCl. The separation between adjacent planes of atoms was measured by Bragg scattering.

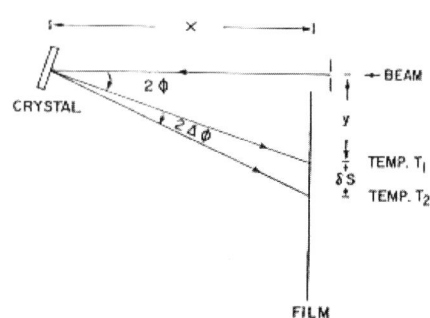

Fig. 12-18. Schematic of a Bragg scattering experiment.

An aside on **Bragg scattering**. *This technique, in use since 1913, involves the reflection of x-rays from a crystal. The intensity of the reflected rays has maxima at certain angles of reflection, and these angles depend on the separation between the layers of atoms in the crystal.* ($\sin \phi = \lambda/2d$ *where ϕ is the reflection angle for a maximum, d is the distance between crystal planes, and λ is the x-ray wavelength.)* **Fig. 12-18.**

In this experiment, the target crystal was placed in a cryostat in which the temperature could be precisely controlled. (**Fig. 12-19**) The whole cryostat could be rocked back and forth so that the angle between the incident x-ray beam and the atomic planes would sweep through the angle at which the Bragg condition was satisfied, giving a sharp line on the x-ray film. The position on the film of the reflected beam gives the angle ϕ and thus the separation between crystal planes.

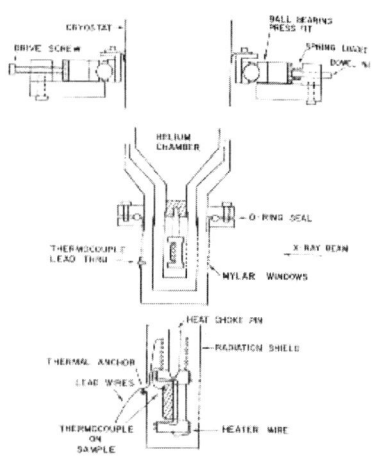

Fig. 12-19. Apparatus for "rocking crystal" Bragg scattering experiment.

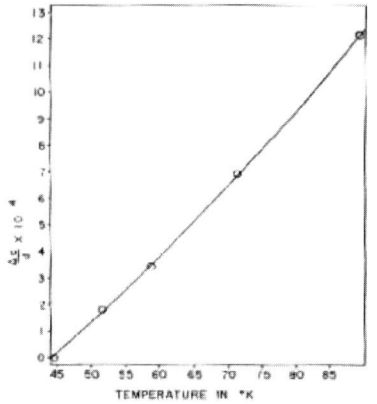

Fig. 12-20. Thermal strain vs temperature.

The results of a typical run are shown in **Fig. 12-20**, where the fractional expansion $\Delta d/d$ of rubidium iodide is plotted against temperature from 45 to 90 K. The resulting linear coefficient of expansion, $\alpha = (\Delta d/d)/\Delta T$, can also be written in terms of the bulk modulus B and the specific heat c_v thusly: $\alpha = \gamma c_v/3B$ where γ is a parameter which connects the volume changes with the vibrational frequencies of the ions in the crystal. This constant of proportionality, γ, is called the Grüneisen parameter. In the theory, γ is expected to approach two limiting values, at low and high temperatures respectively. The experimental data plotted in **Fig. 12-21** confirm this behavior at the high temperature end. (*J. Phys. Chem. Solids* **25** 801 1964)

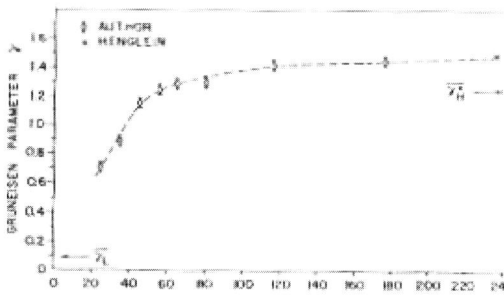

Fig. 12-21. Measured Grüneisen parameter vs. temperature.

Schuele returned to the ultrasound measurements of elastic constants which Smith had initiated. These measurements provided an alternative way to determine the Grüneisen parameter. The velocity, v, of high-frequency sound was measured along the principal axes of NaCl and KCl crystals. Since $v^2 = C/\rho$, where ρ is the density, the elastic constant C can be extracted. The measurement is repeated at various pressures, P, all the way up to 3300 atmospheres, and the dependence of C upon P is determined. According to the prevailing theory, γ may be deduced from the behavior of dC/dP. Thus the authors were thus able to compare the "thermodynamic" γ with the "elasticity" γ. *J. Phys. Chem. Solids* **26** 537 1965; **27** 493 1966; **28** 1225 1967; **30** 589 1969.

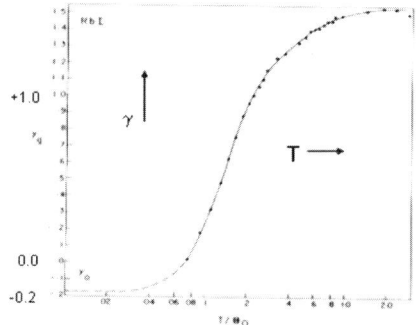

Fig. 12-22. Grüneisen parameter vs. temperature

Ultrasound measurements made on rubidium iodide resulted in the data shown in **Fig. 12-22**. This substance had been shown to have a negative Grüneisen parameter γ at low temperatures. The authors conclude that "comparison between the thermal value and the elastic data is excellent." "Low Temperature Grüneisen Parameter of RbI from Elasticity Data" *J. Phys. Chem. Solids* **31** 647 1970. Pulse-echo measurements were made to determine elastic constants of several other crystal substances. Another variable was introduced as ultra-sound velocities were measured as a function of externally applied stress, up to pressures of 3000 atmospheres. "Third-Order Elastic Constants of Al_2O_3 "*J. Acous. Soc. of Amer.* **48** 190 1970.

Two isotopes of lithium, ^6Li and ^7Li, were chosen for a study of the effect of atomic mass on the thermal and elastic properties of the solid. (These atoms, which differ by a single neutron, were favorites of nuclear physicists as well, as mentioned in Chapter 7.) Because their atomic masses are small, the ratio of the masses of these two isotopes is rather large, and the effects may be correspondingly notable. The velocity of ultrasonic pulses through 3-in long, half-inch diameter single crystals was measured. These ultrasound velocities, measured along various axes of the crystal, along with the densities of the solids, give the desired elasticities. Measurements were made at temperatures ranging from 100 K to 295 K. In addition, the pressure dependence of the elastic constants was measured by applying pressures up to 30,000 psi. (2×10^9 dyne/cm^2) It was found that there was *no discernable difference* between the two isotopes for the elastic constants or their derivatives, and that the ultrasound velocities scaled simply as the inverse square root of the atomic mass. The crystal properties depend presumably on the three electrons in each atom, and not on the masses of the nuclei which tie the electrons together. "Temperature and pressure dependence of the single-crystal elastic constants of ^{6}Li and natural lithium" *Phys. Rev.* **B16** 5173 1977.

Dielectric properties of crystals

Around 1970, Schuele and his students, Carl Andeen and John Fontanella, moved away from thermal and elastic properties to a new area of research: the determination of the *dielectric* properties of crystals. In their introduction to a paper in the *Review of Scientific Instruments,* they note: "A study of the literature on the static dielectric constant of ionic crystals reveals a surprising amount of discrepancy. This is unusual for a field that is over a hundred years old and is particularly disturbing to theoreticians who require accurate values of these constants in order to determine characteristic model parameters as tests of their various hypotheses." "The static dielectric constant ε_s and its variation with temperature and pressure contain important information concerning the constitution of solids. The constant itself provides a measure of the ability of polarizable entities within the solid to respond to an electric field. The variation of ε_s with pressure, then, reflects how these systems are affected by a change in interatomic distance while the temperature dependence of ε_s includes both analogous size effects and those that are intrinsically thermal." "Accurate Determination of the Dielectric Constant by the Method of Substitution" *Rev. Sci. Inst.* **41** 1573 1970.

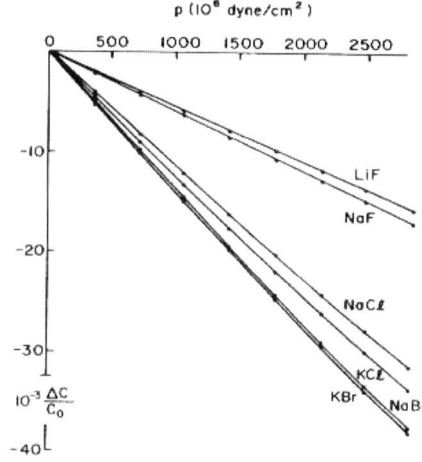

Fig. 12-23. Change in capacitance of halide salt samples with increasing pressure.

In two papers, Schuele reported on what were basically capacitance measurements. The sample dielectric materials were disk-shaped, 25 mm diameter, 1.6 mm thick, at 308 K and at pressures up to 40,000 psi. Each paper tabulates the density dependence of the dielectric constants. **Fig. 12-23** shows an example of the data on pressure *vs* the fractional change in capacitance for six halide salts. As the atoms are squeezed a bit closer together, the dielectric constant, and thus, the capacitance, decreases. "Pressure and Temperature Derivatives of the Low-frequency Dielectric Constants of LiF, NaF, NaCl, NaBr, KCl, and KBr" *Phys. Rev.* **B2** 5068 1970; **B6** 582 1972.

Clearly, these very high pressure measurements require an accurate and precise pressure gauge. Schuele and his students published a paper describing their design for such a device. Since their earlier work established the effects of pressure on dielectric constants, it was obvious that the measurement of a capacitance could be used to determine a pressure. Single crystals of CaF_2, again 25 mm in diameter and 1.6 mm thick, had 2000 Å layers of aluminum plated on their surfaces. Their capacitance was found to change about one percent for every 2.7 kilobar change in pressure. The gauge was calibrated at the National Bureau of Standards, and was found to be stable and accurate at the 0.01% level over the range zero to 2.5 kilobars. "A Capacitive Gauge for the Accurate Measurement of High Pressures" *Rev. Sci. Instr.* **42** 495 1971.

An interesting technique for the determination of a dielectric constant is called "the method of substitution". The capacitances with five different "fillings" are measured: vacuum, fluid #1, fluid #2, fluid #1 surrounding the sample to be measured, and fluid #2 surrounding the sample. It is a nice exercise for the intro-E&M student to show that the dielectric constant of the sample can be written in terms of these five capacitances, independent of the properties and thicknesses of the fluids or sample size. "Low-frequency Dielectric Constants of the Alkaline Earth Fluorides by the Method of Substitution" *J. Appl. Phys.* **42** 2216 1971.

Fig. 12-24.
Donald Schuele.

Measurements of the dielectric constants for a variety of commonly used materials were made with precisions of about one part per thousand. In the case of anisotropic materials, such as quartz and sapphire, measurements were made as a function of the angle between the applied electric field and the major axis of the crystal. The dielectric constant for quartz, for example, is found to vary linearly with the square of the sine of this angle, decreasing about 2% from 0 to 90 degrees. "Low-frequency dielectric constants of α-quartz, sapphire, MgF_2, and MgO" *J. Appl. Phys.* **45** 2852 1974.

In the four year period beginning in 1968, Schuele supervised seven doctoral students, five of them measuring elastic constants and the other two determining dielectric properties. A photograph of Schuele, taken around 1980, is shown in **Fig. 12-24.**

In 1976, Schuele took over from Ken Kowalski as chairman of the department and held that position for five years. In 1981 he was appointed vice-dean and then dean of CIT (that portion of CWRU which included the engineering, science, and mathematics departments.) When the undergraduate college was reorganized into an engineering school and a college of arts and sciences, Schuele resumed his research and teaching.

Polymers: dielectric properties

In 1984 Schuele and grad-student Alfredo Bello tackled the dielectric properties of the same polyvinyl fluoride polymer that Gordon had been studying using x-ray absorption. In the early 1990's, Schuele and Gordon studied the dielectric properties of polymer *liquid crystals*. During the same period, Phil Taylor and Rolfe Petschek were developing theoretical models for polymer liquid crystals as will be described in Chapter 17. Gordon and Schuele used dielectric relaxation spectroscopy to get a handle on how these fascinating groups of atoms arrange themselves along a polymer backbone. The resulting materials combine the good film-forming and mechanical properties of polymers with the optical, electro-optical and magneto-optical properties of liquid crystals. Because Schuele had been studying the dielectric properties of materials, his laboratory was well equipped to take on these exciting new liquid crystal polymers.

In *dielectric relaxation spectroscopy*, the sample to be studied is placed between two electrodes and subjected to an alternating electric field. The magnitude and phase of the resulting current are measured as a function of the frequency of the applied field. The current generally lags behind the field. The "out-of-phase" component is a measure of the delay in the alignment of the various dipole pieces in the sample. The different building blocks (mesogens) of the polymer LC, having different sizes, charges and masses, respond to specific ranges of frequencies. When one plots the size of the "out-of-phase" component (called the "loss") against the frequency, one sees peaks corresponding to the contributions from different mesogens. Further information comes from making the measurements at different temperatures. The typical "relaxation times" decrease exponentially with increasing temperature, i.e. things react faster. Measurements are made over a wide range of frequencies, from 10 Hz to 100 kHz, because of the great variety of dipole structures under study.

A few terms from grad-student Robert Akins' dissertation illustrate some possibilities: side-group-rotations, phenyl ring flips, crankshaft relaxations. It was interesting to look at how the relaxation-time changes as the material underwent phase changes. Because some of the materials have rather large side-chains, they can lose their order and become amorphous as the temperature is increased. It is important to know where these "glass-transitions" occur for various LC polymers. Akins' dissertation specifically addressed this issue: "Dielectric investigation of double glass transitions in polymers". Gordon, Schuele, Akins and another grad-student, Z. Z. Zhong, published five papers on dielectric relaxation measurements of a series of liquid crystal side-chain polymers containing a variety of spines and branches. (*Change the shoes on the millipede and see what tunes she will dance to.*)

Having earned his doctorate at Case, Schuele has always had strong connections with Case alumni organizations. In 1987 he was named the first Albert A. Michelson Professor of Physics. This professorship was created in conjunction with the celebration of the centennial of the Michelson Morley experiment. Schuele retired in 2005.

Hoffman: thin films

We pick up on the work of **Richard W. Hoffman** in 1966 when he published a 60-page review on "The Mechanical Properties of Thin Condensed Films" (*Physics of Thin Films*, **3** 211 1966 Academic Press). This paper, with 121 references, presents an overview of the then-current understanding of films formed by evaporation techniques. Included are some of the ideas described earlier in this chapter on the determination of the surface structures. Methods for measuring stress in films are outlined (e.g. the bending glass slide technique), as well as models for the origin of stress. Finally, techniques for the measurement of the tensile properties of various films are presented. Hoffman had become one of the principal spokesmen for this area of physics.

The creation of very thin (order of 0.1 micron thick) NaCl crystal substrates is described in a brief article. (*J. Vacuum Sci. & Tech.* **6** 65 1968). The goal is to be able to lay down thin metallic films on a substrate which is transparent to the electron beams of an electron microscope. This allows the investigation of the films without the destructive technique of "stripping" them off the substrate.

In 1969, Hoffman addressed a meeting sponsored by the New York Academy of Sciences on the use of thin films in electronics. He concludes: "we see that the important advances in using thin films in electronics have come about by developing careful control of impurities and structural defects in the formation of the films, since the properties of interest are very sensitive to these defects." *Trans. N.Y. Acad. of Sci.* **31** 868 1969.

Magnetization in films

Hoffman and his group experimented with using *Mössbauer spectroscopy* as a way to examine the magnetization of thin iron films. A thin iron film is prepared with the inclusion of about ten parts per million of radioactive cobalt-57. This nucleus decays with a half-life of 271 days by electron capture to an excited form of iron-57, which in turn emits a 14.4 keV gamma ray The outgoing γ's pass through a thin iron absorber and the fraction of γ's absorbed is measured. The energy of the emitted γ depends on the magnetic field in which the emitting atom finds itself, and it is this field which Hoffman wants to explore. In Mössbauer spectroscopy, the absorber is moved slowly toward the emitter so that the effective energy of the γ may be raised, enabling the γ to have just the exact amount of energy to excite an iron nucleus in the absorber. Dips in the fraction of γ's which make it through the absorber occur at particular speeds of approach. Typically, speeds of a few millimeters per second translate to shifts as small as one part in 10^{12} of the energy of the γ. Using this technique, Hoffman found that films of different thicknesses had different internal magnetic fields. "Mössbauer Spectra of Monolayer Iron Films" *Jour. of Vac. Sci. and Tech.* **7** 118 1969. "Interpretation of Mössbauer Spectra in Thin Iron Films" *Jour. of Vac. Sci. and Tech.* **9** 177 1971.

Hoffman then returned to the earlier technique for measuring the magnetization in nickel films, i.e. measuring the torque on a specimen placed in a magnetic field. His apparatus was greatly improved over that used eight years earlier. A schematic drawing of the experiment is shown in the **Fig. 12-25**. In addition to new servo-mechanisms and photometric readout, the new computer technology was brought to bear: "torque values….were transferred to punch cards, and a complete analysis of the data, including a final error analy-

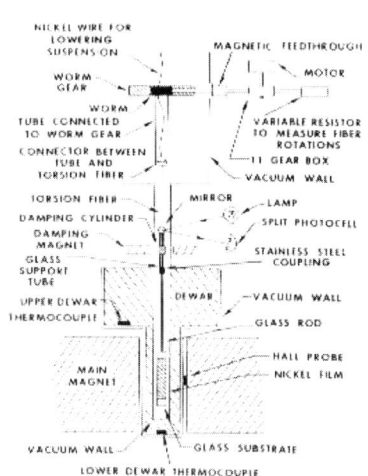

Fig. 12-25. Schematic of device for automated measurement of torque.

sis, was performed by a single computer program". "Saturation Magnetization and Perpendicular Anisotropy of Nickel Films" *Jour. Appl. Phys.* **412** 1623 1970.

Stress in films

Hoffman's group pursued the question of stress in thin films over the course of several years. (We mentioned above the technique of measuring the bending of the substrate caused by stress in the film.) They found that the stress depends strongly on the temperature, on the thickness of the film, and on the preparation (cleaning) of the substrate. "Structure and Intrinsic Stress of Platinum Films" *Jour. of Vac. Sci. and Tech.* **8** 151 1971. "The Origins of Stress in Thin Nickel Films" *Thin Solid Films* **12** 71 1972. The authors concluded that the fall-off of stress with increased temperature is associated with relaxation of grain boundaries and diffusion of substrate atoms into the film. The bending substrate method was later replaced by the use of a microbalance which could measure the stress force directly. "Growth Effects on Stress in Nickel Films" *J. Vac. Sci. Technol.* **10** 238 1973.

Hoffman again turned to Mössbauer measurements to look at what happens when a thin cobalt film is cooled to a crystalline form. "Pure amorphous cobalt films (~100 Å) have been prepared by condensation onto liquid helium-cooled glass substrates at 10^{-10} Torr. The films underwent an abrupt and irreversible fourfold drop in resistivity at about 55 K, characteristic of metallic films during the amorphous to crystalline transition. Analysis of the 6-line Mössbauer spectra indicates the hyperfine field is 1.5% lower and each line is 30% broader compared with the same quantity in the crystalline phase." "In summary, the picture of an amorphous film that is obtained from these results is that of a structure in which disorder produces a striking change in the electrical resistivity but affects very little the atomic bonding, the magnetization direction and the electronic density at the nucleus". "Electrical and Magnetic Properties of Pure Amorphous Cobalt Films" *J. Appl. Phys.* Suppl. **2** 729 1974.

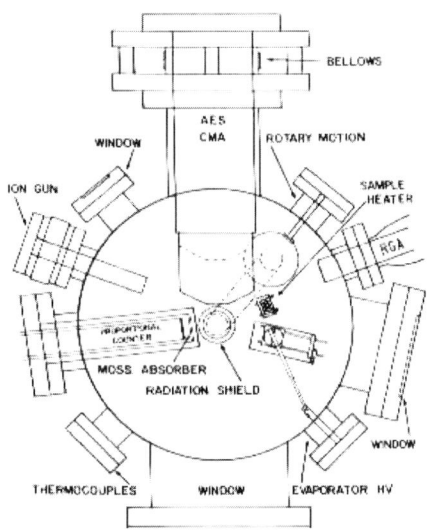

Fig. 12-26. Sketch of typical high-vacuum system for surface research.

New tools for surface science

The high-vacuum systems used by Hoffman and the surface physics group became more and more complex as new devices were placed inside them. **Fig. 12-26** shows a sketch of such a system. It comes from a paper describing Mössbauer effect emission spectroscopy (MEES) for the study of thin films. The sample must be placed in an ultra-high-vacuum (down to 10^{-9} Pa pressure), held at controlled temperatures ranging from 20 K to 1500 K. It must be kept in view by the Mössbauer absorber, by an Auger electron spectrometer (AES), and by the various hot ion sources which create the thin films or lay down the radioactive ^{57}Fe atoms on the sample. "Apparatus for Mössbauer effect emission and AES study of free surfaces" *J. Vac. Sci. Technol.* **16(2)** 466 1979.

*An aside on Auger Electron Spectroscopy (**AES**): A beam of electrons is directed at the surface to be studied. Some of these knock an electron out of low-lying energy levels in atoms stuck to the surface. Subsequently, an electron from a higher level drops down to fill the hole, while simultaneously a **second** electron is ejected from the atom. These "Auger" electrons have well-defined energies for each type of atom, and their detection signals the presence of a particular type of atom on the surface.*

In a different approach to the study of thin films, Hoffman determined their *tensile* properties by measuring how much they stretched when subjected to a known force. Devices to do this, called nanotensilometers, were used not only for thin metallic films, but for polymers and even biological samples. Hoffman and his students, Carl Hagerling and C. G. Andeen, designed and tested a device capable of measuring forces over the range 10^{-8} to 0.5 Newtons, with elongations from 0.1 nm to 1 mm. The forces were applied electromagnetically and positions measured by monitoring the capacitance of the moving parts. "The nanotensilometer – an accurate, sensitive tensile test instrument" *Proc. 7th Int. Vac. Congr. Vienna* p1769 1977.

Some early bio-physics

In a rather interesting departure from his thin film work, Hoffman teamed up with a researcher from the Western Reserve School of Medicine, John H. Bauman, to design and test a device to allow *in vivo* measurement of the iron content of livers. Basically, the device was a set of transformer coils into which the subject animal could be placed. The presence of iron in the liver changes the mutual inductance of the transformer. From the abstract: "Studies using live rats demonstrate that with this technique iron-loaded animals can be distinguished from control animals, because the high hepatic concentration of storage iron in the liver of the experimental group exhibits positive magnetic susceptibility." (*IEEE Transactions on Bio-medical Engineering,* 239 Oct. 1967) When I called this paper to the attention of my colleague, David Farrell (whom we shall meet in Chapter 15), he remarked that Hoffman's 1967 work was indeed strongly related to his (Farrell's) decision to initiate a research program in biomagnetism.

All in all, between 1959 and 1994, Hoffman would oversee the research of 40 doctoral students (a departmental record, by far). The first 16 (before 1978) studied the magnetic and mechanical properties of thin films. The later doctoral topics would include the Mössbauer experiments and some pioneering work using electron energy-loss spectroscopy, x-ray photoemission spectroscopy, and low energy ion bombardment techniques. These methods for studying the physics of atoms on surfaces will be further described a bit later in this chapter.

Exotic films

In the mid-1990's, Hoffman joined John Angus of the CWRU chemistry department in a series of surface studies of diamond-like *carbon films*. Angus had developed techniques for creating such films by chemical vapor deposition. In a paper which included four of Hoffman's students among its authors, a study of the secondary emission of electrons from diamond-like surfaces is described. Gerry Mearini and Isay Krainsky both had been hired by NASA Lewis Research Center, and Chris Zorman and Yaxin Wang were still grad students. They were interested in increasing the yield of emitted electrons. These diamond-like surfaces are used in photomultiplier tubes and similar devices, partly because of their ability to dissipate heat. The surfaces were first coated with a thin layer of cesium iodide and then subjected to an electron beam which preferentially knocked out the iodine atoms, leaving a "cesium-terminated" diamond surface. The result was that the yield of secondary electrons was enhanced by a factor of 5 or more. They conclude: "A material with such high stable secondary yield can be used to construct an electron amplifier with gain several orders of magnitude higher than is presently attainable." *Appl. Phys. Lett.* **66** 242 1995.

Fig. 12-27. A shiny new vacuum system, three grad students, and Dick Hoffman.

Fig. 12-27 shows Hoffman and a few of his students alongside one of their high-vacuum chambers. He was elected chair of the Thin Films Division of the American Vacuum Society in 1968 and AVS president in 1976. Hoffman became the third Ambrose Swasey Professor of Physics in 1981. He was named "Honorary Member of the American Vacuum Society" in 1995. Hoffman remained active in research and teaching until the mid-1990's, even though he was beginning to suffer from a progressive debilitating disease. He died at age 74 in 2002, leaving a legacy of forty doctoral students, most of whom are active in teaching or materials research.

Chottiner: next generation surface physics

Gary S. Chottiner joined the department in 1980, its first "permanent" addition in ten years. Raised in Pittsburgh, he did his undergraduate work at Carnegie-Mellon University. He came to CWRU in with a fresh doctorate from the University of Maryland, where he had worked with R. E. Glover III and Robert Park. His PhD thesis was on the adsorption of oxygen on tin, the first of an extensive list of publications on surface phenomena. (*J. Vac. Sci. & Tech.* **15** 429 1978) Photo **Fig. 12-28**. Chottiner built upon

Hoffman's work in surface studies, but he soon moved into other areas of surface phenomena such as chemical catalysis and fuel-cell technology.

The study of thin layers of materials placed on an underlying surface has become a major area in experimental physics, chemistry, and materials science, as well as a major contributor to technology. For starters, the physics of a two-dimensional array of atoms or molecules is entirely different from that of normal three-dimensional systems. The techniques for the production and analysis of thin films pioneered by Hoffman and his contemporaries have been greatly refined and improved. The following "aside" describes some of them.

The "Many Acronyms of Surface Science" (MASS)

Fig. 12-28. Gary Chottiner.

This is probably a convenient place to describe some of the methods used to study thin layers of atoms and molecules on surfaces. Each technique involves a beam of photons, electrons, ions or molecules which is aimed in an ultra-high vacuum environment at the surface to be analyzed. The goal of these measurements is to collect information on the density, composition, and arrangement of the "absorbate" on the "substrate".

*Auger electron spectroscopy (**AES**) was described earlier in this chapter. Recall: a beam of electrons strikes the surface and atomic electrons are ejected with energies specific to the type of atom. Mössbauer effect emission spectroscopy (**MEES**) was also described above as a way to measure the magnetic fields in thin films.*

*X-ray photoemission spectroscopy (**XPS**) uses a beam of x-rays (around a keV) to knock electrons out of the atoms. The measured kinetic energies of these electrons are subtracted from the incident x-ray energy to give the atomic binding energy, and thus the element may be identified. XPS has the advantage over AES in that there is no charge build-up on the surface. XPS can be sensitive enough to detect the small shifts in the binding energy caused by the atom's chemical environment, e.g. the kind of molecule it might live in. Even more subtle details in these spectra provide information on the electron's interaction with the underlying substrate.*

*Secondary ion mass spectroscopy (**SIMS**) uses a beam of ions or atoms with energies in excess of a keV which knock (or sputter) a whole atom, molecule, or cluster of atoms off the surface. These loose pieces can then be identified by using standard mass spectrometry. SIMS can detect extremely tiny amounts of elements on the surface, as little as one millionth of a monolayer. It is often used to detect impurities on the surface.*

*Low energy electron diffraction (**LEED**) can be used to determine the "crystal structure" of the layer of atoms on the surface. A beam of electrons (with energy in the 20 to 500 eV range) strikes the surface at right angles. These electrons have wavelengths comparable to the size of the atomic spacing in the surface layer and so produce an inter-*

ference pattern similar to the Bragg scattering of x-rays from a crystal. The backscattered electrons are detected and the interference pattern yields information on how the surface atoms are arranged.

*Infra-red reflection-absorption spectroscopy (**IRAS**) is one example of a "vibrational spectroscopy". Monochromatic infra-red radiation strikes the surface at a glancing angle, passes through the adsorbate, reflects off the substrate, travels back through the adsorbate, and is then detected. Sometimes the photon loses energy by exciting a vibrational mode of an adsorbate atom or molecule. Absorption lines in the observed spectra can be used to identify the type and orientation of the adsorbate. In Fourier transform infra-red spectroscopy (**FTIR**), extremely sparse mono-layers of such adsorbates as H or CO or CO_2 can be characterized.*

*High resolution electron energy loss spectroscopy (**HREELS**) is a second technique in which vibrational modes in the adsorbate are excited. Highly monoenergetic electrons are used as probes, rather than infra-red photons.*

*In temperature programmed desorption (**TPD**) the measurement involves slowly raising the temperature of the surface and watching the rate at which the pressure rises in the chamber. At a certain point, a large number of absorbed molecules might break away, giving a sharp rise in the pressure. This provides a measure of the strength of the adsorbate-substrate bond.*

Usually, two or more of these approaches are used in parallel in the study of a given sample. The vacuum chamber contains a variety of sources and detectors of ions, electrons, and photons along with devices used to produce the surface layers in the first place. Equipped with the necessary acronyms, we may now describe Chottiner's work. During his first few years at CWRU, he worked with Dick Hoffman, beginning with a study of conducting polymers. Both Hoffman and Schuele had looked at the properties of thin film polymers, and Chottiner joined an HREELS/AES experiment. "HREELS and Auger Studies of Conducting Polymers" *Appl. Surf. Sci.* **21** 80 1985.

During the 1980's Chottiner worked to equip his Surface Science/Thin Film Laboratory with state-of-the-art ultra-high-vacuum systems in which thin films could be prepared and then analyzed. He published a series of papers describing innovative techniques for handling samples over a wide range of temperatures at very high vacuum. (e.g. "Ultrahigh Vacuum Cryostat and Sample Manipulator for Operation between 5 and 800 K", *Rev. Sci. Inst.* **56** 1799 1985.)

With grad student Wayne Jennings, Chottiner studied the way in which sulfur migrates to the metal-oxide interface when iron becomes oxidized. This was part of a more general study of how oxidation takes place, undertaken with colleagues in the Materials Sciences department. "Sulfur Segregation to the Metal/oxide Interface during the Early Stages in the Oxidation of Iron" *Surf. and Interface Analysis* **II** 377 1988.

Another of Chottiner's students, Steve Eppell, looked at platinum deposited on a carbon surface (more precisely, on highly oriented pyrolytic graphite). In a study which presaged later work on platinum clusters, Eppell determined the correlation between the size and height of the platinum islands and the catalytic properties of the surface. "Scanning Tunneling Microscopy of Platinum Deposits on the Basal Plane of Highly Oriented Pyrolytic Graphite" *Langmuir* **6** 1316 1990. (Eppell is now associate professor of biomedical engineering at CWRU.)

Surfaces for electrocatalysis

In 1990, Chottiner began a productive collaboration with electrochemist Daniel A. Scherson, a professor in the CWRU Department of Chemistry. The study of electrochemistry at CWRU goes back to the 1970's with the pioneering work of Professor Ernest B. Yeager, and continues to this day in the multidisciplinary center named for Yeager and headed by Scherson.

Electrocatalysis is the study of how chemical reactions may be accelerated at the surface of an electrode. This requires that the electrode's surface be modified by the presence of a sub-monolayer adsorbed species of appropriate ions, molecules or metals. Many of the surface science techniques listed above are applicable to the characterization of such surfaces. Of particular interest are electron-transfer processes such as those which take place in energy-storage devices like rechargeable batteries and hydrogen fuel cells. The marriage of the new ultra-high vacuum techniques developed for surface-science with established electrochemical approaches has resulted in an exciting new research area.

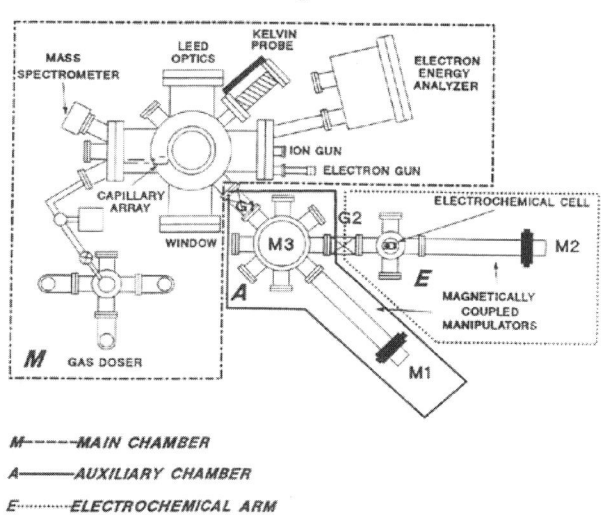

Fig. 12-29. Schematic of transfer system for UHV analysis of electrochemical samples.

In batteries, the ability of electrons to be transferred from one medium to another is critical. The transfer rate can be significantly modified by the presence of absorbed atoms or molecules on the surface. The size and orientation of the absorbed particles as well as the thickness of the layer all affect the electrochemical behavior.

Grad student Kuilong Wang completed a series of experiments in 1992 on "three electrochemically interesting interfaces": potassium hydroxide on nickel, carbon dioxide on potassium-coated silver and a certain solvent, THF, on lithium. Key to this type of experiment is the ability to prepare an interface in an "electrochemical cell" and then to transfer it, without disturbing its microscopic structure, to the usual ultra-high-vacuum chamber for analysis. **Fig. 12-29** is a schematic of a "transfer system" taken from Wang's dissertation. The measurements involved XPS and HREELS and TPD.

The goal of this type of measurement is an understanding of the electron transfer properties (essentially the current-voltage relations) for a given configuration and composition of the interface. The technological goal is an electrode for a battery or fuel cell or catalytic cell, or even a microscopic layer in a microchip, which will function with minimal deterioration. "Ultra-high vacuum transfer system for electrochemical studies" (*Rev. Sci. Instr.* **64** 1993)

Some of the group's experiments studied the bonding between a variety of electrocatalysts and substrates. A typical example resulted in a 1992 paper on the binding of carbon dioxide molecules with a surface of silver. Ordinarily, CO_2 has no affinity for silver, but when the surface is doped with potassium, it was found that the CO_2 is bound to the surface through a bond between the carbon and silver atoms. AES was used to monitor the potassium concentration, and TPD was used to determine the strength of the bond between the CO_2 and the silver substrate. "Activation of Carbon Dioxide on Potassium-Modified Ag(111) Single Crystals" *J. Phys. Chem.* **96** 3788 1992.

By 1995, Chottiner and Scherson, aided by a large number of physics and chemistry post-docs and graduate students, had begun working with metallic lithium, today a well-known component of batteries. They used an extended menu of techniques: AES, IRAS, TPD, XPS and FTIR In an early lithium study, for example, they looked at the reactivity of propylene carbonate (PC) toward metallic lithium. PC is "a viscous, high dielectric-constant solvent with wide applications in lithium based battery technology." Because the lithium electrodes are exposed to the PC, it is important to know how they react with it. "Reactivity of Propylene Carbonate toward Metallic Lithium" *J. Phys. Chem.* **99** 7009 1995.

Between 1990 and 2003, Chottiner and Scherson coauthored over thirty papers in various physical chemistry and materials journals. These involved the experimental characterization of a great variety of electrolytes and electrocatalytic surfaces. They studied not only the electrochemical properties of many combinations of materials, but also the techniques for creation and analysis of the desired thin films.

Recently, Chottiner's work has emphasized the reactions of lithium atoms with molecules they might encounter in a rechargeable battery. Another line of research with great technological potential involves the use of clusters of platinum atoms in catalytic converters and fuel cells. In addition, Chottiner and his colleagues have joined professor of chemical engineering, John Angus, the CWRU "diamond-man", in looking at the surface properties of high temperature materials such as sapphire, silicon carbide, or diamond. These materials might eventually find applications in many areas, from microelectronics to automobile engines. Because of its clear potential applicability to important technology, much of this research has been funded through a grant from DARPA (the Defense Advanced Research Projects Agency).

In 2000, Chottiner took over from Bill Fickinger in the departmental post of Director of Undergraduate Studies. He has worked to continue the development of the physics curriculum with the introduction of new major options and with the expansion of the department's pioneering senior project program.

Chapter 13   Nuclear and Particle Theory

Kowalski,   Nagarajan,   Pearle,   Rix,   Kantor,   Shakin,   Brown
1963-   1964-69   1966-69   1967-70   1967-74   1970-73   1970-

With Foldy, Tobocman and Thaler on board, the Case department had a solid base in nuclear and particle theory. Between 1963 and 1970, *seven* more hires were made in this area. Of these, only two, Kowalski and Brown, would remain in the department longer than five years. Initially, Kowalski would work in nuclear scattering theory and Brown in higher energy particle theory. Like their colleague, Bill Tobocman, both would eventually apply their mathematical talents to other areas of physics.

Kenneth L. Kowalski

Ken Kowalski was born in Chicago in 1932. He earned his bachelors degree at the Illinois Institute of Technology in 1954, and then spent three years at NACA-Lewis (The precursor to NASA-Glenn) in Cleveland where he studied supersonic flow and fluid mechanics. He subsequently enrolled at Brown University. He completed his doctorate there in 1963, working with David Feldman. His dissertation title: "On the Two-Body Formulation of High-Energy Nuclear Scattering Problems". Kowalski's work at Brown was related to theoretical studies being done at Case by Foldy, Tobocman, and Thaler, and by Kisslinger at WRU. The opportunity to work with these senior nuclear theorists played a significant role in his decision to join the Case faculty in 1963. A photo of Kowalski is shown in **Fig. 13-1.**

Fig. 13-1. Ken Kowalski.

Between 1963 and 1983, Kowalski would tackle a series of topics in collision theory, in both the nuclear physics and the high-energy, or hadronic, domains. There would be continual overlapping of these pursuits, so we shall present them roughly by subject, saying a few words and citing a paper or two for each.

"Very" high energy hadronic scattering

This early work addressed certain features of the high-energy, low momentum transfer (i.e. glancing) collisions which were beginning to be studied at the new proton accelerators. In one paper, predictions, derived from very basic mathematical assumptions (analyticity and unitarity), are made for the shape of the angular distribution in the forward direction at very high energies. "Lower bounds on the Shrinking of Diffraction Peaks", *Phys. Rev.* **137B** 1350 1964.

Scattering from the deuteron

What happens when a proton of 10, 20 or 100 MeV scatters elastically from a target deuteron? It sees a proton-neutron combination with a net spin of one unit, rather loosely bound together, with which it interacts through both the Coulomb force and the nuclear force. The collision may involve various (and interfering) values of total angular momentum. Clues for what is happening come from the experimental angular distributions and polarizations of the outgoing protons. A large amount of such data became available in the early 1960's. Starting with the 1963 paper written with Feldman, Kowalski would pursue this problem for over two decades, introducing different calculational techniques and different assumptions. This would mark the start of his interest in "more-than-two-body interactions".

The abstract of the 1963 paper describes the early stages: "Some formal and practical problems concerning the effects of the internal target nucleon motion and of the multiple scattering on the elastic scattering of high-energy nucleons by deuterons are considered. In order to provide a foundation for the examination of these effects, two well-known forms of the impulse approximation are studied within the context of a multiple-scattering formalism…. The direct use of the optical-model approach is shown to be impractical for very light nuclei and, in particular, for the deuteron. An alternative means of obtaining solutions of the multiple-scattering equations (when the number of target nucleons is small) which permits the exact treatment of the ground-state scattering while allowing a systematic treatment of the contributions due to the excited intermediate nuclear states is discussed" "Elastic Nucleon-Deuteron Scattering" *Phys. Rev.* **130**, 276 1963. In one of the subsequent papers, Kowalski demonstrated the sensitivity of the polarization of the outgoing neutron to the D-wave part (i.e. the part with 2 units of angular momentum) of the deuteron wave function. "Two-nucleon interactions, the unitary model, and polarization in elastic nucleon-deuteron scattering" *Phys. Rev.* **C5** 306 1972.

Some terms: The **optical model** *picture is one in which the scattered particle is described by quantum mechanical waves spreading outward in a pattern similar to that formed by a plane optical wave incident on an opaque disk. At a more technical level, the model provides a way to calculate the "optical potential" from the basic nucleon-nucleon interaction and the wave function of the target nucleus.* The **impulse approximation** *describes the interaction as one in which the beam particle scatters from only one of the nucleons in the deuteron, its partner barely aware of the altercation. In the case of* **elastic scattering**, *the final state deuteron is somehow reconstituted and survives the collision. The* **polarization** *of the protons leaving the p-d collision is measured by using an incident beam of polarized protons, and observing the number scattered left and right.*

Off-mass-shell processes

Off-shell scattering refers to the fact that in scattering from a complex target, such as a nucleus, the participating particles can be "virtual" or "off their mass-shell", that is their effective mass-squared differs (briefly) from their energy-squared minus their momentum-squared: $m^2 \neq E^2 - p^2$. In what Kowalski describes as his most widely cited paper, he introduced new techniques for the solution of integral equations which arise in calculating the amplitudes for off-shell scattering. This paper addressed the questions of how to handle the short-range repulsion in the nuclear force and how to separate the off-shell and on-shell contributions. "Off-shell equations for two-particle scattering" *Phys. Rev. Lett.* **15** 798 1965. Because the nuclear force has only a short range, one may ignore "distant" collisions, i.e. collisions with large orbital angular momentum. In a following paper, Kowalski compares the contributions from these states for on-shell and off-shell scattering. "Angular Momentum Dependence of Off-Shell Amplitudes *Phys. Rev.* **163**, 1030-1031 1967

Three-(and more)-particle scattering

The study of the proton-deuteron, three-pion and similar "three-particle" systems resulted in a series of papers. The idea was to improve on earlier approaches in which the effects of three two-particle forces are combined. For over a decade, Kowalski, along with several of his colleagues and post-docs, would investigate techniques for calculating the details of such interactions. The integral equations which arise are enormously more complicated than those which describe two-body collisions. It was found, however, that the quantum mechanics rules for the identical interacting nucleons helped to make the calculations somewhat more manageable.

In the 1960's and 70's dozens of strongly interacting, mostly short-lived, particles were discovered at the high energy accelerators (see for example Chapter 16). Some of these decay into three particles, such as the omega meson which decays into three pions. Kowalski and collaborators applied some of the techniques developed for the 3-nucleon system to the 3-pion system. One study addressed the possibility that the newly discovered mesons which decay into three pions might be described by combinations of two-pion scattering amplitudes. "Minimal 3-to-3 Scattering Amplitudes" *Phys. Rev.* **D7** 2957 1973. "Three-pion Scattering Amplitudes and the K-matrix Formalism" *Phys. Rev.* **D13** 2352 1976. Eventually, all the mesons would be better described as quark-antiquark combinations.

Optical potential scattering

In the early 1980's Kowalski and post-docs Alan Picklesimer (Roy Thaler's collaborator mentioned in Chapter 9) and Rudy Goldflam wrote a series of papers expanding on the use of the optical model. These addressed the application of an optical potential approach to multi-body processes. From an abstract: "dynamical equations for the optical potential are obtained starting from a wide class of N-particle interactions....with arbitrary multiparticle interactions....including all effects of nucleon identity." "Dynamical equations for the optical potential" *Phys. Rev.* **C 23** 597 1981.

Relativistic pion nucleus scattering

As described in Chapter 9, Kowalski joined Roy Thaler and post-doc E. Siciliano to study the scattering of pions by nuclei. This followed the development of pion beams at Los Alamos and their use as probes of nuclei. The multiparticle aspect of these interactions was clearly within Kowalski's area of expertise. This time, however, the pions were relativistic, and the calculations required relativistic quantum field theory. "Connected-kernal multifermion Bethe-Salpeter equations" *Phys. Rev.* **D20** 2526 1979. "Composite-particle structure of pion-nucleus amplitudes" *Phys. Rev.* **C19** 1843 1979.

In 1982, Kowalski joined in some work which his colleague Bob Brown had been doing on the origins of zeros predicted to appear in the angular distributions of gauge boson scattering. "Classical radiation zeros in gauge theory amplitudes" *Phys. Rev.* **D28** 624 1983. We shall take a look at this in the section on Brown later in this chapter.

Thermal field theories

It was a distinctly different branch of many-particle physics which interested Kowalski in the mid-1980's, in particular during his sabbatical at Argonne National Laboratory. Thermal field theories, or "finite temperature field theories", are a combination of statistical mechanics and quantum field theories. They have applications to particle physics, cosmology, and condensed matter physics. Among his papers on this topic is "Real-time Fermion thermal field theories", *Phys. Rev.* **D35** 2415 1987.

In 1988 Ken organized the first truly international conference hosted by the CWRU physics department: *The Workshop on Thermal Field Theories and their Applications*. The 558-page proceedings were published in their entirety in a major European journal: *Physica* **A158** 1 1989. In light of the wide interest in this developing field, a report on this workshop was included in the popular voice of particle physics, the *CERN Courier* **29** 21 1989. This event was the first in an ongoing series of international conferences on this topic.

Non-linear Optics

Kowalski entered into a very different partnership within the department, this time with experimentalist Ken Singer. Singer (Chapter 18), who arrived in 1990, studies non-linear optical materials and has advised a significant cadre of graduate students. It was much appreciated, therefore, when Kowalski, who had many times taught the graduate course in electromagnetic theory, became interested in the theory of non-linear optics and developed a course for the graduate students. Furthermore, Kowalski, Singer and grad student Jim Andrews published a theory paper on non-linear optics. "Pair correlations, cascading, and local-field effects in nonlinear optical susceptibilities" *Phys. Rev.* **A46**, 4172 1992.

Experimental Physics at the Tevatron

In 1988, a young particle theorist joined the department. Cyrus Taylor's interest in some theoretical aspects of very high energy scattering led him, in 1993, to participate in a proposal for an experiment at the Tevatron proton accelerator at Fermilab. We shall describe Taylor's work in Chapter 18. Ken Kowalski joined Taylor and his collaborators on the proposal, and participated in the subsequent experiment. The experiment, named

MiniMax, was run at the Tevatron in 1995 and 1996. Kowalski would, in fact, spend the fall semester of 1993 at Fermilab. The principal purpose of MiniMax was a search for "disordered chiral condensate". Basically, this effect is related to the relative rates for the production of charged and neutral pions in very high energy collisions. The experiment required the detection and identification of charged and neutral particles produced at low momentum transfers (i.e. at very small angles) in proton-antiproton collisions at 1.8 TeV center of mass energy. We shall write a bit more about MiniMax in Chapter 18. "Search for disoriented chiral condensate at the Fermilab Tevatron" *Phys. Rev.* **D61** 32003 2000.

More recently, Kowalski has become interested in theoretical particle astrophysics, specifically the origins and properties of cosmological magnetic fields. Research in theoretical cosmology by several members of the current department will be the subject of Chapter 18.

The **funding** for research in theoretical nuclear and particle physics came from grants by the Atomic Energy Commission and later the National Science Foundation. Typically, these funds provided for faculty summer salary (2/9 of academic year salary), salary for post-docs, stipends for graduate students, travel expenses for meetings and conferences, computer costs, sometimes a bit toward secretarial services, and up to 40 or 50% overhead to the university. This would add up to about $50K to $75K per year per faculty theorist.

Kowalski was chairman of the CWRU department from 1971 until 1976. He took advantage of two year-long sabbaticals, the first in 1968 as visiting professor at Leuven (Louvain) in Belgium and the second in 1986 at Argonne National Laboratory. In 1987, Ken was co-chairman, with Bill Fickinger, of the "Modern Physics in America Symposium", part of the celebration of the centennial of the Michelson Morley experiment. The proceedings of that event, co-edited by Kowalski and Fickinger, were published as AIP Conference Proceedings No. **169**, 1988. (A list of speakers and titles appears in Appendix D.) In 1991, Ken published a book on scattering theory which brought together much of his earlier work. The book was co-authored by S. K. Adhikari, who was, at the time, on the faculty of the Federal University of Pernambuco in Brazil. "Dynamical Collision Theory and Its Applications" Academic Press 1991 494 pp.

Five more theorists

Nagarajan: nuclear models

Mangalam A. Nagarajan came to the Case department as a research assistant in 1962 and two years later was appointed to the faculty. He was born in India in 1933 and awarded his doctorate at the University of Calcutta in 1962. His principal interest was the theo`ry of nuclear structure and reaction mechanisms. Among his publications while at Case were three co-authored with Bill Tobocman (Chapter 9). "Equivalence of Elementary and Composite Particles" *Phys. Rev.* **137** 1236 1965. "Boundary Condition Constraints for the Shell Model: A Method for Nuclear Structure and Nuclear Reactions"

Phys. Rev. **138** B1351 1965. "Test of the Boundary Condition Constraint Method for Nuclear Reactions" *Phys. Rev.* **140** B63 1965.

The *shell model* gives the best description of the energy levels in nuclei. *This model for loading protons and neutrons into nuclei is similar to the Bohr picture of an atom, in which electrons are forced to fill successive shells in order to have unique quantum numbers (thus satisfying the Pauli principle). The shell model predicts energy levels and quantum numbers for the nucleons in both the ground and excited states of the nuclei.* Nagarajan joined the many nuclear theorists who were developing and applying the shell model. "Separability of Center-of-Mass Motion in the Nuclear Shell Model" *Phys. Rev.* **135** B34 1964. "A Shell Model Calculation for the Reaction $N^{15}(p,n)O^{15}$" *Nucl. Phys.* **A113** 412 1968.

Fig. 13-2. M. A. Nagarajan.

Nagarajan also worked on stripping reactions in which an incoming deuteron projectile leaves its proton behind in a target nucleus. We described this process in the section on Tobocman in Chapter 9. "The Study of $B^{10}(d,n)C^{11}$ and $B^{11}(d,n)C^{12}$ Reactions" *Nucl. Phys.* **A93** 190 1966. Nagarajan remained at Case for a total of seven years. Interest in nuclear theory was waning as subnuclear theory was waxing. Nevertheless, Nagarajan would contribute to the understanding of medium energy (d,n), (t,p) and similar stripping processes in nuclei for at least two more decades. He has been part of theory-experiment collaborations at Lawrence Berkeley Lab, Saclay, and Oak Ridge. His principal base has been the Daresbury Lab in England.

Pearle: relativity, philosophy

Fig. 13-3. Phillip Pearle.

Another young theorist hired by Case just before the 1967 creation of CWRU was **Phillip Pearle**. **Fig. 13-3.** Born in New York City in 1936, Pearle did both his BS and PhD (1963) at MIT. He then took a position as instructor at Harvard where he wrote a paper whose title suggests an independent approach to physics: "Alternative to the Orthodox Interpretation of Quantum Theory" *Amer. J. Phys.* **35** 742 1967.

Pearle's principal interests were relativistic classical mechanics and the philosophical aspects of quantum mechanics. As an assistant professor (and an extremely popular teacher), Pearle published several papers in these areas, including "Construction of an Invariance from a Conservation Law, and Vice Versa, in Classical Mechanics" *J. Math. Phys.* **9** 1092 1968. "Relativistic Classical Mechanics with Time as a Dynamical Variable" *Phys. Rev.* **168** 1429 1968. Pearle remained at the newly merged CWRU for only four years, joining the others who were squeezed out of the oversized department. Pearle has been at Hamilton College for the past 35 years, where he has continued to publish regularly on similar topics.

Rix: hadronic collisions

In 1965, the Case department added a fresh PhD to its growing group of particle theorists. **John Rix** started as a post-doc and was promoted to assistant professor in 1967. **Fig. 13-4**. Rix was born in 1938 in the Cleveland suburb of Bay Village, did his BS at Cal Tech (1960) and his PhD at Harvard (1965). Working under Walter Gilbert, he wrote his thesis on dispersion relations. In his letter of application, he described his research as "dispersion relations for charged particle scattering with particular emphasis on Coulombic competition with short range forces." This work was published after his arrival at Case in a paper coauthored by M. B. Halpern of UC Berkeley: "Exact Solution of the One-Photon-Exchange N / D Equations" *Phys. Rev.* **147**, 984 1966.

Fig. 13-4. John Rix..

Rix then worked with Roy Thaler (Chapter 9) on essentially the same topic: interference between the Coulomb and strong interactions. The authors developed a technique which uses the experimentally determined Coulomb scattering amplitudes in conjunction with the data from the scattering of charged particles by nuclei to deduce the "pure" nuclear (i.e. strong) scattering amplitudes. They include in their prescription modifications for relativity and for spin. "Separation of Strong and Electromagnetic Effects in Charged-Particle Scattering" *Phys. Rev.* **152**, 1357 1966.

The mass of the neutron is about a tenth of a percent, i.e. 1.29 MeV, greater than that of the proton. After the discovery of the π meson which clearly played a role in the nuclear force, it was not unreasonable to look at a model in which the neutron is a bound state of a negative pion and a proton. Rix took a look at a variation of this idea in which the neutron is not a simple atom-like bound state, but a dynamic system in which virtual pions are emitted into a "pion-cloud" and reabsorbed. "Remarks on the Nucleon Mass Difference" *Phys. Rev.* **158**, 1600 1967. Not very long after Rix's efforts, the quark-gluon picture of the nucleons would emerge as a better model for hadrons.

Rix supervised the work of one doctoral student, Gee-Yin Chow. In the 1960's, a large amount of data on "high" energy scattering was being generated at the accelerators of CERN, Brookhaven and Berkeley. The elastic collisions of nucleons and mesons were found to be dominated by small angle diffractive scattering and the size, shape, and width of the small angle differential cross sections were becoming available. Rix and Chow developed a consistent theoretical description of these data, including calculations for the charged particle multiplicities. "Elastic Diffraction Scattering of Hadrons at High Energies" *Phys. Rev.* **184** 1714 1969. "Distribution over the Number of Prongs in Inelastic

Hadronic Collisions" *Phys. Rev.* **D2** 139 1970. John Rix left the newly formed (and oversized) combined department in 1970.

Kantor: high energy scattering

Paul B. Kantor earned his PhD under Sam Treiman at Princeton in 1963. He then spent two years each at Brookhaven and SUNY Stonybrook before joining the CWRU department, at age 29, in 1967. He had been recommended to Chandrasekhar by Leonard Kisslinger (as had another Brookhaven physicist, Keith Robinson). Kantor's earlier work concerned nucleon-nucleon scattering in the intermediate (a few hundred MeV) energy range. Of special interest was the spin-dependence and predictions of polarization in high energy scattering.

Fig. 13-5. Paul Kantor.

At CWRU, Kantor investigated polarization in the scattering of K mesons by nucleons. Experimental data were becoming available as kaon beams were created at the particle accelerators. "Polarization in K-Nucleon Scattering" *Nuovo Cim.* **84** 353 1969. Another topic having its origins in accelerator experiments was the production of intermediate short-lived resonant particles. Kantor looked into various potential sources of symmetry-breaking such as that observed in the decay into three pions of the η meson: $\eta \to \pi^+ \pi^- \pi^0$. He and grad student Keith Taggert looked into the role of interference with competing intermediate channels (e.g. $\pi^- p \to \eta n \to 3\pi n$ which gives the same final state as $\pi^- p \to \Delta^- 2\pi \to 3\pi n$).

Kantor was promoted to associate professor in 1969. In 1972, he became interested in communication theory and information systems. He decided to join the faculty of the newly formed Complex Systems Institute in the CWRU School of Library Science, while retaining an adjunct position in the physics department. In 1976, Paul left CWRU to create an information retrieval "think tank", Tantalus Inc., which he heads to this day. He has been at Rutgers University since 1991 where, as Distinguished Professor, he is active in information retrieval technology and related areas.

Shakin: interactions with nuclei

The fifth addition to the nuclear theory group was **Carl M. Shakin** who arrived in 1970. **Fig. 13-6.** Born in 1934, Shakin did his BS in 1955 at NYU, followed by a Fulbright at Manchester and his PhD at Harvard in 1961. After a post-doc position at ITP in Copenhagen, Carl was a member of the faculty at MIT for five years before being appointed associate professor at CWRU.

In the course of the following four years, Shakin worked largely with Roy Thaler on a wide range of nuclear scattering

Fig. 13-6. Carl Shakin.

problems. An early paper, coauthored with both Thaler and Kowalski, explores techniques for calculating nuclear transitions induced in energetic collisions. "Off shell continuations of the two particle transition matrix with bound states" *Phys Rev.* **C3** 1146 1971. This was followed, during the next three years, by no fewer than thirty-two publications with his new colleagues. Among these were the following examples, in which the general theme is the investigation of nuclear structure by scattering experiments using beams of photons, pions, kaons, or nucleons. "Theory of Analog Resonances" *Rev. Mod. Phys.* **44** 48 1972. "Nuclear γ Rays Following K⁻ Capture" *Phys. Rev.* **C5** 238 1972. "Intermediate Structure and the Photodisintegration of O^{16}" *Phys. Rev.* **C5** 1898 1972. "Charge-Dependent Effects in the Photodisintegration of ^{4}He." *Phys. Rev. Lett.* **28** 1729 1972. "Elastic Scattering of Nucleons from Correlated Nuclei" *Phys. Rev.* **C7** 494 1973. "Off-shell Effects in Nucleon-nucleus Scattering" *Phys. Rev.* **C7** 2346 1973. "Off-shell Effects in Elastic Pion-nucleus Scattering" *Phys. Rev.* **C9** 1370 1974. "Unitarity and off-shell Effect in the Impulse Approximation" *Phys. Rev.* **C9** 1374 1974.

Shakin took a leave of absence in the 73-74 academic year, continuing his research at Brooklyn College of the City University of New York. Even though he had just been promoted to full professor at Case, he decided to remain at CUNY. For the past thirty years, Shakin has continued to contribute prolifically to particle theory. His departure was a major loss for the CWRU theory program.

Robert W. Brown

Fig. 13-7. Robert Brown.

In 1970, 28 year-old Robert W. Brown joined Foldy, Tobocman, Thaler and Kowalski in the CWRU theoretical particle physics program. Brown was born in St. Paul, Minnesota, completed his BS in physics at the University of Minnesota in 1963, and was awarded his PhD at MIT in 1968. He was a student of John Bronzan. His thesis was entitled the "Comparison of the Scattering of Electrons and Positrons from Protons". **Fig. 13-7.** Before coming to CWRU, he spent two years as a post-doc doing particle theory at Brookhaven. The principal work he performed in that period focused on investigations into "Ward Identity Anomalies". These represented the breaking of classical symmetries by quantum effects.

Brown also expanded upon the work he had done for his Ph.D. degree, the comparison of electron and positron scattering by protons. The leading terms in the expression for the interaction describe an incoming lepton which sees a point target, surrounded by an electric field. Smaller correction terms appear which describe how the internal structure of the proton begins to affect the scattering. He pointed out that data from the SLAC accelerator on e^+p and e^-p small angle elastic scattering were sensitive enough to test his calculations. "Comparison of the scattering of electrons and positrons from protons at small angles" *Phys. Rev.* **D1** 1432 1970.

Intermediate Vector Bosons: $W^{\pm}$ and Z^0

Another paper written at Brookhaven in the summer of 1970 was a collaboration with A. K. Mann of U. Pennsylvania and J. Smith of SUNY at Stony Brook. It, too, was directly related to potential accelerator experiments: a comparison between the cross-sections for muon and neutrino induced production of the W boson. The existence of this particle, the proposed mediator of the weak interaction, had been assumed for many years. No one knew what its mass might be, but it was hoped that the energies available at Fermilab would be enough to produce a free W. Brown and his partners calculated cross sections as a function of W mass (1 to 10 GeV) and of incident energy (10 to 1000 GeV), based on the dissociation of the incoming lepton in the Coulomb field of the target proton. They reported that the neutrino induced reaction would be up to five hundred times more probable than the muon induced reaction. "Neutrinos Versus Muons in W-Boson Production" *Phys. Rev. Lett.* **25** 257 1970. It turns out that the W is much more massive than anything that could be produced at Fermilab. It was not discovered until years later, in proton-antiproton collisions at CERN, and it has a mass of 83 GeV.

Over the next two years, Brown, then at CWRU, followed this initial paper with a three-part sequence of longer publications, with Smith and R. H. Hobbs of MIT. The first paper extended the W-production cross-section calculations to interactions of leptons with *nuclei* (rather than just protons). They included corrections for the Pauli principle, for nucleon Fermi motion, for W magnetic moments, etc. The second paper looks at the angular distributions of outgoing muons from these same reactions, assuming prompt decay of the W. The final paper of the series takes up the question of the role of the W (or its neutral counterpart, the Z), as virtual particles, in reactions of the type lepton + nucleus gives three leptons + nucleus. If the available energy in such an interaction is too small to produce a "free" W or Z, the outgoing leptons may still provide some clue of its intervention. (Reactions mediated by the exchange of virtual particles had indeed been observed, and these "off-mass-shell" calculations were all in vogue. Some of the experiments described in Chapter 16, for example, provide evidence for this sort of interaction.) Brown's papers on the subject: "Intermediate Boson. I. Theoretical Production Cross Sections in High-Energy Neutrino and Muon Experiments" *Phys. Rev.* **D3** 207 1971. "Intermediate Boson. II. Theoretical Muon Spectra in High-Energy Neutrino Experiments" *Phys. Rev.* **D4** 794 1971. "Intermediate Boson. III. Virtual-Boson Effects in Neutrino Trident Production" *Phys. Rev.* **D6** 3273 1972.

Brown also expanded his earlier work on "anomalies" with a paper that cleared up discrepancies in the literature concerning the remarkable fact that there are no higher-order corrections to the original anomaly calculated by Stephen Adler, who along with Young and Wong was his collaborator. "Absence of Second Order Correction to the Triangle Anomaly in Quantum Electrodynamics" *Phys. Rev.* **D4** 348 1971.

Photons and electrons

A strictly electromagnetic interaction, the production of $e^+e^-e^+e^-$ by photons incident on virtual photons in the electric field of nuclei, was tackled by Brown and collaborators from Brookhaven and U. Michigan. It was proposed as a way to get at the photon-photon interaction. "Role of $\gamma\gamma \to e^+e^-e^+e^-$ in the High-Energy Cross Section for $\gamma Z \to Z\, e^+e^-e^+e^-$" *Phys. Rev. Lett.* **28** 123 1972. A related work on virtual photon processes in e^-e^- or e^-e^+ colliding beams experiments discussed the possibility of measuring the real photon-photon *hadron* production cross sections. "Study of photon-photon interactions via electron-electron and electron-positron colliding beams" *Phys. Rev.* **D4** 1496 1971.

All these papers, calculations of cross-sections from basic principles, were at least in part meant to guide the experimentalist. In fact, the authors often discuss competing backgrounds and other potential drawbacks which would influence the design of an experiment. "Electromagnetic Background in the Search for Neutral Weak Currents via $e^+e^- \to \mu^+\mu^-$" *Phys. Lett.* **43B** 403 1973. The same considerations were extended to other experiments in other theoretical work. "Non-resonant asymmetry in $e^+e^- \to \pi^+\pi^-$" *Lett. al Nuovo Cim.* **10** 305 1974. "Possible effects of weakly coupled neutral currents in $pp \to \ell^+\ell^- + X$" *Nucl. Phys.* **B75** 12 1974.

Work on the production of two electron pairs was extended to higher energies, and for the first time, Brown (with post-doc Karnig O. Mikaelian) would discuss some of the cosmological ramifications of this process. Their calculations showed that at very high energies (10^{21} eV), cosmic photons bouncing along through the cosmic background radiation would disappear in 10^{26} cm or so, or less than a billion light years. "Role of $\gamma\gamma \to e^+e^-e^+e^-$ in Photoproduction, Colliding Beams, and Cosmic Photon Absorption" *Phys. Rev.* **D8** 3083 1973. "Absorption of High-energy Cosmic Photons through Double-pair Production in Photon-photon Collisions" *Astrophys. Lett.* **14** 203 1973.

Neutral Currents

As experimental evidence for weak "neutral currents" began to become convincing, it seemed that the charged W's mentioned above should be accompanied by an uncharged partner, the Z^o. Neutral currents are required to explain a group of weak interactions in which the participants do not exchange their electric charges. An electrically neutral propagator of the weak field must be exchanged, for example, in elastic scattering of neutrinos by electrons: $\nu e \to \nu e$. In a series of papers similar to the W papers described above, Brown, Mikaelian (then at U. Penn) and grad-student Leon Gordon presented cross sections for Z production as a function of incident energy (up to 1000 GeV) and Z mass (up to 30 GeV). The first paper looked at collisions with protons. "Production of neutral weak bosons in high energy electron and muon experiments" *Phys. Rev. Lett.* **33** 1119 1974. The next set of calculations looked at Z^o production in the fields of nuclei, and included predictions of polarizations and angular distributions of all out-going particles. "Theoretical spectra and polarization in neutral-weak-boson production" *Phys. Rev.* **D12** 2851 1975. "Theoretical estimates for photoproduction and leptoproduction of neutral vector bosons" *Phys. Rev.* **D13** 1856 1976. Most of this work was done while Brown was on sabbatical at SUNY Stonybrook. As in the case of the W, these calculations are easily extended to higher masses and are appropriate for the experimental mass values. The Z was discovered a decade later and it had a mass of 94 GeV.

This series of papers includes calculations of the expected *cross-sections* for W and Z production. Brown and collaborators consider several possible ways to produce intermediate bosons, each associated with the capabilities of existing or planned accelerator facilities. At that time, the Fermilab accelerator produced 200 GeV proton beams, the Stanford Linear accelerator had colliding beams of 20 GeV electrons, and the Brookhaven proton-proton collider, Isabelle, was on the drawing boards. (The Isabelle project was eventually cancelled). In addition, plans were in the offing for the proton-antiproton machines in CERN and Fermilab.

As the experimentalists were pushing their accelerators and detectors to the limits in an effort to see the carriers of the weak force, Brown *et al.* were providing their best estimates of the cross sections for single W or Z, as well as W^+W^- and $Z^\circ Z^\circ$ pair production and, finally, WZ° and $W\gamma$ production. They considered all available lepton beams and colliding beams. For example, they advise the experimenters that: "It is emphasized that (1) the rate of production of Z° pairs is comparable to that of W pairs and that (2) W-pair production with colliding proton beams may be the best way to see high-energy cancellations in cross sections, the hallmark of renormalizability in gauge theories." "W^+W^- and $Z^\circ Z^\circ$ pair production in e+e-, pp and pbar p colliding beams" *Phys. Rev.* **D19** 922 1979. "$W^\pm Z^\circ$ and $W^\pm \gamma$ pair production in ve, pp, and pbar p collisions" *Phys. Rev.* **D20** 1164 1979.

Five years later, after the W and Z had finally been discovered in colliding beams experiments at CERN, Brown returned to the subject with two of his PhD students, Cynthia L. Bilchak and John D. Stroughair. Their goal was to predict angular distributions and polarizations in the various W and Z production channels which would soon be explored at the accelerators. "$W^\pm$ and Z° polarization in pair production: Dominant helicities" *Phys. Rev.* **D29** 375 1984. A decade later, pairs of W's and Z's were produced experimentally. Analysis of these data could set constraints on gauge couplings in the reactions studied by Brown and his collaborators.

A Little Cosmology

Cosmologists have long been intrigued by the fact that the universe seems to have more matter than antimatter and that the galaxies we see today are simply the left-over excess. They have turned to particle physics hoping to identify the source of the asymmetry. An alternate scheme is that there are indeed equal amounts of coexisting matter and antimatter, but they are found in widely separated domains and have not yet had a chance to annihilate. In 1978, Bob Brown teamed up with Floyd W. Stecker, a theoretical astrophysicist at the Laboratory for High Energy Astrophysics at NASA Goddard in Maryland. They had met, through Fred Reines, at a conference on proposed large neutrino detectors. The collaboration of the particle-field-theorist with the cosmic-ray-astrophysicist resulted in a paper which proposed that some of the ideas current in elementary particle theory, such as CP non-conservation and proton decay, could lead one to consider a matter-antimatter *symmetric* universe, but one containing enormous and widely separated domains which are dominated by one form or the other. "We suggest that grand unified field theories with spontaneous symmetry breaking in the very early big bang can lead more naturally to a baryon-symmetric cosmology with a domain structure than to a totally baryon-asymmetric cosmology." "Cosmological Baryon-Number Domain Structure from Symmetry Breaking in Grand Unified Field Theories" *Phys. Rev. Lett.* **43** 315 1979.

Brown and Stecker collaborated once again, this time on a paper which combined Brown's earlier interests in IVB's (intermediate vector bosons, *e.g.* W and Z) with their joint interest in cosmic antimatter. At the time, there were proposals that heavier relatives of the W and Z might exist. The authors suggested that the new and planned very large cosmic neutrino detectors might generate evidence for IVB's too massive to be produced at accelerators. Furthermore, the new detectors could differentiate between incoming neutrinos and anti-neutrinos, thus providing another handle on the question of the amount of antimatter in the universe. "Cosmic-ray-neutrino tests for heavier weak bosons and cosmic antimatter" *Phys. Rev.* **D26** 373 1982.

In the early 1980's, data from experiments at the DESY electron collider in Hamburg indicated a lack of symmetry in the production of muon pairs. In the $e^-e^+ \to \mu^-\mu^+$ reaction, about 10% more negative μ's came out backward in the center of mass (i.e. opposite the incoming e^-) than came out forward. This reaction provided a nice test of electroweak theory, in that it should be mediated by a mix of photon and Z^o exchange. There had been earlier calculations of the expected asymmetry, but as the data were becoming more precise, it was useful to undertake more detailed theoretical calculations. Brown and two colleagues at the University of Dortmund, R. Decker and E. A. Paschos, refined the calculations and compared the experimental results with the calculated asymmetries. Typical values of the asymmetry were about 10%; their theory agreed with the measurements to about 0.2%. "Weak corrections to the $e^\pm e^- \to \mu^\pm\mu^-$ asymmetry" *Phys. Rev. Lett.* **52** 1192 1984.

Predicting Zeros

Most often, theories are developed with the intention of explaining what is observed in experiments. But sometimes the theorist discovers an unexpected feature which jumps out of the theory and which can be tested experimentally. Then, if the effect is observed, the theorist is on the right track. An example is described in a series of papers which Brown coauthored with grad student Deshdeep Sahdev, Mikaelian, Kowalski, and later with S. J. Brodsky of SLAC. The subject is electromagnetic radiation emitted in relativistic collisions. In particular, they noted that interference effects between the incoming and outgoing particle waves give rise to "zeros" in the differential cross-sections. As a result, there are regions in angle and energy where the experimentalist should observe **no** outgoing photons. Because of experimental difficulties and backgrounds (as Brown remarked: "it is hard to see nothing"), it would be many years before the predictions would finally be verified when they were observed in quark-level interactions. "Zeros in amplitudes: gauge theory and radiation interference" *Phys. Rev. Lett.* **49** 966 1982. 'Classical radiations zeros in gauge-theory amplitudes" *Phys. Rev.* **D8** 624 1983; *Phys. Rev.* **D29** 2100 1984.

The subject of these zeros in photon production reappeared ten years later when Brown and his grad student Mary Convery expanded the study to multi-photon production channels. "Radiation tree amplitudes: Zeroing in on more photons and gluons" *Phys. Rev.* **D49** 2290 1994. Meanwhile, theorists at the Brooklyn and Staten Island campuses of the City University of New York had been working on bremsstrahlung (literally "braking radiation"), i.e. photon production in the electromagnetic field of the colliding particles. They were looking at nucleon-nucleon collisions and Brown joined them in a paper on the subject. "Coplanar and non-coplanar nucleon-nucleon bremsstrahlung calculation: A study of pseudoscaler and pseudovector π N couplings" *Phys. Rev.* **C52** R2346 1995.

In the early 1990's, Brown joined forces with Cyrus Taylor (to be introduced in Chapter 18) and research fellow Shmaryu Shvartsman, to look at a problem in quantum field theory. In quantization, there is always a dilemma revolving around whether one has identified all the degrees of freedom, and the associated question of how to quantize each of them. "Role of zero modes in the canonical quantization of heavy-fermion QED in light-cone coordinates" *Phys. Rev.* **D48** 5873 1993.

Solitons, Strings and Undergrads

Bob Brown is, certainly among the theorists in the department, the one who has most often involved undergraduate physics students in research. For many years, he has taught the immensely popular "honors" course for freshman, an introductory course into which he inserts exciting topics in modern theoretical physics. Each student in the course participates in projects, many computer-based, which illustrate cutting-edge applications of the physics taught in the usual introductory course. In addition, Brown has regularly worked with the senior majors in helping them organize their preparation for their GRE's (graduate record exams) and their selection of and application to graduate school. These interactions with undergrads have often led to collaborative published work in which the students were major players.

In the 1980's, undergraduate students joined Brown in making discoveries in classical nonlinear physics. The first, with one undergraduate and two graduate student co-workers, addressed proving the full stability of underwater solitary waves in the presence of large (nonlinear) disturbances. The second concerned research on cosmic strings. These hypothetical huge, massive, but vanishingly thin strands, stemming from phase transitions in the early universe, may account for some features of galaxy formation. In one paper Brown and his grad student David DeLaney reported their discovery of a convenient way to represent the Fourier series for a vector which promised to be useful in string theory. "Product representation for the harmonic series of a unit vector: A string application" *Phys. Rev. Lett.* **63** 474 1989.

A sophomore undergraduate joined Brown in a general mathematical formulation of this solution, and another, even younger undergraduate worked with Cyrus Taylor and Brown in presenting an elegant spinor representation of the general harmonic string solution. Additional solutions were also studied featuring kinks representing intersections of strings with each other or with themselves. "Closed strings with low harmonics and kinks" with M.E. Convery, S.A. Hotes, M.G. Knepley, and L.S. Petropoulos *Phys. Rev.* **D48** 2548 1993.

In 2004, Brown was recognized for his dedication and originality in the teaching of undergraduate physics by being awarded the national AAPT (American Association of Physics Teachers) Excellence in Introductory College Physics Teaching Award. From the department website: "This is the highest award given for introductory physics teaching in the country. This award reflects the incredible job Bob has done over the past decades in innovative teaching, including using the web well before anyone else and developing peer teaching techniques, as well as inspiring scores of undergraduate students to excel." In 2005, Brown won the equally prestigious Cherry National Teaching Award.

A Problem in Jackson

Bob Brown has taught the graduate course on the theory of electromagnetism many times over. This is one of the few absolutely required courses for PhD students, pretty much worldwide. And the textbook first written by J. D. Jackson in 1962 continues to be "*the* authority". A wonderful feature of "Jackson" is the many problems it includes, problems which have challenged grad students (and their teachers) for two generations.

In 1991, Brown, three of his graduate students, Mike Martens, Labros Petropoulos, Jim Andrews, and two former students, M.A. Morich and John Patrick, published a paper on the design of a magnet system which qualified as a rather advanced "Jackson problem". This paper would mark the beginning of a new and different line of research for Brown. In it, the authors describe a technique for calculating the spatial distribution of electric currents which would produce the optimum magnetic fields for a medical imaging device, an "inverse problem". The application of classical electromagnetic theory to the technology of magnetic resonance imaging is a far cry from theoretical quantum particle physics. Brown's entry into the field of medical imaging follows by a hundred years the pioneering x-ray work of Dayton Miller described in Chapter 4. At the same time as Brown's MRI work, Bill Tobocman was developing a parallel program in the application of scattering theory to medical ultrasound imaging (Chapter 9).

An aside on the physics of nuclear magnetic resonance imaging *(now called MRI, because the* Nuclear *part of NMR was scaring people away). The object of the game is to determine how many hydrogen atoms there are in a given region of space and what their environment is - without touching them. This information can be used to learn about the interiors of biological systems without cutting into them. The physics of the game is based on what happens to the proton nucleus of the hydrogen atom when it finds itself in a magnetic field. This is the only place where "non-classical" physics comes in. Quantum mechanically, the proton has an intrinsic spin and an associated magnetic di-*

pole moment. If the proton is sitting in an external magnetic field, B_o, and the proton spin is "tipped over" by a short magnetic pulse, its magnetic moment will precess with a specific frequency f_o around the direction of B_o. Quantitatively, the proton precession frequency f_o in MHz equals 42.6 times B_o in Tesla. The precession gives rise to a changing magnetic field which can be detected. That's the end of the physics.

Now the trick of MRI: if one puts a gradient on B_o, that is, B_o varies smoothly along one dimension in the sample, the frequency of the detected signal will depend on where the protons are. Thus one can identify which slice of the sample the signal is coming from. This insight led to the explosion of MRI applications described almost daily in the news media and the Nobel Prize in Medicine for the inventors of the gradient-field idea in the Fall of 2003.

Now the technology: one must design coils to produce the fields (for B_o, for the gradient ΔB_o, and for the short excitation pulse) and the receiver coils to detect the precession signal.

Some details: it turns out that the room temperature thermal energy is far greater than the energy associated with the precessing protons, but there are so many protons involved that a measurable signal survives. The rate of precession depends slightly on the environment surrounding the proton because of interactions with neighboring atoms. Furthermore, the rate at which the precession signal decays also depends on the environment. Consequently, frequency and time-dependent amplitude measurements provide additional useful information. The bottom line: one can learn where the protons are (in three dimensions), and what their chemical environment is.

The 1991 paper by Brown and his students was on the design of the coils which produce the gradient, ΔB_o, for an MRI scanner, but with minimized inductance. It is desirable to reduce coil inertia in order to be able to turn such coils on and off as rapidly as possible, and thereby speed up the imaging process. **Fig. 13-8** shows the calculated spatial distribution of the currents. The group actually built a ½-scale model, with all the wires held in position on a plywood sheet. They then measured the resulting B field in the region where the imaging would be done. **Fig. 13-9** shows the comparison between the calculated and measured fields. The agreement was better than 5%. Two of the authors were already employed by Picker International which is now owned by Philips, which, along with General Electric and Siemens, is a leader in MRI technology. "Insertable biplanar gradient coil for MR imaging," *Rev. Sci. Instr.* **62** 2639 1991. Two years later, the same authors would tackle another geometry, one especially relevant to the human

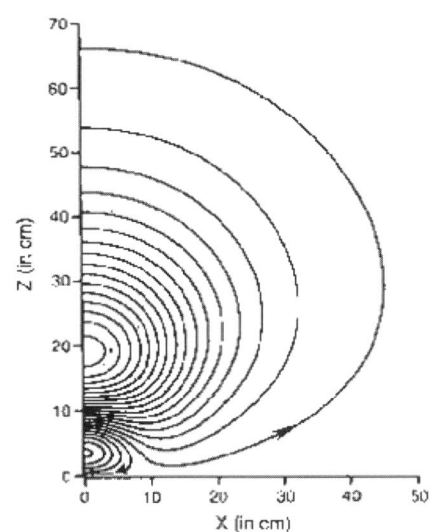

Fig. 13-8. Current directions in two dimensional model of an MRI gradient coil..

profile. "An MRI elliptical coil with minimum inductance," *Meas. Sci. and Tech.* **4** 349 1993.

Brown and his group continued to improve upon the design of MRI magnet coils, with various goals in mind. Among these are uniform gradients to allow high resolution, low inductance to enhance quick response, large openings to allow easy access, and shielding coils to reduce stray fields outside the device. The group also moved into related MRI computational electromagnetics analysis in the design of the main magnets that produce the large Tesla-level field and the rf coils which detect the proton spin fields. A new short-coil design with a paradigm shift to include a central bundle of wires was found. "An inverse approach to the design of MRI main magnets," *IEEE Trans. Mag.* **30** 108 1994. The same "inverse" procedure used in gradient design was applied to rf coil development in a later paper. "A hybrid inverse approach to the design of lumped-element rf coils," *IEEE Trans. Biomed. Eng.*, **46** 353 1999.

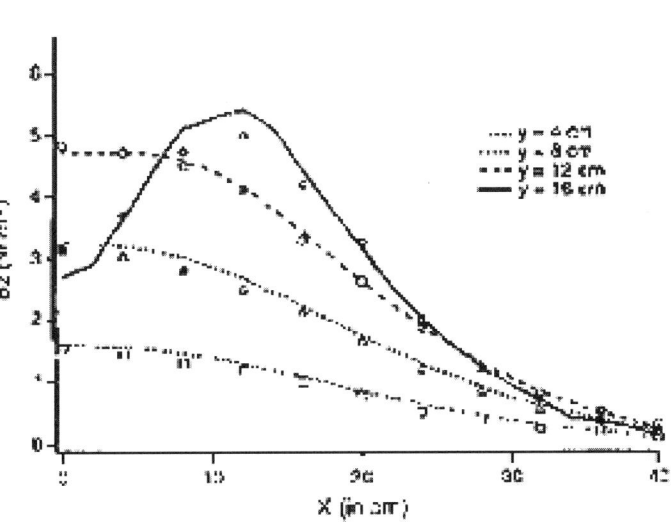

Fig. 13-9. Comparison between calculated and measured fields in MRI model configuration.

The team also modeled the *temperature* distribution in surrounding tissue in medical procedures where MR-guided rf probes are used for ablating tumors. Good agreement was found between the simulations and the measured ablation volume in comparisons with clinical work carried out at the neighboring University Hospitals. "Calculated rf electric field and temperature distributions in rf thermal ablations: comparison with gel experiments and liver imaging," *JMRI* **8** 70 1998.

In 1999, Brown and his colleague, research fellow Shmaryu Shvartsman, published a paper titled: "Supershielding: Confinement of Magnetic Fields". *Phys. Rev. Lett.* **83** 1946 1999. The fact that

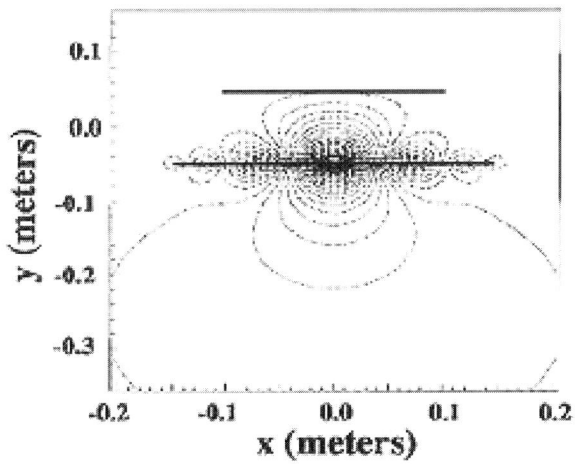

Fig. 13-10. Magnetic field lines confined by supershielding. Note the absence of field above the upper current sheet.

the paper was published in a flagship AIP journal rather than in a technical MRI or medical journal indicates that the topic is of general interest to physicists, in the J. D. Jackson spirit. The work develops a theoretical limit on how good the shielding can be for open MRI coils. Shielding is essential to kill the production of eddy currents in the surrounding main magnet. The field due to these currents degrades the image and the currents themselves experience a Lorentz force from the main field that leads to additional noise. In this classic paper, a "shielding error function" is defined and an equation is given for it. The succinct result is that minimizing this function guarantees the best shielding available, subject to other constraints and design goals.

The technique described in the paper is based on the placement of two strips of conductors, each divided into narrow sub-strips, each of which carries a specified amount of current. **Figs. 13-10** shows how the appropriate current distributions can reduce the field above the upper strip to zero. **Fig. 13-11** shows the current distributions in the upper and lower strips which result in the shielded field.

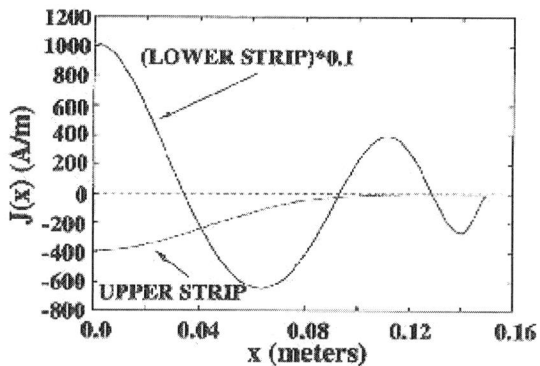

Fig. 13-11. Current distributions in the two strips in a SUSHI system.

A large effort with a correspondingly large group of graduate students and faculty continues to be made in follow-up work that has led to smaller and smaller error functions and thus better and better shielding in the standard MR system. "Application of the SUSHI method to the design of gradient coils" *Mag. Res. Med.* 45 147 2001 and *IEEE Trans. Mag.* 37 3116 2001. (This tasty contraction of SUper-SHIelding is easy to remember.)

In 1999, Brown and his longtime collaborator, Mark Haacke, and their former students, Mike Thompson, and R. Venkatesan, published a comprehensive book on the subject of MRI. They describe the book as a text for a two-semester course for graduate students or advanced undergrads in physics or engineering. The 914-page book includes chapters on the related physics, mathematics, biology, and engineering. The 27th and last chapter is an introduction to MRI coils and magnets. This is where the Jackson – and Morse and Feshbach – influence can be found. The text has been described variously as the "green MRI bible" and the "daily companion of the MRI scientist," by researchers in the field. "Magnetic Resonance Imaging: Physical Principles and Sequence Design", E. Mark Haacke (Washington University), R. W. Brown, M.R. Thompson (Picker International), and Ramesh Venkatesan (GE Medical Systems, Bangalore), 914 pp., John Wiley & Sons, New York 1999.

A problem in topology and another in fluid analysis

Visitors often come to the Rockefeller building with a question which a physicist may or may not be able to answer. One such visitor was a surgeon from nearby University Hospitals who was seeking advice on how to make quantitative and reproducible measurements of surgical modifications. Specifically, she wanted to know what easily measured parameters would allow one to determine the volume of a woman's breast, before and after surgery. She was referred (by your author) to Bob Brown who had experience with MRI and the associated measurements on patients. The meeting led eventually to a paper coauthored with post-doc Y. C. Norman Cheng and the surgeon, M. Kurtay. "A Formula for Surgical Modifications of the Breast" *Plastic and Reconstructive Surgery* **106** 1342 2000.

Another of Brown's contributions to technology involved work done with departmental lab director Bill Condit, colleague Don Schuele, Norman Cheng, and several students. The challenge was to develop an easy way to quantify the level of contamination of lubricating oils by wear particles and other impurities. The solution was based on the measurement of the capacitance of a capacitor cup filled with the oil. Recall that Schuele is an expert at such measurements as described in Chapter 12. The measured effective dielectric constant can be related to the concentration of impurities. Furthermore, the application of a voltage to the capacitor can be used to produce current discharges. The critical voltage for these is related to the amount of iron wear particles.

A summary, of sorts

The following are two excerpts from the description of Brown's teaching and research which appears on the department website.

"My basic research has been supported by the NSF for 20 years, including REU (Research Experience for Undergraduates) awards for 26 undergraduates, 15 published papers in which the students were authors or coauthors, and 12 NSF graduate fellowships. There have been 15 Ph.D.s under my advisory."

"In much of the applied research, I have worked for a number of years in the search for ways in which elegant and powerful mathematical tools can be used to solve practical problems, tools such as variational calculus with constraints. … This is analogous to field theory work in basic particle theory. Our problem-solving group, consisting of students and faculty, has worked out a large number of designs that have led to publications, industrial products, and … patents. The emerging lesson is that more and more products in the business world, including financial investment instruments, can be effectively mathematically modeled and their qualities thereby mathematically optimized."

Chapter 14   WRU Experiment, Round 2

B. Robinson,	Casper,	McGervey,	Jha,	Reichert,	Huang
1960-70	1960-67	1960-99	1966-69	1966-70	1966-74

Western Reserve chairman John Major took charge of the rapid expansion of the WRU department. In the decade following 1955 he brought in several theorists, including Stefan Machlup, Joseph Weinberg and Paul Zilsel. Their work was described in Chapter 11. During the same period, he added experimentalists Berol Robinson, Karl Casper, John McGervey, Ben Green and B.S. Chandrasekhar. The work in experimental low temperature physics done by Green and Chandra will be described in Chapter 15. In the current chapter, we shall look at the research of Robinson, Casper, and McGervey, as well as that of three other experimenters who joined WRU shortly before the 1967 federation.

Nuclei

Robinson and Casper studied nuclear excited states. Their experiments made use of radioactive materials produced at nuclear reactor facilities or at particle accelerators. Typically, if the sources had lifetimes long enough, the experiments could be performed in a modestly equipped laboratory, such as those found in many universities. Particle detectors including scintillation counters and the newer solid-state devices, along with associated electronic pulse height and timing circuits, were adequate to elucidate many details of nuclear structure. Thousands of measurements made at laboratories around the world contributed to the compilation of extensive tables of nuclear properties, listing the energies, spins and parities of each ground state and excited level. Each entry in this data bank required the measurements of lifetimes, decay branching ratios, and angular distributions of the γ's, electrons and positrons emitted by the unstable nuclei. Some of this information has been applied to nuclear technology, for example in nuclear reactor or weapons design, but most of it has been used to develop, test and substantiate theoretical nuclear models. This was "the hot topic" in experimental and theoretical research in the first few decades of the "nuclear age".

Robinson: nuclear spectroscopy

Fig. 14-1. Berol Robinson

Berol Robinson was born in Detroit in 1924. His photo is shown in **Fig. 14-1**. He served as a radar officer in the Air Force during World War II, completed his A.B. at Harvard, and went on to receive a PhD at Johns Hopkins in 1953. At Hopkins, Robinson was one of Leon Madansky's first students, working on angular and polarization correlations of gamma rays emitted sequentially from radioactive cesium. He spent three years on the faculty of the University of Arkansas where, with radio-chemist Richard W. Fink, he published a review of experiments on electron capture. "Recent experimental results on orbital electron capture" *Rev. Mod. Phys.* **32** 117 1960. *(Electron capture is the process whereby an orbiting electron is captured by the atom's nucleus, transforming a proton into a neutron. The resulting new nucleus is most often in an excited state and subsequently emits one or more photons.)*

Robinson joined the WRU department in 1956 at age 32. He brought with him the unfinished vacuum-tube multichannel-analyzer he had begun to construct at Arkansas. He continued his work in experimental nuclear physics, eventually joining Mike Kalvius and S. Jha in the application of the Mössbauer effect to the determination of very short lifetimes of nuclear states.

Aside on nuclear lifetimes. The promptness with which an excited nucleus emits a γ depends on the quantum numbers of the initial and final nuclear states. Consequently, the measurement of the transition probabilities (or equivalently the lifetimes) of the excited states provides information on the spin and parity of each level. Lifetimes of long-lived nuclear states can be measured by measuring the radioactivity as a function of time. For sequential decays in which a second emission takes place very quickly after the first, the time interval between them can be measured electronically by "delayed coincidence" circuitry. For states which decay extremely quickly, one can estimate the lifetime by measuring the "width" of the state. The outgoing particle has a spread in energies, partly due to the finite resolution of the equipment, partly due to a Doppler-shift caused by thermal motion of the emitting nucleus, and partly due to the Heisenberg uncertainty principle. This last contribution to the spread in energy (i.e., the width of the bump in the energy plot) is correlated with the lifetime by the uncertainty relation: $\Delta E\, \Delta t = h/2\pi$. A width of one eV implies a lifetime of about 10^{-15} sec; one milli-eV implies 10^{-12} sec, i.e. one picosecond.

Lifetimes by Mössbauer

Robinson and his colleagues specialized in the measurement of extremely short lifetimes by exploiting the Mössbauer effect. Mössbauer's 1958 discovery (1961 Nobel prize) was all the rage in the 1960's. In Chapter 12 we described how Dick Hoffman used this effect to measure tiny shifts in the energy of gammas emitted by ^{57}Fe nuclei in order to determine the magnetism in thin iron films. Later in this chapter, **Jonathan Reichert** will be another member of the Mössbauer club.

*A further aside on **Mössbauer spectroscopy**. In Chapter 12 we described how Hoffman used a moving absorber to measure precisely the energy levels in iron atoms in thin films. A related technique for minimizing the energy stolen by the recoiling nucleus is called "recoilless emission". In this case, the emitting nucleus sits in a crystal lattice at low temperature so that when the γ is emitted the recoil momentum is taken up by the whole crystal. As a result, the γ gets almost all the energy from the transition. If the emission is nearly recoilless, then one can measure the energy spectrum with enough precision to determine the natural width and lifetime of the state.*

An important aspect of Robinson's work was the preparation of the radioactive nuclear sources. As an example, to study two excited states in osmium-189, he and his colleague **S. Jha** (to be introduced shortly), bombarded an iridium-191 target with 26 MeV protons at the University of Colorado cyclotron. This sometimes produced platinum-189 which decayed into iridium-189 which decayed into excited osmium-189 which decayed to the osmium ground state:

$$_{77}Ir^{191} \ (p,3n) \ _{78}Pt^{189}$$
$$_{78}Pt^{189} \ (10 \text{ hours by e-capture}) \rightarrow \ _{77}Ir^{189}$$
$$_{77}Ir^{189} \ (13 \text{ days by e-capture}) \rightarrow \ _{76}Os^{189}* \rightarrow \ _{76}Os^{189} \ \gamma$$

The star in the last line indicates that the osmium nucleus is in an excited state. An important number is the "13 days", long enough to get home from Colorado with a radioactive sample. An alternative procedure, which replaces the first two steps above, was used at the cyclotron of the nearby NASA-Lewis Research Center. It involved bombardment of a rhenium target by alpha particles:

$$_{75}Re^{187} \ (\alpha,2n) \ _{77}Ir^{189}$$

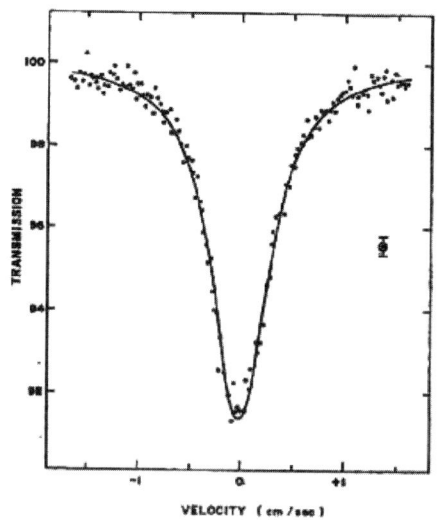

Fig. 14-2. Mössbauer spectrum.

The Mössbauer velocity spectrum for the 69.6 keV γ's coming from the excited osmium is shown in **Fig. 14-2**. The transmission is plotted against the speed of the absorber, from -2 to +2 cm/sec. The source was the bombarded rhenium foil and the absorber was osmium metal (enriched in the A=189 isotope). Both source and absorber were at liquid helium temperature to reduce thermal motions. Because the emission is recoilless, the shape of this curve reflects the true energy spread of the nuclear state. The width of the observed dip was (5.34 ± 0.08) mm/s, which corresponds to a lifetime of (2.35 ± 0.06) nanoseconds. Note how *slowly* the absorber must be moved (half a centimeter per second) to map out the width of the state. "Studies of Osmium-189: Gamma Rays, Lifetimes, and Mössbauer Effect" *Phys. Rev.* **180** 1158 1969.

Robinson initiated a small program of research on positron annihilation physics which was taken over and expanded brilliantly by John McGervey and his students. This work will be described later in this chapter. Robinson took a year's leave in 1961-62 on a fellowship from the Israel Atomic Energy Commission. He worked at the "Atoms for Peace" swimming-pool reactor at the Soreq Research Laboratory. He kept in constant contact with WRU chair, John Major, as they planned the coming year when the department would move into the brand new Science Center (now named in honor of WRU president John S. Millis).

Robinson and Jha and their several graduate students published a series of papers on the properties of nuclear states, for example "Gamma Rays in the Decay of Barium-131" *Phys. Rev.* **101** 149 1956. "Precision Determination of the Energy of the Gamma Ray of Potassium-40" *Phys. Rev.* **B134** 506 1964. "Intensity of the E3 Transition in Argon-38" *Phys. Rev.* **B140** 1529 1965.

Their Mössbauer work resulted in papers such as "The Mössbauer effect in ^{191}Ir and ^{189}Os" *Phys. Lett.* **25B** 115 1967. "Gyromagnetic Ratio of the 129-keV State in Iridium-191" *Phys. Rev.* **185** 1555 1969. "Mössbauer-Effect Studies in Hafnium-Metal Single Crystals" *Phys. Rev.* **187** 475 1969.

In 1967, Robinson was appointed vice-chair of the WRU department, thus allowing then-chairman Chandrasekhar to concentrate on the preparations for the merger with Case. Robinson was an extremely popular teacher and was successful in devising ways to make undergraduate laboratories more effective. In fact, Robinson and Weinberg were awarded first prize in the 1966 AAPT teaching apparatus competition for two "up-to-date" demonstration experiments: one on the Mössbauer effect, the other on positron annihilation.

His promotion to full professor was strongly supported by his chairman and department in 1967. But, in view of the merger and related uncertainties, it was not approved by the acting-president. Two years later, after Harvey Willard had taken over as chairman of the combined department, Robinson took a year's leave of absence. He accepted a position at the Education Research Center of MIT, where he joined an experimental program in undergraduate education, including research in alternative learning methods and undergrad research participation. The time spent at MIT would be a turning point in Robinson's career. He was asked to prepare the "educational component" of the "U. S. Metric Study". This assignment would pique his interest in science in the public arena. The report, mandated by Congress, concluded that the US should "go metric" by 1980. (A similar recommendation was made before World War II – and it too was ignored.)

Robinson resigned from the CWRU department in 1970, joining a dozen other members of the WRU department, tenured and non-tenured, who left for one reason or another between 1965 and 1970. He joined the Paris headquarters of the United Nations Educational, Scientific, and Cultural Organization (UNESCO). His responsibilities in the Division of Science Teaching led him to work on educational programs in Brazil, Africa, and all over the Middle East. He retired from UNESCO in 1985. Still based in France, Robinson is (2004) a very active spokesman for the Association of Environmentalists for Nuclear Energy, as an advocate for the use of nuclear power for energy production. He is the founder and president of the American affiliate.

Karl Casper: nuclear beta decay

Karl J. Casper was a member of the WRU department from 1960 to 1967. He had completed his PhD at Ohio State in 1960, having worked in experimental nuclear physics with P. S. Jastram. His work at WRU was mainly concerned with the development and use of solid state detectors to make precision measurements of the energy spectra of beta rays (i.e. electrons and positrons) emitted by radioactive nuclei.

An aside on "beta decay". If the energy of a nucleus would be reduced by the transformation of a proton to a neutron with the emission of a positron or by the transformation of a neutron to a proton with the emission of an electron, the transformation will take place eventually, along with the emission of a neutrino: that is $p \rightarrow n\ e^+\ \nu$ or $n \rightarrow p\ e^-\ \bar{\nu}$ The probability for this happening depends on the available energy and on the spin and parity of the initial and final states. The energy of the electron or positron can have any value from zero up to a specific maximum available kinetic energy. This maximum energy gives information about the mass-energy of the initial and final nuclide (nuclide is shorthand for any nuclear state). The shape of the electron's energy spectrum gives information about the quantum numbers of the initial and final nuclides. Energy spectra for beta decay are usually displayed in a special way, in a graph called a Kurie plot. The variables plotted are chosen so that a Kurie plot takes the form of a straight line for the least complicated transitions. The place where the Kurie plot goes to zero gives the "end point" energy, i.e. the maximum energy which the electron can have for the given transition.

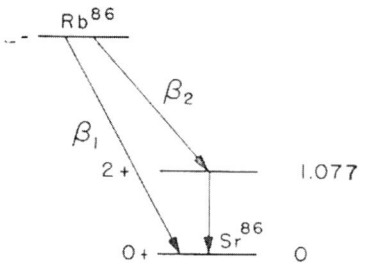

Fig. 14-3. Beta decay of rubidium.

As an example of Casper's work, **Fig. 14-3** shows the energy level diagram for the beta decay of rubidium to strontium, where transitions take place to two levels in the strontium nucleus. **Fig. 14-4** is a Kurie plot for the betas from the β_2 branch. Its shape indicates that the transition is "allowed", i.e. having no change in angular momentum. Casper and grad student Richard Thompson, had developed particularly sensitive "lithium drifted silicon detectors" which had the required resolution. This experiment's contributions to the nuclear tables were the energies of the two transitions and the spin and parity of the excited state in strontium. The data were analyzed on the Case Univac 1107 computer. "Beta Decay of ^{86}Rb" *Nucl. Phys.* **72** 106 1965. "Energy of the First Excited State of Sr^{86}" *Nucl. Phys.* **47** 443 1963.

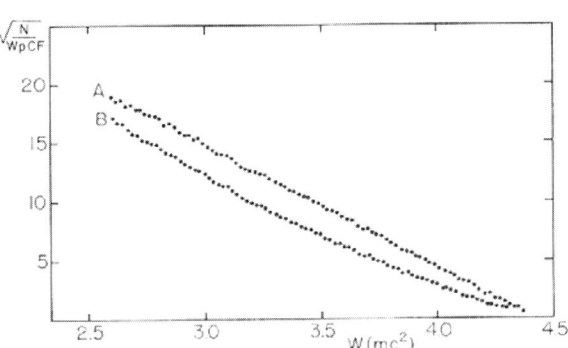

Fig. 14-4. Kurie plots. (curve A has been corrected for detector response.

In a somewhat different type of measurement, Casper determined the spin and parity of several nuclides by observing the angular correlation between pairs of sequentially emitted gammas. For example, **Fig. 14-5** shows the decay scheme for rubidium Rb^{84}. This nucleus changes a neutron into a proton and can arrive at any of three levels in the resulting Kr^{84} nucleus. The ten-microcurie Rb^{84} source (half life 33 days) was created at the 86-in cyclotron at Oak Ridge National Laboratory. Lithium-drifted germanium detectors were used to catch the gammas coming from the excited states at 1901 and 883 keV. The main result of the "table-top" measurement was the determination of the spin and parity of the 1901 keV level (2^+). "Angular Correlation of Gamma Rays in the Decay of Rb^{84}" *Phys. Rev.* **138** B1378 1965.

Casper and Robinson collaborated on a study of the effects of random background and finite efficiencies on coincidence logic. "Analysis of Chance Coincidences in Fast-Slow Coincidence Systems" *Nucl. Instr. & Meth.* **24** 482 1963. The development of the high resolution detectors warranted a technological paper: "Fabrication Methods for Lithium Drifted Surface Barrier Silicon Detectors" *Nucl. Inst. & Meth.* **40** 330 1966.

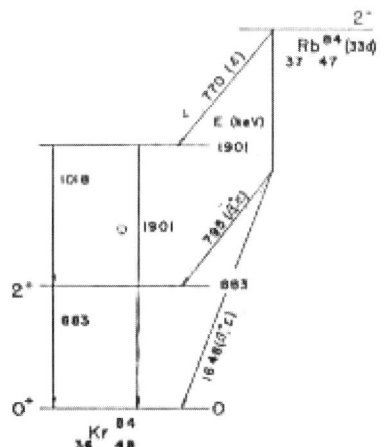

Fig. 14-5. Decay scheme for rubidium-84.

One limitation on the detection of the beta rays in a nuclear decay is the finite solid angle presented by the detectors. The solid-state detectors of the time were quite small. The effective solid angle can be increased by placing the radioactive source and detectors in a strong magnetic field. Casper built a magnetic spectrometer to achieve this. He used a geometry in which two 2-inch-diameter detectors were placed at the ends of a six-inch-long cylinder with the source at the middle, all lying in a superconducting, "commercially available", 30 kiloGauss solenoid. The betas leaving the source in nearly all directions get caught-up by the magnetic field and spiral down the cylinder to the detectors, sort of like incoming cosmic rays which wind around the earth's magnetic field lines. "Superconducting Magnet Beta-Ray Spectrometer" *Rev. Sci. Inst.* **38** 1110 1967.

Karl Casper left WRU in 1967, the year of the merger with Case Tech. In a letter from chairman Chandrasekhar, Casper was informed that support for the program in low energy experimental nuclear physics could not be guaranteed. Casper accepted a position at Cleveland State University where he is now professor emeritus.

S. Jha: γ's from nuclei

Shacheenatha Jha was born in Bihar, India in 1918 and was 48 years old when he joined the WRU department in 1966. He had done his BS and MS (1941) at Patna University where he remained as part of the staff for an additional five years. He completed his PhD under N. Feather at the University of Edinburgh in 1950, and returned to

Fig. 14-6. S. Jha

India to do experimental nuclear physics research at Patna and the Tata Institute in Mumbai for eleven years. In 1961 he came to America to become assistant professor at the Carnegie Institute of Technology. He joined the WRU department in 1966 as associate professor. His photo is in **Fig. 14-6**.

At WRU, Jha worked with Berol Robinson. Several of their joint papers on Mössbauer studies of nuclear levels were described earlier in this chapter. In a paper with two colleagues at NASA Lewis, Jha reported a study of the isomers of xenon-130. They detected γ's emitted by xenon nuclei which had been produced in the beta decay of two neighboring nuclides: $_{53}I^{130} \rightarrow \beta^- \, _{54}Xe^{130}$ and $_{55}Cs^{130} \rightarrow \beta^+ \, _{54}Xe^{130}$ (each with neutrinos, of course). This was an interesting way to get at some of the xenon levels because the iodine-130 has spin 5 and the cesium-130 has spin 1. As a result, they decay into different selections of xenon levels. The iodine source was prepared by neutron capture and the cesium source by alpha bombardment at the NASA cyclotron. Forty different transitions were observed and eighteen levels identified. "Levels of ^{130}Xe Populated in β Decay of ^{130}I and ^{130}Cs" *Phys. Rev.* **174** 1472 1968.

In a paper with his graduate student, Peter Bond, Jha tackled another nucleus: molybdenum-95. The sample was prepared at NASA by alpha bombardment of niobium which produced technecium-95 which decays to the isomers of Mo^{95}. *(Isomers are excited states which take a fairly long time to decay.)*

$$_{41}Nb^{93} \, (\alpha,2n) \, _{43}Tc^{95} \rightarrow (20 \text{ hour e capture}) \, _{42}Mo^{95}*$$

A similar, but alternate route from Nb to Tc to Mo passes through a metastable state in Tc, one with a 61 day mean-life. The Tc ground state and this metastable state have different quantum numbers, (9/2+ vs 1/2-), so different levels in the molybdenum are fed from each. One set of transitions decays away in the first few days. The experimenters also looked at the angular correlations between γ's coming from sequential decays, another tool in tying down the parameters of the molybdenum levels. The resulting level-scheme was compared with several theoretical models for this almost spherical Mo nucleus, including a model proposed by Jha's colleague, **Leonard Kisslinger**. "Nuclear-Structure and Hyperfine-Field Studies with Mo^{95}" *Phys. Rev.* **C2** 1887 1970.

Jha teamed up with colleague **John McGervey** and Peter Bond to devise an experimental setup which allowed the accurate measurement of the time interval between two γ-rays. This was used to determine the lifetimes of excited states in ten different radioactive nuclei. The measured lifetimes were in the 1 to 10 nanosecond range, and the experimental uncertainty about 3%. An example of a measured state is the 127 keV level in ^{101}Ru. A sample of radioactive ^{101}Rh (prepared at NASA), decays spontaneously to a state of ^{101}Ru which is (198+127) keV above the ground state. This emits a 198 keV γ when it drops to the 127 keV level. It then subsequently emits a 127 keV γ as the nucleus returns to its ground state. The time difference between the detection of the 198 and the 127 keV γ's gives the lifetime of the 127 keV level. The ten different nuclei studied came from the decays of ten different radioactive sources, prepared by commercial suppliers or at NASA or at Oak Ridge National Laboratory. "Measurements of Some Nuclear Lifetimes in the Nanosecond Region" *Nucl. Phys.* **A163** 571 1971. Jha took a faculty position at the University of Cincinnati in 1969, not long after the merger of the WRU and CIT departments.

McGervey: positrons

John D. McGervey received his PhD in 1961 from Carnegie Institute of Technology. He was a student of Sergio DeBenedetti; his dissertation was titled "Mean Lives of Positrons in Oxidizing Solutions" (according to the abstract: "an attempt to detect oxidation of positronium (Ps) by positive ions"). McGervey was hired that same year by John Major as an assistant professor in the Western Reserve department. His picture is in **Fig. 14-7**.

Fig. 14-7. John McGervey.

We have seen in Chapter 11 how the Case researchers, Gordon, Eck, and Schuele, were investigating the behavior of electrons in crystals and their interaction with the lattice. They did so by looking at such things as the effect of external fields, temperature, and pressure on electrical and magnetic properties. McGervey had worked with positrons at Carnegie Tech and he would use them as a complementary probe of solids.

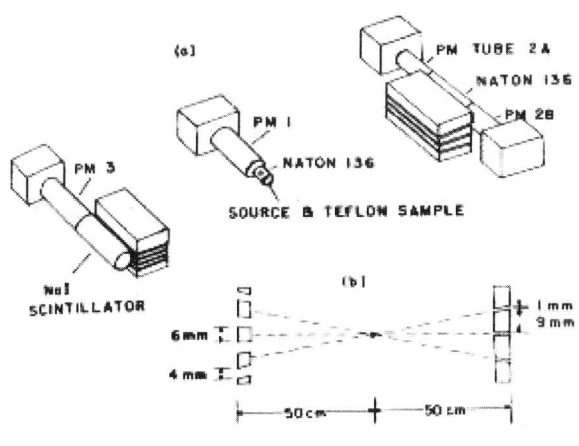

Fig. 14-8. Positron annihilation telescope.

When a positron from a radioactive source finds itself inside a crystal or other material, it most often finds an electron with which it forms a mini-hydrogenlike-atom, called positronium (Ps). If the electron and positron have opposite spins, the atom (called parapositroniuim or p-Ps) has zero angular momentum and decays in about 10^{-10} s into two γ's. If the spins are parallel, the "orthopositronium or o-Ps has total angular momentum unity and must decay into three γ's. This is more complicated and it takes a thousand times longer: about 10^{-7} s. For the 2γ decay, the angle between the outgoing γ's gives a measure of the momentum of the positronium (it'd be 180 degrees if the Ps were at rest), and provides therefore a probe of the momentum distribution of the electrons in the material.

With graduate student Virginia Walters (the first woman physics PhD student at Western Reserve), McGervey did a series of experiments in which positrons were introduced into a Teflon sample, and their survival time was measured as a function of the angle between the two outgoing γ's. **Fig. 14-8** shows a schematic of the experiment, with the Na^{22} positron source and Teflon sitting inside a hole in the scintillator which is looked at by photomultiplier #1. When a positron is emitted, it is quickly followed by a 1.3 MeV γ. The detection of this gamma marks the time at which the positron enters the Teflon. Two sets of lead collimators were placed 50 cm from and on opposite sides of the Teflon. Narrow slits in the lead allowed the selection of well-defined ranges of the angle between the two γ's.

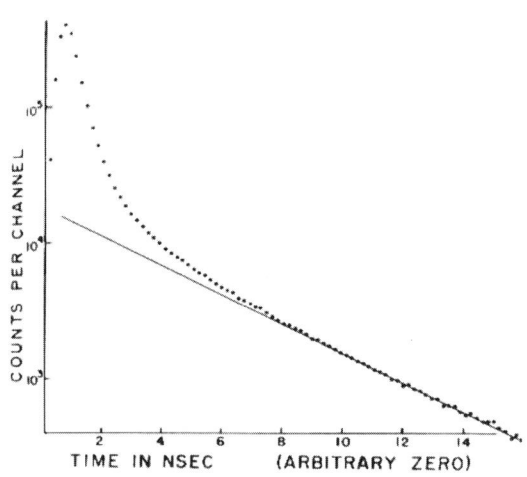

Fig. 14-9. e$^+$ annihilations in Teflon: 3 different lifetimes.

Three ranges of angles (i.e. away from 180 degrees) were studied, 0 to 2 mrad, 4 to 12 mrad, and 8 to 12 mrad. Scintillators and photomultipliers #2 and #3 were placed beyond the slits. The lifetime of the positron in the material, as determined by the delay between counter #1 and #2, was recorded in a multichannel analyzer. A chi-square fit to the lifetime distribution, shown in **Fig. 14-9**, indicated that it was a superposition of three different mean lifetimes: 0.33, 1.05, and 4.06 ns. (Note that this is a logarithmic plot, covering three orders of magnitude.) That means that the positron found at least three different things to do in the Teflon. It could form spin zero para-Ps, or spin one ortho-Ps, or perhaps form some sort of bound-state with the Teflon. The fractions of the annihilations associated with each lifetime could also be determined: roughly 67, 17, and 16%. The long-life time components were found to be associated with the larger angle annihilations, as is evident in **Fig. 14-10** where the small and large angle data are plotted separately. Fits to the combined angle-and-lifetime data "support the hypothesis of a bound-state between the Teflon molecule and the positron". "Correlation of Positron Lifetime with the Angle between the Annihilation Gamma Rays" *Phys. Rev. Lett.* **13** 408 1964; also *Nucl. Instr. and Methods* **25** 219

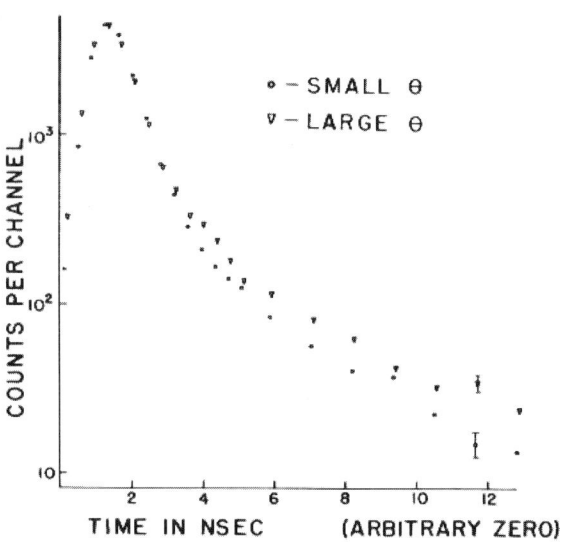

Fig. 14-10. e$^+$ survival times for two γ separation angle selections.

1964; and later "Correlation between Lifetime and Momentum for Positron Annihilations in Teflon" *Phys. Rev.* **B2** 2421 1970.

Positrons as probes of crystal structure

McGervey next used positron annihilations in silicon and germanium to look at electron momentum distributions along three different orientations of the crystals. The data were compared with theoretical predictions for each orientation, and found to be in good agreement for only one of the three orientations. "Electron Momentum Distribution in Silicon and Germanium by Positron Annihilation", *Phys. Rev.* **151** 615 1966.

In McGervey's later experiments, various alloys of copper and aluminum or copper and nickel were tested. A bit of radioactive ^{64}Cu was introduced into the mix to provide an *in situ* source of positrons, and the emerging γ's were detected after passing through 3 mm wide slits 4 meters from the crystal. The crystal was aligned so that the measured electron momentum could be related to the lattice orientation. (Recall that the angle between the two annihilation γ's gives the momentum of the electron before it got involved with the positron.) This in turn gives information about the Fermi surface for the crystal (we described Fermi surfaces in Chapter 12). McGervey was able to identify "necks" in the Fermi surface (i.e. preferential directions in which the electron momenta are large) and to determine the effects of the alloy composition. "Positron Annihilation in Copper Alloys" *Phys. Rev. Lett.* **24** 9 1970.

While on sabbatical in Jülich in Germany, McGervey worked with W. Trifthäuser on another experiment which used positron annihilation to probe the properties of metal crystals. The object was to determine the vacancy formation energies (how much energy does it take to kick an electron out of its place in the crystal?). The experimental technique involved measuring the rate of positron annihilation as a function of the temperature of the crystal. For "zero angle" events, where the two annihilation photons emerge back to back, the rate is expected to be proportional to $\exp(-E_V/kT)$, where E_V is the desired vacancy formation energy. Measurements were made from near 0 to 1000 K, and the data indicated the expected exponential form between 400 and 800 K. A value, $E_V = 0.98 \pm 0.07$ eV, was determined for both silver and copper. "Vacancy-formation Energies in Copper and Silver from Positron Annihilation" *Phys. Lett.* **44A** 53 1973.

Positrons to detect voids

A second paper written with his German collaborators concerned the detection of voids in aluminum caused by neutron irradiation. The authors explain: "The creation of voids in neutron-irradiated materials, and their deleterious effects on mechanical properties, pose serious problems in reactor technology. In the case of irradiated aluminum it is puzzling that, even during annealing studies, no voids smaller than about 100 Å diameter are seen in the electron microscope. Hence, additional experimental techniques known to be sensitive to very small voids have been sought." Here, then, was a challenge for the positron-annihilation approach. Single crystals of pure aluminum were exposed to high neutron fluxes at Oak Ridge National Laboratory. These were then exposed, in the Jülich laboratory, to a source of positrons, and the two 0.51 MeV photons resulting from each annihilation in the aluminum were measured using a two-narrow-slit arrangement as described above. The photons travel approximately in opposite directions, and, as mentioned above, the deviation from co-linearity is a measure of the momentum of the positronium, and indirectly of the "target" electron. McGervey and his colleagues found that photons coming from the irradiated sample (i.e. the one with tiny voids) were significantly more co-linear than those emitted from the non-irradiated sample. This means that the electrons' higher momentum component is suppressed. The author's suggest, in explanation, that the positrons are trapped at the boundary of the voids, and encounter mostly low-momentum valence electrons. "Positron-annihilation studies of voids in neutron-irradiated aluminum single crystals." *Phys. Rev.* **B9** 3321 1974. *Phys. Rev.* **B9** 2402 1974. **Fig. 14-11** is a plot of positron meanlives in irradiated aluminum versus the annealing temperature shows how abruptly the voids are cooked out of the crystal. *Phil. Mag.* **36** 117 1977.

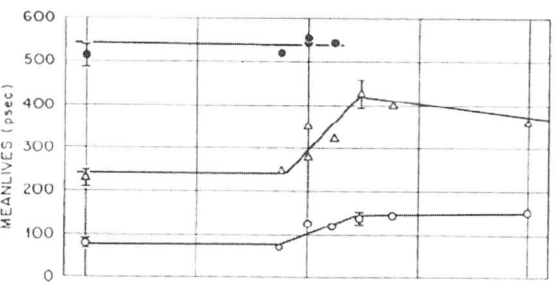

Fig. 14-11. Positron survival times vs. annealing temperature in irradiated aluminum.

McGervey joined collaborators at the University of Guelph and Queen's University in Ontario to look at positron survival times in cadmium as a function of temperature and pressure. One would expect that lowering the temperature or raising the pressure would increase the density of the host material, and thus decrease the lifetime of the positron. The experimenters measured the positron lifetime at atmospheric pressure both at 77 K and at 296 K, finding 170 ± 2 and 190 ± 2 picoseconds respectively, as expected. They then increased the pressure at 296 K until the cadmium was squeezed down to the density it had at 77 K and one atmosphere, i.e. 9.8 kilobar (98 thousand atmospheres). The positron lifetime dropped to 178 ps, but not all the way down to 170 ps. The conclusion was that the fate of the positron depends not only on how close it gets to the atomic electrons, but perhaps also on how vigorously the atoms are jiggling around. "Temperature and pressure dependence of positron mean lives in cadmium" *J. Phys. F: Metal Phys.* **7** L255 1977.

McGervey had become an expert in the technique of using positron annihilations to probe electrons in matter. He joined his colleague, **Arnold Dahm**, in an experiment to chase down the electrons in gaseous, liquid and solid helium. Dahm had been studying electrons in helium, and the positron annihilation technique provided yet another tool. Because the physics of this work is more "helium" than it is "positronium", I shall describe it in the section on Dahm's work in Chapter 15, rather than here.

McGervey experimented with a superconducting material, V_3Si, to examine the connection between defects and positron lifetimes. His colleagues Farrell and Chandrasekhar had been studying a certain class of materials in which it was believed that the presence of defects radically changed the temperature at which the materials became superconducting. (Their work is described along with Dahm's in chapter 15.) McGervey measured the lifetime of positrons in V_3Si as a function of pressure up to 20 thousand atmospheres, with the expectation that the voids associated with the defects might be squeezed out by high pressure. They found no change in the positron survival time over this wide range of pressure. "It is tempting to conclude that the postulated vacancies in V_3Si do not exist. However, it is also possible that vacancies are present, but that they do not produce a sufficiently deep potential well to trap a positron." "Pressure dependence of positron lifetimes in V_3Si" *Phys. Lett.* **63A** 393 1977.

John McGervey authored a very popular paperback called *Probabilities in Everyday Life* (Ballantine Books 1986) in which he analyzed, for example, the odds in cardplaying, in horseracing, in the stock market, in surviving tobacco smoking, in life insurance, in driving without seatbelts, and a number of other fascinating areas in which people try to beat the odds. On the internet, he is most often cited as being the debunker who compiled the astrological signs of some 17,000 scientists and found them evenly distributed. He was a born-campaigner: for railroad travel, for arms control, for any number of (in my opinion) enlightened causes. He published two undergraduate texts: *Introduction to Modern Physics* (Academic Press 1983) and *Quantum Mechanics* (Academic Press 1995). For many years John directed a program to assist primary and secondary school teachers of science. He died only two years after his retirement in 1999.

Reichert: hyperfine structure in solids

Jonathan F. Reichert earned his BS in physics at Case in 1953. He completed his PhD at Washington University St. Louis in 1962, where he worked with Jonathan Townsend on the magnetic interaction between electrons and the lattice nuclei in solid metals. "Dynamic Nuclear Enhancement in Metallic Sodium" *Phys. Rev.* **137** A476 1965. He continued in this area of experimental condensed matter physics for two years as a post-doc at Harvard. "Electron-nucleus Double-Resonance Studies of F Centers in KCl: Electric Field Effects" *Phys. Rev. Lett.* **15** 780 1965. In 1965, the 34 year old Reichert accepted an offer of an assistant professorship from John Major's successor as chairman, Gerald Tauber. Reichert's photo is in **Fig. 14-12**.

Fig. 14-12. Jonathan Reichert.

Electron spin resonance in crystals

Reichert measured hyperfine interactions in atoms. These result from the interplay between the magnetic moment associated with the electron's spin and orbital motion with the magnetic moment of the nucleus. The atomic energy levels are consequently split into two or more closely spaced levels. But Reichert and his collaborators were not studying atomic physics. They were using hyperfine structure to study the interaction between electrons and nuclei in a crystalline structure. By subjecting a sample to high frequency electromagnetic radiation, they could simultaneously induce transitions associated with the nuclei and with single electrons trapped in vacancies in the crystal. **Fig. 14-13** shows such a trapped electron in a sodium chloride crystal. Their technique was called ENDOR (electron-nucleus-double-resonance) spectroscopy. (It was thus a combination of electron-spin-resonance, ESR, and nuclear-magnetic-resonance, NMR.) In these experiments, the samples were further subjected to extremely strong steady electric fields, up to 50 kV/cm. The idea was to pull the electron around in its trap so it could cozy up to one neighboring nucleus or another. "Electric Field Effects in Electron-Nuclear Double-Resonance Spectroscopy of F Centers" *Phys. Rev.* **180** 482 1969.

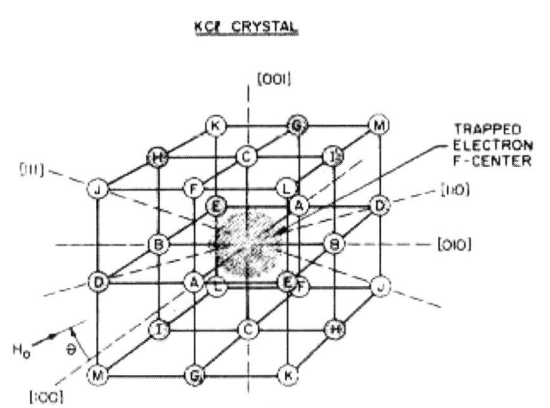

Fig. 14-13. An electron in a NaCl cage.

This was followed by a second paper, with his student Zahiruddin Usmani, which elaborated on the associated theory: *Phys. Rev.* **B1** 2078 1970. A variation on this experiment was performed using lithium fluoride as the medium. The twist here was the use of a 50-50 mixture of the two lithium isotopes: Li^6 and Li^7. The object was to find out what happens when the electron lies between two nuclei with differing quantum numbers. ("Isotopic Substitution for Observing Linear Electric Field Effects" . II. Theoretical Considerations", *Phys. Rev. Lett.* **24** 709 1970.

Reichert also worked on a related experiment with his colleague Chao-Yuan Huang whose work will be described later in this chapter. They measured the hyperfine structure associated with manganese ions in fluoride crystals, studying how the coupling with the lattice changes with increasing temperature. "Temperature-dependent Hyperfine Coupling Constants of Mn^{2+} in Fluorides" *Phys. Lett.* **26A** 219 1968.

In another application of electron-spin-resonance, Reichert got together with Arnold Dahm to look at spin-flip transitions of electrons injected into liquid helium. This work will be described in Dahm's section of Chapter 15.

Finally, we mention an experiment in which the ESR spectroscopy was combined with Mössbauer spectroscopy. Reichert would thus follow Hoffman and Robinson in using this powerful tool for the precise measurement of energy values. He and grad-student James Lock were looking at magnetic transitions in ^{57}Fe when the iron nucleus resides in a nonmetallic host lattice. (As in experiments described earlier, a cobalt-57 source provided the γ-rays to be absorbed by the iron nucleus.) The Mössbauer spectrum of the γ-rays shifted when the sample was subjected to external radio-frequency radiation. "This article reports what we believe is the first observation of an electron spin-Mössbauer double resonance effect of dilute paramagnetic ions in a nonmetallic host lattice." What they were seeing was a change in the energy at which the γ-rays are absorbed by an iron nucleus when a microwave magnetic field is simultaneously producing transitions in the electronic levels in the iron atom. "Mossbauer-Electronic Double Resonance in $NH_4(^{57}Fe,Al)(SO_4)_2 \cdot 12\ H_2O$" *J. Magnetic Resonance* **7** 74 1972.

Jonathan Reichert came up for promotion in 1969; unfortunately it was a time when the newly combined department was clearly too large. The department was moving toward more particle physics and the vote was not in his favor. Reichert was an extremely popular teacher and the students reacted strongly to the department's action. The headline of the newly established Observer, CWRU's student newspaper, was "Dr. Reichert canned". The Cleveland Plain Dealer picked up on the story with front-page coverage. The following month the students organized a Teach-In on the issue, but to no avail. Reichert moved on to SUNY at Buffalo where he continued his research and teaching.

Chao-Yuan Huang: foreign atoms in solids

Chao-Yuan Huang joined the Western Reserve physics department in 1966 during the period of the negotiations which led to federation with Case. Huang was born in Taiwan in 1935; he completed his doctorate in applied physics at Harvard in 1964. At WRU, Huang joined the Condensed State Center which had been set up by Chandrasekhar and colleagues.

Huang's principal interest was the study of what happens to a foreign atom when it is placed within a crystal lattice. This was quite similar to what Reichert was doing. In the simplest terms, the atom's electrons feel the presence of its neighboring lattice atoms and its energy levels are slightly shifted and their widths changed. Huang observed these changes experimentally by "electron paramagnetic resonance" (EPR) spectroscopy. The sample, held at very low temperatures (4 K), is placed in a strong magnetic field (3 to 10 Tesla) which can be varied. The field splits the energy levels in two, one with electron spin up, the other with spin down. The amount of this "Zeeman" splitting is proportional to the strength of the applied field.

Fig. 14-14. C-Y Huang.

One can flip the spin of the electron by applying microwave radiation to the crystal (in the 10 GHz or 3 cm wavelength range). When the microwave energy matches the difference in the energies of the two levels, the electrons flip over and the radiation is strongly absorbed. The location and shape of the bump in the plot of the absorption vs. magnetic field gives an accurate measure of the electronic energy levels. Huang made EPR studies of a large number of rare-earth ions which have unpaired electrons and permanent magnetic moments. The ions were placed in a variety of crystal structures and the temperature dependence of the electronic levels was measured. From this, the coupling between the phonons of the crystal lattice and the ion's electrons can be determined. "Temperature-dependent hyperfine interactions of Mn^{2+} in alkali halides *Phys. Rev.* **158** 280 1967. "Phonon-induced spectral linewidths in crystals" *Phys. Lett.* **24A** 740 1967. The role of the magnetic moment of the ion's nucleus was studied by using different isotopes. "Temperature-dependent isotope shifts and phonon-induced zero-field splitting" *Phys. Lett.* **27A** 437 1968.

In the early 1970's, Huang supervised the research of two graduate students: Kazushi Sugawara and Frederic Rachford. While at CWRU, Huang published over three dozen experimental and theoretical papers, many with Sugawara, on the interactions between selected foreign ions and the host crystal. In 1974, he became interested in a very different line of research: superconducting materials. In a series of papers with Rachford, Huang reported measurements, for example, of current flow through small "microbridges" in superconducting tin. This work was related to the new "superconducting quantum interference devices" (SQUID's). "Limiting flux passage time in narrow superconductors" *Phys. Rev. Lett.* **35** 305 1975.

Huang moved to the Los Alamos National Lab in 1974 while maintaining his connection with CWRU as adjunct associate professor. He continued collaborative studies with his CWRU colleagues for several years. He would remain at LANL for eleven years, working on superconductivity and on the use of μ-mesons, essentially heavy electrons, as probes of materials. After a four-year stint as senior scientist at the Lockheed Research Laboratory, he returned, after more than thirty years in the US, to Taiwan to assume a professorship at National Taiwan University.

Chapter 15   Matter at Low Temperatures

Green,   Chandrasekhar,   Sparlin,   Adler,   Farrell,   Dahm
1961-68   1963-87   1964-68   1965-68   1966-   1968-2001

Much of the experimental physics research done at Case during the 1950's and 1960's was concerned with the properties of solids. The work on crystals done by Chuck Smith and that done by Eugene Crittenden on thin metallic films (described in Chapter 7) led to the later research of Schuele, Hoffman, Gordon and Eck (Chapter 12). Many of these studies were done on materials held at a few degrees above absolute zero. The subtle interactions between the electrons and the atoms of the crystal become more transparent as the temperature is lowered and the thermal motions of the players become less important. In the middle 1960's, a new line of work was begun at the new "Center for the Condensed State" on the Western Reserve side: the physics of superconductors and superfluidity. These studies, also done at very low temperatures, generally went by the name "low temperature physics". They concern materials which exhibit extraordinary electrical or mechanical properties below well-defined temperatures, and whose study has lead to significant advances in the applications of quantum mechanics. Nowadays, the more inclusive term "condensed matter physics" embraces "low temperature physics" as well as what was once called "solid state physics".

Ben A. Green, Jr. - electrons in matter

Fig. 15-1. Ben Green (in the 1964 yearbook)

The first experimental low temperature research at WRU was done by **Ben A. Green, Jr.** Green was born in Alabama and earned his BS at the University of Alabama. He received his PhD from Johns Hopkins in 1956 and spent four years at Union Carbide Metals Co. before joining the WRU faculty in 1961. His work at Union Carbide was on determining the lifetime of positrons stopped in lead, above and below the temperature at which lead becomes superconducting. He used a ^{22}Na source which emits a positron in coincidence with a 1.3 MeV photon. The positron annihilates with an electron in the lead, resulting in two 0.511 MeV photons. The delay between the first photon and the latter photons was measured at 10 K and 4.2 K. The expectation was that the probability that the positron would find and annihilate with an electron would be different in the superconducting lead. Contrary to some earlier published predictions, no difference in lifetime greater than $2\ 10^{-10}$ s was observed. "Lifetime of Positrons in Superconductors" (*Phys. Rev.* **102** 1014 1956.)

At WRU, Green was the first to conduct research on the properties of materials at liquid helium temperatures. He began with a study of the effect of selected impurities on the electrical resistivity of liquid sodium. The idea is that the impurities provide scattering centers which the electrons must cope with, and that the change of resistivity will depend on the size of the foreign atom. His experimental results agreed reasonably well with the rather simple model proposed in the paper. "Size effect in electrical resistivity of alloys" *Phys. Rev.* **126** 1402 1962.

Between 1962 and 1968, Green, with the help of two doctoral students and three MS students, published six papers in the *Physical Review* on the low temperature specific heats of various metal alloys. The object of this type of measurement was to determine the separate contributions to the specific heat of the electrons and of the lattice. The density of electrons could be varied by changing the concentration of the solute. Samples of the alloys were placed in a novel calorimeter designed by Green. The amount of time it took for a constant-rate heat source to heat the samples was measured. The resulting specific heats were fitted to $c = \gamma T + AT^3 + BT^5$, where the first term is associated with the electron contribution and the other two terms with that of the lattice. The values of γ, A and B were tabulated as a function of tin concentration in the gold-tin alloy. The measurements were made at temperatures between 2K and 4K. **Fig. 15-2** shows, for example, the value of γ as a function of e/a (the number of valence electrons per atom, which varies as the tin content is changed). "Specific Heats of Au and Au/Sn". (*Phys. Rev.* **150** 519 1966). "Rigid Band Behavior in Aluminum-Based Alloys – Electronic Specific Heat" (*Phys. Rev.* **153** 800 1967.) The experimenters found that γ, the electron contribution to the specific heat increases with e/a at a rate three times faster than predicted by theory. This sort of result always sends the theorists back to work.

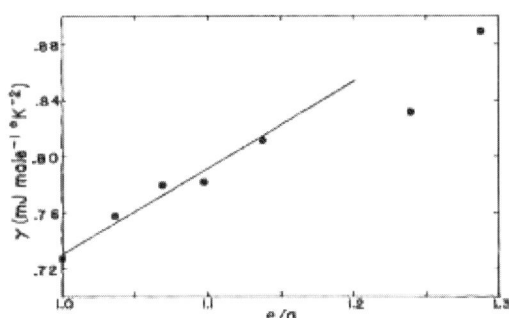

Fig. 15-2. Electron contribution to alloy specific heat vs. electron concentration.

Green became very interested in research on the teaching of introductory physics and he spent the 67-68 year on leave at the University of Maryland as staff physicist with the Commission on College Physics. He subsequently decided to move on to the Education Research Center at MIT and resigned from the newly merged CWRU department.

Chandrasekhar

We now back up four years to when Green joined forces in 1963 with a new colleague who was also a "low temperature" experimentalist. **Bellur S. Chandrasekhar** was well-established in the field of superconductivity research. "Chandra" would eventually become chair of the WRU department and later dean of the CWRU College of Arts and Sciences during his 25 years at the university. He was born in 1928 in Bangalore, India. He completed his B.Sc. at Bangalore, his M.Sc. at Delhi, and D.Phil. as a Rhodes scholar at Oxford in 1952. At that time, there was but one Rhodes scholarship allotted to India. At Oxford he did research under F. Simon and K. Mendelssohn on the properties of He II superfluid films. ("Pressure Measurement in Superflow") He then took a post-doc position at the University of Illinois working with Dillon Mapother. Chandra was employed for eight years by Westinghouse Laboratories where his interests turned first to electrons in metals (in work similar to the Fermi surface measurements done by Gordon and Eck and the ultrasonic elastic constants measurements done by Schuele as described in Chapter 12). He then switched to the study of superconductivity and the design of superconducting magnets. He was a member of the group which built one of the first very high field (68 kilogauss) superconducting magnets. Chandra accepted a position at WRU in 1963. He is shown in the foreground in the photo (**Fig. 15-3**).

Fig. 15-3. Chandrasekhar (foreground).

In an essay describing his years at CWRU, Chandra tells how in 1961 he met John Major (then chair at WRU, Chapter 10) when Major was on sabbatical leave in Paris. Major described how he was building a new and competitive research department and he invited Chandra to join in. Chandra had enjoyed his time as a post-doctoral researcher in 1953 at Illinois and thought often about returning to the academic world. When he visited WRU in 1962 he was especially impressed by the spirit of cooperation among the WRU faculties of physics, chemistry, and biology, most notably in the persons of John Major, Ernest Yeager, and Howard Schneiderman, respectively. These men were planning several exciting interdisciplinary research programs for WRU and they had the enthusiastic support of WRU president, physicist John S. Millis.

Chandra predicts oscillatory magnetostriction

Chandra's first publication after his arrival at WRU was a letter titled "A Note on the Possibility of Observing deHaas-vanAlphen Oscillations in Magnetostriction". (*Phys. Lett.* **6** 27 1963.) We have looked at the DHVA effect as studied by Gordon and Eck in Chapter 12. We saw there that when metal crystals are placed in a magnetic field, the electron orbits within the lattice are perturbed, producing periodic variations in such properties as the magnetization and the resistance as the applied field is changed. In his letter, Chandra predicted similar effects for the lattice spacing (or equivalently the length of the crystal), an effect known as magnetostriction. He remarks, "such an oscillatory magnetostriction not only provides an interesting new technique for the study of the electronic structure of such metals and semimetals, but also can play a vital role, hitherto overlooked, in the analysis of other oscillatory phenomena such as oscillations of acoustic velocity in a magnetic field." The one-page letter outlines a preliminary theory and ends by saying that "attempts to observe the effect experimentally in bismuth are underway."

About eight weeks later, Chandra and his new colleague, Ben Green, published a letter: "Observation of Oscillatory Magnetostriction in Bismuth at 4.2 °K". (*Phys. Rev. Lett.* **11** 331 1963.) A 5 cm long, 6 mm diameter, 99.999% pure sample of bismuth was placed within a 20 cm long superconducting solenoid. This provided the intense magnetic field, variable up to 25 kilogauss. The sample's length was monitored by measuring the capacitance formed by a flat electrode placed at its end 70 microns away from a parallel fixed electrode. At 25 kGauss magnetic field strength, the change in length of the sample was only about one part in a million. **Fig. 15-4** shows the change in length plotted against the applied magnetic field. The oscillations are anything but subtle, and just as predicted in the earlier letter.

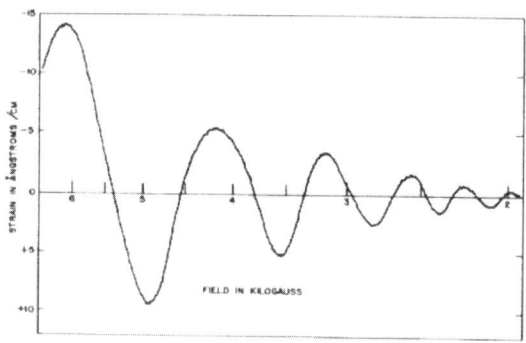

Fig. 15-4. Experimental observation of magnetostriction.

Chandrasekhar worked for a time with another young colleague, assistant professor **Don M. Sparlin**. Sparlin, who did his PhD at Northwestern, would spend only three years in the WRU department. For his dissertation he had worked on the de Haas-van Alphen effect in various metals. At WRU he joined Chandra in 1964 in the study of magnetostriction. He left the department just after the formation of CWRU, moving on to the University of Missouri at Rolla, where he would continue his research and teaching for several decades.

Josephson effect in liquid helium

Chandra's first graduate student, Brij M. Khorana, studied the ac Josephson effect in superfluid helium in an experiment reminiscent of Chandra's own doctoral research fifteen years earlier. Superfluid helium can flow through a narrow orifice without any resistance, an effect explained quantum mechanically by the passage of quantized vortices through the constriction. If two reservoirs of liquid helium are placed one above the other and connected by a thin tube, the level of the upper reservoir increases incrementally in time. In a carefully crafted experiment, Chandra and Khorana were able to show that the rate dn/dt at which the vortices pass through the tube is proportional to the gravitational potential difference between the two levels: dn/dt = mgz/h where m is the mass of the helium atom, h Planck's constant, and z the difference in height between the two levels. *Phys. Rev. Lett.* **18** 230 1967.   Chandra later pointed out that work by Pobell suggested that at least a part of these observations were due to acoustic resonances in the experimental chamber. ("ac Josephson Effect in Superfluid Helium" *Phys. Rev. Lett.* **18**, 230 1967.) We shall describe Chandra's further research on superfluidity later in this chapter.

Superconductivity work begins

In 1964 Chandra organized an international conference on Type II superconductivity at Western Reserve, for which he edited the three volumes of proceedings. This event, sponsored by the National Science Foundation, marked a high point in the story of the then recently discovered *high field* superconductors. As Chandra recalls, "everyone who was anyone in the field came, some 200 of them as I remember: Tinkham, de Gennes, Werthamer, Geballe, Gorter, Goodman, etc., etc."

In 1965, Chandra began work with a new member of the department, **John G. Adler**. Adler (born in Budapest in 1935) had completed his doctorate at the University of Alberta in 1963. There he had studied the tunneling of electrons from a normal metal through a thin insulating layer into a superconducting metal. He and Chandra began their collaboration with a similar experiment.

*A few comments about **superconductivity**: When I first wrote this section, I oversimplified the explanation of superconductivity by repeating what I had long taught in introductory modern physics courses: in a nutshell, in a superconducting material, below a certain low temperature, the conduction electrons somehow pair up to produce entities with zero spin. These boson "Cooper pairs" all crowd into the same low energy level (in marked contrast to the fermion unpaired electrons in "normal" materials which are required to occupy different levels). When this happens, the pairs move through the metal without losing or gaining energy. As a result, an electric current flows through the material without any "ohmic" losses, i.e. with no resistance.*

BSC on BCS

I showed this description to Chandra in May 2002 when he, coincidentally, was visiting Cleveland from his home in Bavaria. He kindly let me know that my description was somewhat naive and suggested that I refer to his splendid little book, "Why Things Are the Way They Are" (Cambridge University Press 1998). There, in Chapter XII, the Bardeen, Cooper, Schreiffer (BCS) theory is described. Chandra explains that each conduction electron interacts with a second electron which has equal and opposite momentum (i.e. at the opposite extreme of the Fermi surface) by exchanging a phonon. (The picture of particles interacting via the exchange of a boson has been key to understanding electromagnetism and particle physics.) The phonons are quanta of energy associated with the mechanical vibrations in the crystal lattice. The phonon exchange neutralizes the strong electric repulsion between the two electrons. Because the energy of the paired electrons is lower than that of unpaired electrons, they bind together. Each "Cooper pair", having zero net momentum, is coupled to all the other Cooper pairs and when a superconducting current is flowing in the metal, they all move together with the same velocity. They cannot lose energy by interacting with the lattice, and thus meet no resistance to their flow.

Superconductivity has been known since 1911. The earliest known superconducting materials, called Type I, lose their superconductivity when placed in even a fairly weak magnetic field. Resistance to current flow occurs when the magnetic field penetrates the superconductor, forming discrete bundles of magnetic flux lines. This makes Type 1 materials essentially useless for most applications. Eventually, materials which remain superconducting in very high magnetic fields were found. The work by Chandra and his younger colleagues on these "Type II" superconductors will be described shortly.

Electron tunneling

*Now some remarks about **tunneling:** (remember, we were talking about Chandra's and Adler's work on the passage of electrons through an insulating barrier): If a particle is to enter a region in which its potential energy is increased, it must have enough energy of some kind to do so. A ball rolling inside a bowl will not escape from the bowl unless it has enough kinetic energy to jump over the "potential barrier". However, in quantum mechanics, particles behave as waves, and it is possible for them to be, some of the time, in a region where they do not belong, i.e. where they could never get classically. When an electron arrives at a thin layer of material which would be a barrier according to classical physics, its wave function can penetrate the barrier and there is a finite probability that it can "tunnel" through the barrier – essentially violating conservation of energy for a very short time. This effect is the key to the functioning of many semi-conductor devices. (The same process allows alpha particles to escape from radioactive nuclei.)*

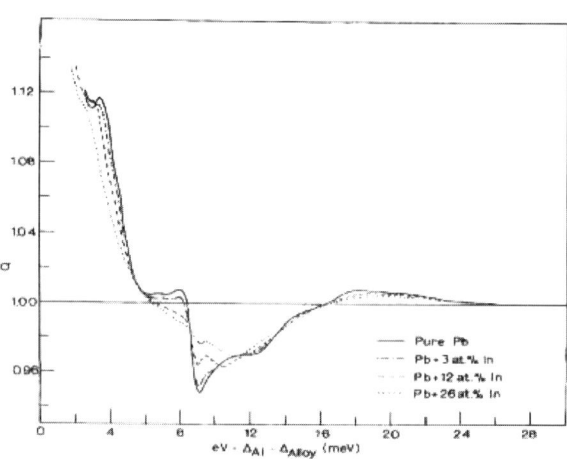

Fig. 15-5. Conductivity vs. applied voltage for Pb/In alloys.

Back now to Adler's and Chandra's experiments. As the potential difference between the two sides of the metal/insulator/superconductor sandwich is increased, the rate of flow of electrons through it shows a lot of interesting structure. These rapid variations in the conductivity through the sandwich provide clues to the interaction among the electrons, the phonons, and the crystal lattice. At Alberta, Adler had looked at a sandwich of aluminum oxide, superconducting indium, and aluminum. (*Phys. Rev. Lett.* **10** 217 1963)

In a *Physical Review Letter*, Adler and Chandra showed evidence for sharp changes in the conductivity through sandwiches in which the superconducting layer was pure lead, or lead alloyed with various concentrations of indium. **Fig. 15-5** shows the conductivity versus the energy difference (proportional to the applied voltage) across the sandwich for four indium concentrations. Note how, at certain energies, the electrons find it more difficult to pass through the sandwich. This indicates their increased interaction with the phonon population. "Effect of Alloying on the Phonon Spectrum of Lead-Indium Alloys" (*Phys. Rev. Lett.* **16** 53 1966)

Adler continued this work with a more comprehensive survey of electron tunneling through other superconducting layers, mainly lead with varying concentrations of bismuth, tellurium, or indium. The main import of these studies was to establish to what extent the electron-phonon interaction could be elucidated in tunneling experiments. "Electron and phonon effects in superconducting fcc lead-based alloys", *Phys. Lett.* **24A** 407 1967 and "System for observing small nonlinearities in tunnel junctions", *Rev. Sci. Instr.* **37** 1049 1966. Adler left CWRU in 1968 to accept a position as associate professor at the University of Alberta.

In 1969, Chandra wrote the first chapter of a book on superconductivity. (*Superconductivity*, editor R.D. Parks, published by Marcel Dekker, New York) Chandra's 49-page introduction to the subject provided a clear and comprehensive review of the early experiments and theories leading up to the BCS theory. In the last chapter, Philip Anderson looked into the future, and in-between 29 other authors described how BCS had, in the 12 years since it appeared, explained practically everything about superconductivity. The two volumes were a classic in the field for many years.

Chandra's research on electronic structure and tunneling provided excellent opportunities for his graduate students. In an experiment, similar to that done with Adler, grad student Alan Geiger looked at a sandwich of aluminum and lead separated by a thin layer of aluminum oxide. As the voltage across the interface was varied, well-defined peaks in d^2I/dV^2 (sort of a derivative of the conductivity) were observed. **Fig. 15-6.** It was deduced that this structure is caused by the excitation of vibrational states in impurities in the aluminum oxide. Studies of this type are important to the technologies based on such junctions. ("Inelastic electron tunneling in Al-Al oxide-metal systems" Phys. Rev. **188**, 1130 1969.) Adler, then at Alberta, was a co-author.

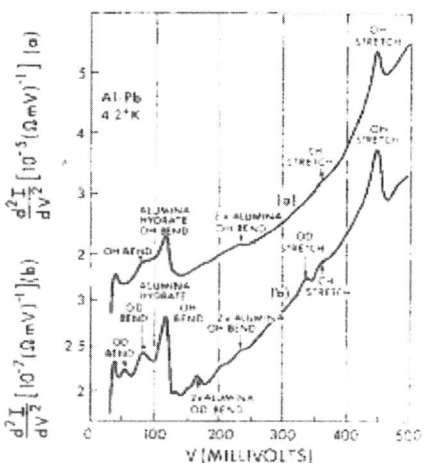

Fig. 15-6. Structure in conductivity through thin aluminum oxide layer.

Three other students, Jerome Jackson, Brij Raj Sood, and Ashok Gupta, investigated properties of superconducting lead-indium alloys in similar experiments. The goal was to understand the effects of alloying on the phonon spectrum (essentially the vibrations of the lattice). Measurements were made for various concentrations of indium, up to eight percent. The results showed again that the conductivity through the barrier has interesting structure as a function of applied voltage and that this is related to excitations of impurities – in this case the indium. Similar measurements were made in other materials, e.g. molybdenum and bismuth.

In a later project, Chandra returned to the subject of magnetostriction. With grad-student Murray Finkelstein, he did DHVA measurements on a sample of tin in a very high (55 kilogauss) magnetic field. The goal was to determine the response of the Fermi surface to the deformation of the lattice caused by magnetostriction. This was followed by a long series of similar studies on other materials: Be, Bi, Cu, Pb, Te. Much of the analysis was done on the department's PDP-9 computer.

In 1971, Chandra co-authored with E. Fawcett of the University of Toronto an extensive review of magnetostriction in metals and how it can be used to deduce the Fermi surfaces of metals and semiconductors. *Advances in Phys.* **20** 775 1971.

Research on superconductivity and other low temperature phenomena expanded as Chandra was joined by two new experimentalists, David Farrell and Arnold Dahm. His work with them will be described on the following pages.

David Farrell joins Chandra

In 1964, Chandra invited 25-year-old **David E. Farrell** to join his research program as a post-doc. Farrell had completed his doctorate at Imperial College London that same year, working with J. G. Park and B. R. Coles. His dissertation explored what was then a rather esoteric (and neglected) aspect of superconductors, namely their anisotropy. He did this indirectly, by measuring how the transition temperature (T_c) of a particular set of dilute alloys depended on the average number of valence electrons per atom.. "Effects of Electron Concentration and Mean Free Path on the Superconducting Transition Temperatures of Zinc Alloys" *Phys. Rev. Lett.* **13** 328 1964. Anisotropy has remained a central theme of Farrell's research, both because of its fundamental interest, and because its understanding and control later turned out to be critical to the realization of "high-T_c" superconductivity.

Fig. 15-7. David Farrell.

At WRU, Farrell was promoted to assistant professor in 1968. **Fig. 15-7.** At CWRU, over the next four decades, he has pursued an extensive research program probing the issue of superconducting anisotropy and related issues. In addition, he has pioneered the use of superconductivity in medicine, including, in the late 1990's, the use of magnetic diagnostic devices based on the newly-discovered high-T_c superconductors.

Type II Superconductors

An aside on Type II superconductors: According to the now well-verified theory of Nobelist (2003) Alexei Abrikosov, in the presence of a magnetic field the fields and currents in Type II superconductors are fundamentally inhomogenous. The magnetic flux is not distributed uniformly but rather is arranged in a lattice made up of bundles of magnetic flux lines. Because electric currents circle the flux bundles, they are referred to as "vortices." As the field is increased, the density of vortices increases until they over-

lap. *This occurs at the "upper critical field, H_{c2}". At this point, a phase transition occurs, the entire sample reverting to the normal state.*

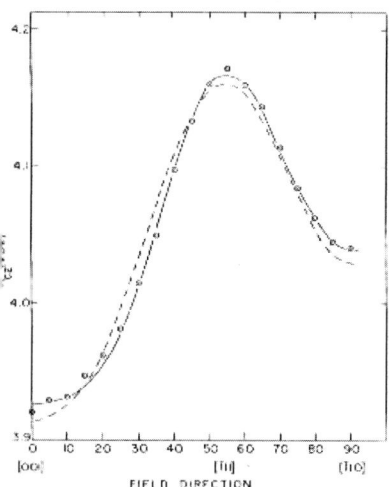

Fig. 15-8. Critical field vs. field direction relative to crystal axis.

Exploring the implications of Abrikosov's theory, Farrell and Chandra measured the value of the critical field for niobium as a function of the angle between the applied field and the crystallographic axes, focusing on its anisotropy. How does H_{c2} depend on the field direction relative to the crystal's axes? (Abrikosov's theory in its simplest form predicts that there should be no anisotropy.) To determine H_{c2}, they placed the sample in an adjustable magnetic field. Adjacent to the sample were two vibrating coils which measured its magnetization. The external field was increased until the magnetic moment of the crystal niobium disk vanished. A typical plot of critical field versus direction of the applied field relative to the lattice is shown in **Fig. 15-8**. They found that the value of H_{c2} changes by as much as 10% as the field direction is changed, a fact that required "non-local" corrections to the Abrikosov formalism. "Precision Measurement of Anisotropy in the Upper Critical Field of Superconducting Niobium" (*Phys. Rev.* **176** 562 1968).

Landau domains: moving islands of superconductivity

By the early seventies, Abrikosov's flux lines had been observed directly in Type II superconductors, and it was widely accepted that resistance in these technologically important materials was due to the motion of flux lines. However, such motion had never been directly observed. In 1972 Farrell turned his attention to the non-quantized flux inhomogeneities called "domains" that are realized in Type I superconductors. Compared with the microscopic flux vortices which form in Type II materials, domains can be of macroscopic (~1mm) size. Farrell asked if it might be possible to produce regular motion and observe it directly.

When a magnetic field is applied to a thin sample of Type I superconducting material, the domains consist of regions in which the sample is either completely superconducting or completely normal. As the field is increased, the superconducting regions shrink and the normal regions grow until the entire sample is normal. Generally, the domain structure is irregular, but, by exploiting geometrical anisotropy, a highly regular array of alternating superconducting and normal domains can be formed. Using a simple point-contact resistance measurement, Farrell was able to show that, in the presence of an electrical current, these macroscopic domains retain their size and shape and move with constant speed across the sample, new domains being created at one edge and disappearing at the other. This work helped to establish the now-accepted physical picture in which *all* resistance in superconductors is ascribed to the motion of flux inhomogeneities. "Experimental Realization of Highly Regular Motion of the Landau Domain Structure"

(*Phys. Rev. Lett.* **28** 154 1972.) "Highly Regular Domain Motion in the Dynamic Intermediate State of Superconducting Tin" (*Phys. Rev.* **B5** 3523 1972).

Passing an electrical current along the sample is not the only way that flux inhomogeneities can be driven into continuous motion. Motion can also be produced by applying a temperature gradient. In the early seventies there were theories for this effect but they lacked any experimental verification. Farrell provided this using the same detection method as he had used previously for electrically driven motion. The abstract of the letter written with grad student, Ashok K. Gupta, summarizes the work: *"Direct evidence is presented for thermally induced flux motion in a superconductor. The method consists of setting up the macroscopic Landau domain structure in a thin single crystal of tin and imposing a temperature gradient along the sample. Motion is monitored with the Sharvin point-contact technique, enabling one to deduce an experimental value for the actual velocity of this motion."*

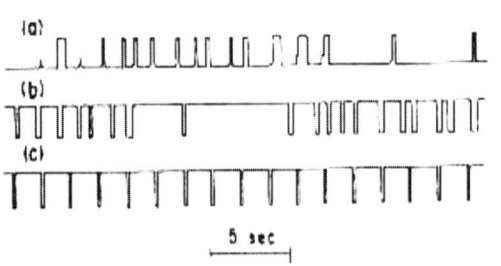

FIG. 4. Sample 1, $T = 2.96$ K, $\beta = 17°$. (a) $I = 4.5$ A, $C_N \sim 0.1$; (b) $I = 2.0$ A, $C_N = 0.80$; (c) $I = 3.0$ A, $C_N = 0.90$.

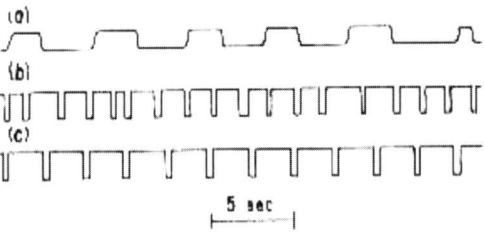

FIG. 5. Sample 2, dimensions $21 \times 11 \times 0.5$ mm, $T = 2.52$ K, $\beta = 8°$; (a) $I = 3.5$ A, $C_N = 0.36$; (b) $I = 2.3$ A, $C_N = 0.73$; (c) $I = 1.5$ A, $C_N = 0.86$.

Fig. 15-9. Resistance between two fixed contacts vs. time as domains are driven across the samples.

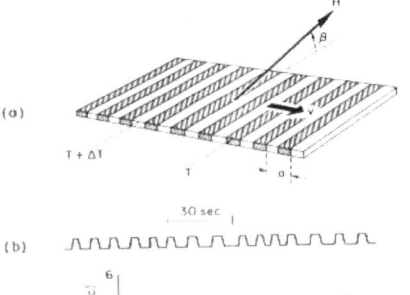

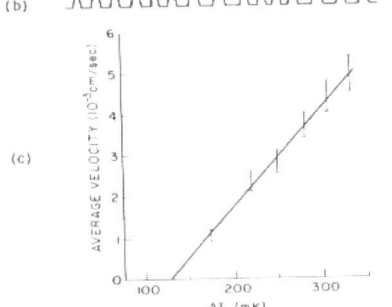

Fig. 15-10. Top: schematic of moving domains; bottom: domain speed vs. temperature gradient.

In the experiment, a single crystal of very pure tin 40 mm long was placed in a magnetic field directed at a small angle relative to the surface of the crystal. This set up the Landau structure. A thermal gradient of a few hundred mK across 1.5 cm was applied, and a fine copper wire fixed to the center of the crystal monitored the contact resistance. The resistance rises and falls as regions of superconducting and normal tin move down the crystal, driven by the temperature gradient. **Fig. 15-9** shows the signal produced in the single probe for various values of the "driving" current and of the fraction of normal material (C_n). The top part of **Fig. 15-10** shows the experimental arrangement and the plot below it shows a clearly linear dependence of the velocity of the domains on the temperature gradient. "Thermally Induced Flux Motion in Superconducting Tin" (*Phys. Rev.* **B7** 3037 1973.)

Farrell subsequently published three related papers on Landau domain structure. "Landau domain structure. I. Theory" (*Phys. Rev.* **B9** 2894 1974.) "Landau domain structure. II. Experiment". (*Phys. Rev.* **B9** 2902 1974.) "Landau Domain Structure. III. Normal Domain Broadening" (*Phys. Rev.* **B11** 4344 1975.) This work brought Farrell international recognition and led to semester-long Visiting Professorships at both Oxford and Cambridge as well as at the Institute for Physical Problems in Moscow. The latter has been home for many leaders in condensed matter physics including Kapitza, Landau, and Landau's best-known student, Abrikosov.

During a stay at Argonne National Laboratory in Illinois, Farrell continued his study of Landau magnetic domains in superconducting tin, using an optical method. In this case, the domains were observed by reflecting light off a thin magneto-optically-active film which was in contact with the surface of the metal crystal. The sample was maintained at 1.6 K. The plane of polarization of the light is rotated upon reflection and the amount of rotation depends on the degree of magnetization of the crystal just below the film. When a magnetic field is applied, perpendicular to the face being observed, Landau domains are produced. The sizes of the domains depend on the strength of the applied field and its angle relative to the normal to the surface. The rather remarkable photographs in **Fig. 15-11** illustrate that one can produce ever larger domains by changing the strength (or angle) of the applied field. "A New Type of Domain Structure in the Superconducting Intermediate State" *Phys. Stat. Solidi* **20** 419 1973.

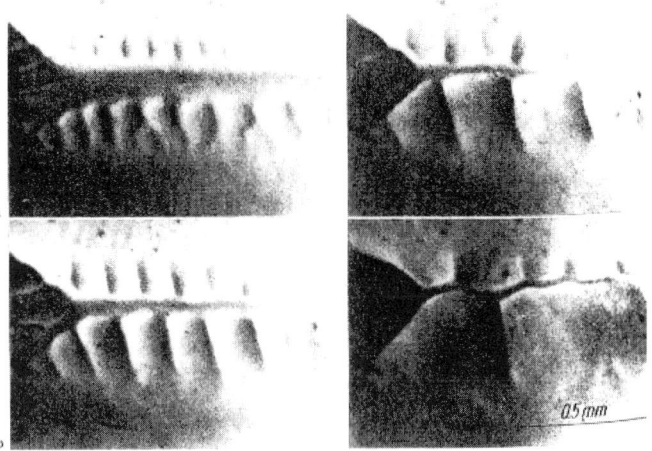

Fig. 15-11. "Photographs" of magnetic domains showing how they grow as the applied field is increased.

Defects in superconductors

Farrell and Chandrasekhar collaborated in the development of an anisotropy-related idea to explain the lowering of T_c of certain superconducting materials produced by the introduction of defects. From their abstract: "A simple explanation is suggested for the recently discovered universal behavior of defected A-15 superconductors. Qualitatively and quantitatively, this behavior can be understood as being the direct result of an extremely large anisotropy of the superconducting energy gap." They showed that for a variety of substances, the fractional decrease in the temperature at which superconductivity occurs is a universal function of the density of defects. The defects were caused by bombardment of the sample by ^{4}He (alpha) particles. **Fig. 15-12** shows the experimental situation, where the change in T_c is plotted for four different materials as a function of the amount of exposure to 2 MeV alpha particles. An intriguing comment is made at the end of the paper to the effect that if *intrinsic* defects could be reduced, some materials might be made to become superconducting at temperatures as high as 35 K, well above any T_c's known at the time. "Defect State in A-15 Superconductors" *Phys. Rev. Lett.* **38** 788 1977.

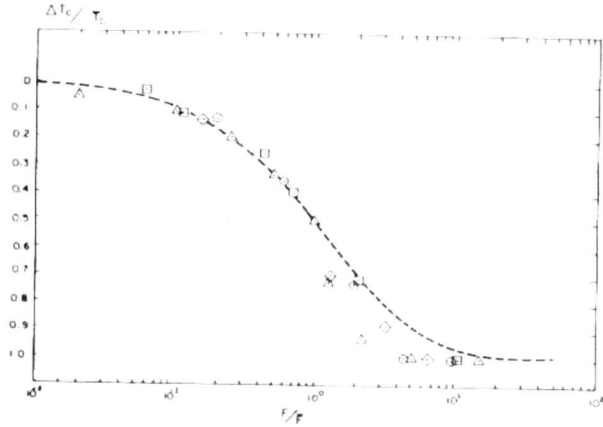

Fig. 15-12. Fractional change in T_c vs. density of defects for four materials.

During the same year, Farrell and his colleague, John McGervey, collaborated on a positron annihilation experiment on one of these A-15 superconducting materials. Specifically, they studied V_3Si in an effort to examine further the consequences of defects. This experiment was described in Chapter 14. A few years later, Chandra and Farrell joined colleagues at nearby NASA Lewis in studying superconductivity in the same material. This time the sample was exposed to a beam of 35 MeV protons from the NASA synchrotron. Among the NASA participants was Chandra's former student, Edward Haugland. As the amount of irradiation increases, defects are produced, the resistivity goes up, and the critical temperature goes down. This type of data helps in the determination of such parameters as the electron-phonon coupling and band structure. "Superconductivity of proton-irradiated V_3Si" *Phys. Rev.* **B24** 90 1981.

Superconductivity in medical diagnostics

In the late 1970's, Farrell was approached by John W. Harris and Gary M. Brittenham of the Cleveland Metropolitan General Hospital. These physician researchers were interested in the non-invasive determination of the iron content in the human liver. Such information provides an important diagnostic tool in cases of systemic iron overload caused by frequent transfusions or certain genetic abnormalities.

Iron in the body is predominantly stored in molecules which are paramagnetic. Paramagnetic materials, when placed in an external magnetic field, become slightly magnetized in the *same* direction as the applied field. Normal livers are slightly diamagnetic, that is, they become slightly magnetized in the direction *opposite* to the applied field. The net induced field therefore is the sum of two comparable, but opposing, contributions, and the presence of an iron-overload can be deduced by comparison of the measured susceptibility with that of a normal liver.

What was needed was a technique for the precise measurement of the induced magnetism in actual patients. This is what Farrell and his group provided by developing a novel way to measure the iron content *in situ*. Superconducting coils inside a liquid helium filled Dewar are located above the patient to provide a constant magnetic field. This field causes a tiny magnetization of the paramagnetic molecules in the liver. The patient is then moved away from the coils, taking his magnetic liver with him. The region formerly occupied by liver is then occupied by water. **Fig. 15-13** shows how an expanding water-filled bellows follows the torso as it is lowered. The resulting minute change in magnetic field induces a current in a second set of superconducting coils in the Dewar. The detector coil current is then monitored by a SQUID (Superconducting Quantum Interference Device). "Magnetic measurement of human iron stores" *IEEE Trans. Magnetics Mag.* **16** 818 1980. "Magnetic susceptibility measurements of human iron stores", *New Eng. J. Med.* **307** 1671 1982.

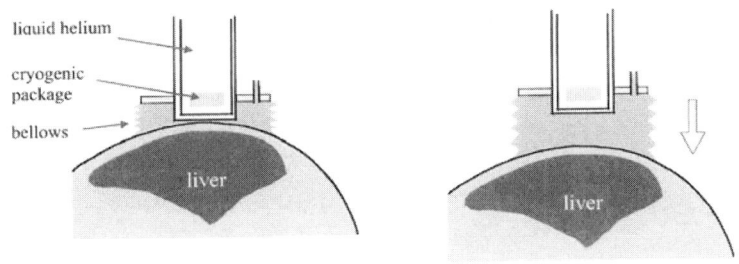

Fig. 15-13. Measuring iron content in the liver.

The feasibility of this "biological susceptometry" was clearly demonstrated by systematic measurements on large numbers of patients with normal and abnormal iron loading. Comparisons with biopsy measurements showed that the technique is reproducible and sensitive enough to provide clinically useful results. As explained in the next section, Farrell and his engineering and medical colleagues are still improving and extending this powerful technique.

In 1982, Farrell hired a brilliant young engineer, Chris Allen, who had just completed his BS in biomedical engineering at Case. Allen began as a junior engineer in the CWRU physics department, working at Metro General. He relocated to the main campus in 1998. In a continuously funded series of NIH projects, Allen has been responsible for developing the myriad array of technical procedures that are needed to transform sensitive low-temperature research instrumentation into useful clinical devices.

By the mid-1980's a commercial version of Farrell's SQUID-based superconducting magnetic susceptometer was in use at a number of diagnostic centers around the world. However, its cost, and the need to use liquid helium, put the instrumentation beyond the reach of most clinicians. Then, in 1986, the superconductivity landscape would change dramatically.

High-T_c superconductors – properties

At the 1987 March meeting of the APS in New York City, details of the discovery of superconductivity in several exotic materials were reported. Transition temperatures as high as 90K had been observed and confirmed. Almost overnight, a whole gamut of new superconductivity applications became possible - in principle, at least. The meeting has been described as the "Woodstock of Physics". As can be imagined, it changed the research careers of many of its attendees, including Farrell's and Chandra's.

The new materials turned out to much more anisotropic than their low-T_c counterparts and it took nearly a decade to understand and control this property and to fabricate flexible conductors. Farrell, who had explored the effect of anisotropy in low-T_c materials in his thesis work, was well positioned to contribute to these fundamental studies. One of the principle difficulties that faced researchers was that *single* crystals were needed to study anisotropy, while growing a crystal greater than a few microns in size was impossible. Within a few months, Farrell and Chandra, together with colleagues from the CWRU engineering school and NASA, published a paper which gave a practical solution to the problem. Tiny (2 to 10 micron) crystallites of yttrium barium copper oxide (nicknamed YBCO, or more precisely $Y_1Ba_2Cu_3O_{7-\delta}$) were set in liquid epoxy which was then cured in a strong magnetic field at room temperature. The field oriented the crystallites so that, magnetically at least, the sample acted like one large single crystal. This was a significant breakthrough and "Superconducting Properties of Aligned Crystalline Grains of $Y_1Ba_2Cu_3O_{7-\delta}$" (*Phys. Rev.* **B36** 4025 1987) became Farrell's most-referenced paper.

This work initiated a decade of basic research into "high-T_c" superconductors by Farrell and his co-workers at NASA-Lewis, Ames Laboratory (University of Iowa), and the University of Illinois. They looked at anisotropies in several high-T_c materials: *e.g.* copper oxides containing yttrium or lanthanum or thallium (referred to generically as "cuprates"). Using epoxy embedding and SQUID technology, they measured "hysteresis curves", i.e. the tendency of the samples to retain their magnetism. This was done as a function of temperature and orientation of the applied field relative to the crystal planes. "Critical current anisotropy in high-T_c superconductors". *Phys. Rev.* **B39** 718 1989.

Farrell's most important contribution to the investigation of superconducting anisotropy was to devise an accurate method for measuring it. The technique, known as "torque magnetometry", involves suspending the sample by a thin tungsten fiber and using a *null* technique to measure the torque it experiences in a uniform field. (A small coil near the top of the torsion apparatus provided counter torque so that the angular displacement of the fiber was zero at all times.) In a typical experiment, the torque on a single crystal of YBCO (50 x 50 x 250 microns) was measured as a function of the angle θ between the applied one-Tesla field and the normal to the planes of the crystal. If the material had no anisotropy, the peak of the torque curve would occur at $\theta=45$ degrees. With increasing anisotropy, the peak moves to higher angles and the anisotropy itself can be deduced directly from the position of the peak and appropriate theory. Farrell made a systematic study of several high-T_c materials, some of which had torque curves that peaked within a few tenths of a degree of 90 degrees! These measurements provided a sound fundamental basis for understanding anisotropy and its impact on a wide range of properties in the superconducting state. "Torque magnetometry: A new probe of dimensionality in high-T_c superconductors" *Phil. Mag.* **B65** 1373 1992.

Throughout this period, a succession of undergraduate physics majors enjoyed the opportunity to become involved in "real" physics research in Farrell's lab. The students were full partners in the work on anisotropy in high-T_c superconductors and were co-authors on many of the papers published by the group.

High-T_c Superconductors to the rescue

By the end of the 1990's, control of the anisotropy of high-T_c materials had advanced to the point that it was finally possible to produce flexible lengths of high-T_c conductors that carried significant electrical currents at liquid *nitrogen* temperatures (80 K). Consequently, the magnetic susceptibility measurements described above can be done without cooling the coils all the way down to the 4K liquid *helium* temperature. The principle goal of Farrell's research since 2000 has been to bring magnetic susceptometry into regular clinical practice by exploiting the new materials. His team was awarded a 5-year NIH contract to design and build a high-T_c instrument that would be an order of magnitude less expensive, simpler to use, and more accurate than the early low-T_c device. In 2004 final technical tests were conducted prior to the installation of the instrument at Columbia University for clinical tests.

David Farrell points out that there are other exciting prospects for high-T_c based susceptometry in medicine, beyond the detection of iron in the liver. Because most biological tissue is diamagnetic with susceptibility proportional to its density, the technique may eventually be suitable for the non-invasive detection of a variety of abnormalities. The "basic" research undertaken by Chandra and Farrell to understand the fundamentals of superconductivity, whether of the low-T_c or the high-T_c variety, has been essential to the exploitation of these materials in a wide range of technologies.

Fig. 15-14. Arnold Dahm.

Dahm: cold helium

A new member of the "low temperature" research group joined the CWRU department in 1968. Thirty-six year old **Arnold J. Dahm (Fig. 15-14)** had completed his doctorate at the University of Minnesota three years earlier. His thesis title was "Effective Mass of Ions in Liquid Helium II"; it was completed under advisor Michael T. Sanders. Dahm spent two years as a post-doc with Donald Langenberg at the University of Pennsylvania, where he worked on the Josephson effect. (This concerns the passage of current across an insulating gap between two superconductors. It is the basis for modern techniques which allow the precise determination of magnetic fields.) "Study of the Josephson Plasma Resonance" *Phys. Rev. Lett.* **20** 859 1968. "Linewidth of the Radiation Emitted by a Josephson Junction" *Phys. Rev. Lett.* **22** 1416 1969.

Solid helium

Dahm has told me that he chose to join the newly formed CWRU department because of the opportunity to work with Chandrasekhar. He began with the measurement of the electrical properties of solid helium. ^{4}He solidifies only at pressures above 25 atmospheres. Dahm was interested in how electric charge is transported in a uniform crystal of solid helium. The experiment was quite straightforward in concept but technically sophisticated. Crystals of helium were grown in a gap (from one to a few millimeters across) between two electrodes, and the current across the diode was measured as a function of the applied voltage. A radioactive source of electrons produces both positive and negative "ions" which travel through the crystal to the oppositely charged electrode. Measurements were made at temperatures from just below the melting point, around 2.5 K down to 1.1 K, and with voltages up to 2000 V. It was found that the current depends linearly on the square of the applied voltage. The mobilities (μ) of the charge carriers were then determined from the slope of the $I^{1/2}$ versus V plots. (μ = a constant times I/V^2 where the constant contains only the electrode separation and area.) The measured mobilities, μ, shown in **Fig. 15-15** as a function of T^{-1} rise precipitously as the melting point is approached. The goal of the measurement was to identify these charge carriers. The negative ions were found to have mobilities 2 or 3 orders of magnitude greater than those for positive ions. "Ionic Mobilities in Solid Helium" *Phys. Rev. Lett.* **28** 1244 1972.

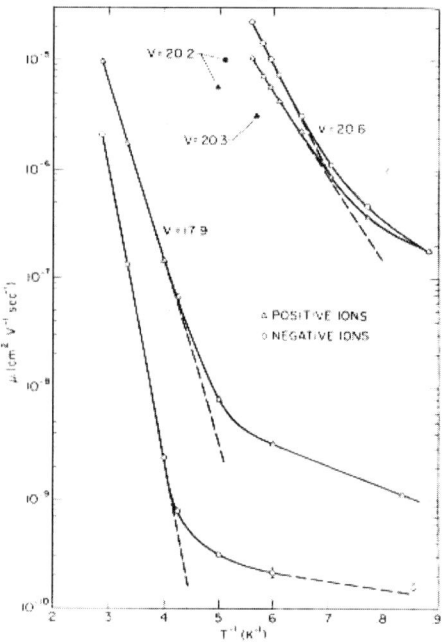

Fig. 15-15. Mobilities of ions moving in solid helium vs. reciprocal temperature.

Dahm reported on the results of this work in a paper given a few years later in Ukraine. The introduction to that paper describes the lattice of solid helium and the motion of ions through it. These are not the chemist's "charged-atom" ions. The negative ions are electrons in a void and the positive ions are regions of enhanced density surrounding a hole (i.e. a missing electron). From the paper: "Helium crystals differ from most other crystals in that the atoms execute large oscillations about their equilibrium positions. This large rms displacement, which is as large as one-fourth of the lattice spacing .... is a direct result of the weakness of the attractive interaction between helium atoms and of the small atomic mass." This situation results in a high rate of atomic exchange and of vacancy hopping. It was hoped that the properties of this rather unique material could shed some light on charge transport in more complex solids. "Ion Motion in Solid Helium" *Fisika Nizkik Temperatur* (Ukr. Acad. of Sci.) **1** 593 1975. The transport of charge in a variety of materials and by an assortment of carriers has remained at the center of Dahm's research for many years.

Dahm, Chandra and liquid helium

Dahm then joined Chandrasekhar in a study of *superfluid* helium. Helium atoms at low temperatures behave similarly to the electrons in a superconductor in that they also undergo Bose condensation. They all crowd into the lowest energy level, and are described by a single wave function. Both phenomena display macroscopic quantum effects. Just as superconducting domains grow with decreasing magnetic field, the superfluid fraction in a sample of liquid helium grows with decreasing temperature. Just as the Cooper electron pairs flow without resistance through the crystal lattice, the zero-spin helium atoms flow through the tiniest orifices without viscosity.

The helium superfluid ground state can have excitations in the form of quantized vortices in the liquid. They are analogous to the angular momentum states in the Bohr atom. They can be produced by the introduction of energy into the system. Chandra had worked on helium superfluidity at Oxford and had published work on the production of vortices by acoustic transducers. In the work done with Dahm, the vortex rings were produced by accelerating ions from a radioactive source across a fixed voltage within the fluid. The radius of each resulting charged vortex ring is proportional to its energy, and thus to the accelerating voltage. The vortices pass through the accelerating grid, travel through the remaining liquid, and impinge on an orifice connecting the helium chamber with a second chamber. A schematic of the arrangement is shown in **Fig. 15-16**. The level of helium in the second chamber is observed to rise an amount proportional to the rate at which the vortex rings arrive at and overlap the orifice.

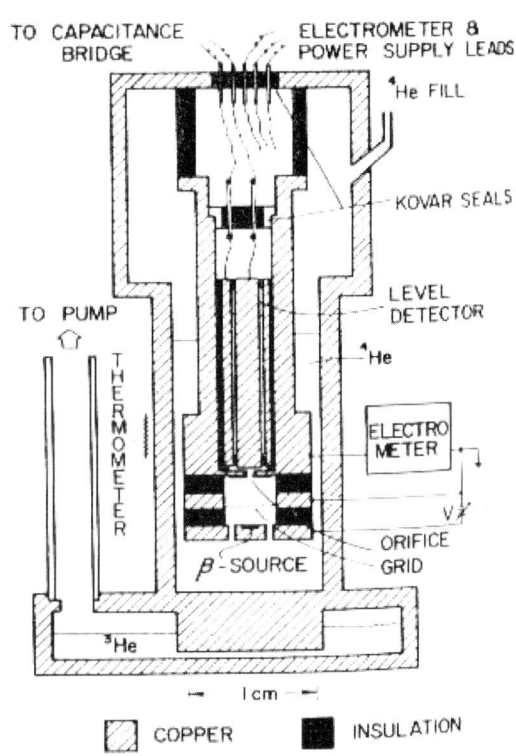

Fig. 15-16. Apparatus to measure flow of charged vortex rings in liquid helium.

A simple model predicts that the level increase per vortex current (for example, microns per pico-amp) should rise quadratically with the accelerating voltage. *(More precisely, the theory states that the phase of the superfluid helium's wavefunction is related to the chemical potential of the bath. When a vortex ring overlaps an orifice connecting two different baths, the relative phase between the two baths changes, along with the chemical potential, and a difference in height is established.)* The experimental data agree qualitatively with the model, especially in that the onset of the level-rise occurs when the vortex ring radius gets to be half the orifice radius, as predicted. "Vortex-ring-Generated Level Differences in Liquid Helium" *Phys. Rev. Lett.* **31** 873 1973.

Electrons, ions, vortices, voids and bizarre molecules in liquid helium

Dahm joined colleague **Jonathan Reichert** to look at electrons trapped in voids in liquid helium. They injected electrons into the liquid by field-emission from iron or tungsten tips. A sketch of the setup is shown in **Fig. 15-17**. The sample was placed in a resonant cavity and subjected to electromagnetic radiation at 13.56 GHz in the presence of a magnetic field. (Reichert had been working in electron spin resonance to study electrons in crystals. Chapter 14.) The field was swept through resonance, that is, through the value at which the difference between the energy of a spin-up electron and a spin-down electron matched the energy of the 13.56 GHz radiation. The observed absorption of power signaled the flipping of the electron spin. "Observation of Electron Spin Resonance of Negative Ions in Liquid Helium" *Phys. Rev. Lett.* **32** 271 1974.

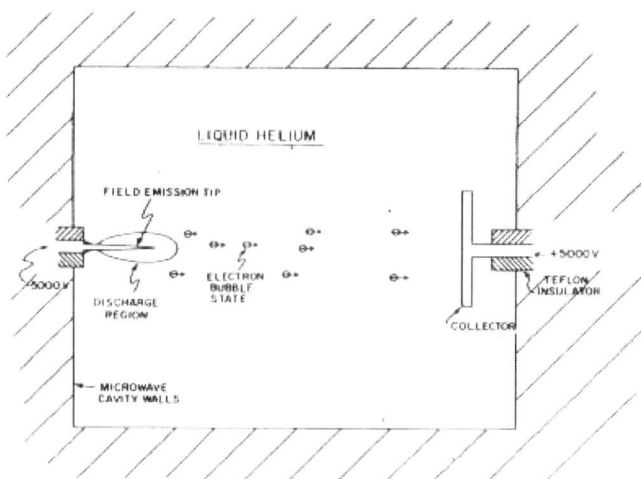

Fig. 15-17. Experiment to observe electron spin resonance of ions in liquid helium.

The continuation of this work was published three years later. By this time, co-author Reichert had moved on to SUNY at Buffalo. The authors describe the unique system of an electron trapped inside a void in liquid helium: "It is now well established that such electrons form a stable microscopically large configuration consisting of a single electron trapped inside a spherical void in the liquid of about 17 Å radius." "In this paper, we shall present the results of an extensive study of this unique system. The negative ion in helium is one of the simplest bound electronic systems in nature, possibly the simplest. That is, of all bound electronic configurations in condensed matter, the electron in the helium void would be expected to have the weakest interactions with its surroundings. The measured magnetic-resonance parameters can be used to verify this theoretical expectation. This paper reports the measurements of the linewidth, the electron g-value, and signal intensity as a function of pressure and temperature."

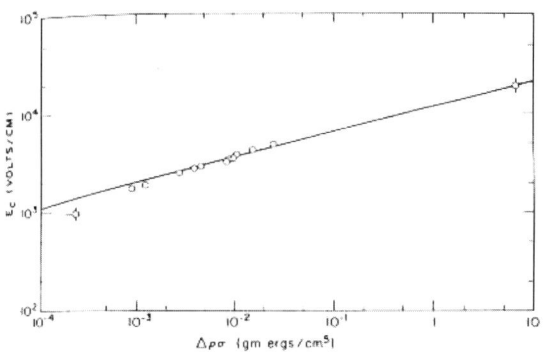

Fig. 15-18. Electric field required to pull charged droplets out of liquid helium surface.

The measured g-value was the same, within errors, as that of the free electron, indicating the weakness of the coupling of these electrons with their environment. *(The g-value relates the magnetic moment of the electron with its spin, and the energy needed to flip the spin is proportional to the magnetic moment.)* "Study of the electron spin resonance of negative ions field emitted into liquid helium" *Phys. Rev.* **B15** 2630 1977.

In an experiment related to the measurement of the mobility of ions through liquid and solid helium, Dahm investigated how charge can be pulled through the *surface* of a liquid into the vapor above. As in the earlier experiments, ions are produced in the liquid by a radioactive source and collected by the plates of a diode, but in this case, one plate lies parallel to and above the surface, the other lies within the liquid. The main experimental result was that the onset of the flow of charged droplets occurred at a critical applied electric field, $E_c = (12.3 \pm 0.8)(\Delta\rho\,\sigma)^{0.28\pm0.03}$ kilovolts per centimeter, where $\Delta\rho$ is the difference in density between liquid and gas and σ is the surface tension. A theoretical derivation of this result is presented. **Fig. 15-18.** When the electric field was configured so as to concentrate the charge in a limited area on the surface, it was found that a portion of the surface would be pulled upward by the field so that bursts of charged droplets would escape; the surface would then fall back to its original level. The resulting oscillatory signal is shown in **Fig. 15-19**. Similar results were found with both liquid helium and nitrogen. This implies that the underlying physics was *classical* mechanics, and not quantum physics, i.e. it has nothing to do with superfluidity. "Extraction of Charged Droplets from Charged Surfaces of Liquid Dielectrics" *J. Low Temp. Phys.* **23** 477 1975.

Dahm continued the study of ion mobility in liquid helium by looking at the way that these ions interact with the quantized vortices which can form in a mixture of ^{3}He and ^{4}He. In a theoretical paper, he and post-doc Whittack Huang proposed a model for the way in which positive and negative ions travel along vortex lines in rotating liquid helium. "In this paper we calculate the ^{3}He atomic scattering contribution to the positive-ion drag and discuss the qualitative difference of this contribution to the drag on the negative ion." "Mobilities of Ions Trapped on Vortex Lines in Dilute ^{3}He-^{4}He Solutions" *Phys. Rev. Lett.* **36** 1466 1976.

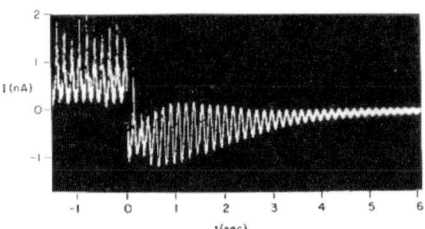

Fig. 15-19. Bouncing of liquid helium surface as bursts of droplets are extracted.

and neutral currents

Apparently, electrons and positive ions are not the only things which can flow through liquid helium. When an alpha source is submerged in liquid helium, or when a high-voltage tungsten-tip is placed in the liquid, some sort of *uncharged* entities are formed which then move slowly through the liquid. "These excitations, which were shown not to be photons, traveled in straight lines for distances of the order of 1 cm at temperatures below 0.45 K, did not respond to electric fields of the order of 10^5 V/cm, and produced He$_2^+$ ions and electrons at the free liquid surface." The experimenters measured the delay, (typically 1 or 2 seconds), between the excitation of the tungsten-tip and the appearance of ions at the surface. They conclude, "Our data suggest that the neutral excitations observed in this experiment are injected into the liquid and drift with the local fluid velocity to the detector, and that the fluid is driven by the electric force acting on the ionic charges." "We conclude that the neutral excitations are metastable $a^3\Sigma_\mu^+$ helium molecules." (Dahm tells me that this is a molecule made of one helium atom in the ground state and another with one electron in an excited state.) "A Study of the Neutral Excitation Current in Liquid ^{4}He Above 1 K" *J. Low Temp. Phys.* **36** 47 1979.

Dahm was the recipient of two Fulbright grants, one to work at the University of Sussex in 1977-78 and a second, at the University of Mainz in 1983-84. These two sabbatical leaves allowed him to explore new areas of low temperature research which he would subsequently pursue in his lab at CWRU.

Positrons in liquid helium

As both Farrell and Chandra had done earlier, Dahm worked with **John McGervey** (Chapter 14) to learn more about voids and bubbles in solid and liquid helium. With grad student James Smith, they would apply McGervey's positron annihilation technique to helium. "We undertook this work to gain a better understanding of the overlap of the positronium wave function with the liquid, the pickoff rate in the liquid, and the bubble model as applied to positronium." If the presence of a positronium atom repels the liquid surrounding it, forming a void or a bubble, the probability that the positron will "pick off" an electron in the liquid will be lessened and the survival time of the positronium will be increased. The use of positrons to sample the electron distribution in matter is discussed in Chapter 14 in the section on McGervey's research.

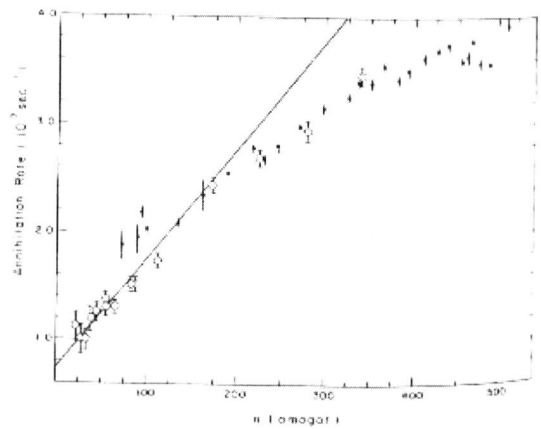

Fig. 15-20. Positron annihilation rate as function of helium gas pressure.

Positrons from the decay of a radioactive ^{22}Na source were introduced into the helium. As described above in the section on Ben Green's work with positrons, this nucleus decays by emitting a positron followed within a few nanoseconds by a γ. The time between the detection of this γ and those from the positron annihilation gives the desired lifetime. The counters could detect only the slow (10^{-7} s) ortho-positronium decay, the one in which three γ's are produced in the annihilation. The measured annihilation rates in *gaseous* helium (i.e. the reciprocals of the mean life), are shown in **Fig. 15-20** as a function of the pressure. The decay rate for the positronium increases linearly as the pressure rises, as expected, but at high pressures (and high densities) it flattens out. The authors are not sure why this happens, but they suggest that at high densities, bubbles might form and the positrons might take refuge in them, away from the electrons. "Orthopositronium decay in gaseous, liquid and solid helium" *Phys. Rev.* **B15** 1378 1977.

Ions moving in solid helium

Dahm continued his work on the motion of ions through *solid* helium. This bizarre material which exists only at very high pressures has a density only about one fifth that of water and is remarkably compressible. In this experiment, with grad student Shing-choy Lau, the mobilities of both positive and negative ions were measured in hexagonal close packed (hcp) crystals. It was found, not surprisingly, that the mobility depended on the direction, relative to the lattice axes, in which the ions are pushed through the crystal by the applied electric field. Some of the measurements were made in samples which had been strained by melting and re-solidifying small portions of the crystal. These crystals were found to permit the flow of positive ions in periodic bursts, an effect thought to be associated with "the simultaneous release of charges from many traps ... with a recharging of the traps as the current decays to steady state." "Motion of charged low-angle grain boundaries in solid helium" *Phys. Rev.* **B23** 1139 1981.

Why would anyone be interested in the motion of electric charge, by whatever means, through a crystal structure? For starters, the whole science of semi-conductors and their applications to electronics depends on just that. In a conductor, every atom contributes one or more electrons which are free to jump from atom to atom with only the slightest encouragement. In a semi-conductor, the electrons need a significant energy boost to get away from their atom. This can come from thermal agitation or from a rearrangement of the energy levels caused by impurities or lattice defects. The simplicity of the solid helium crystal makes it an interesting "laboratory" for the study of charge transport.

Sheets of electrons

In a departure from the study of what happens to ions in bulk liquid or solid helium, Dahm discovered that some very interesting physics can take place on the two-dimensional *surface* of liquid helium. When electrons are placed on the liquid helium surface and held in place by suitable electric fields, they arrange themselves to form a two-dimensional crystal, called a Wigner lattice. It is fascinating to imagine this regular array of free electrons – all pushing on one another, but at the same time being held in place by their attraction to the underlying helium surface. The system is attractive experimentally because it is relatively easy to prepare and manipulate and theoretically because of its "simple interparticle interactions". Dahm became interested in this system of electrons and looked into experiments in which phonons could be used as probes of its properties. Electrons on the surface of liquid helium would be central to Dahm's research for the next two decades.

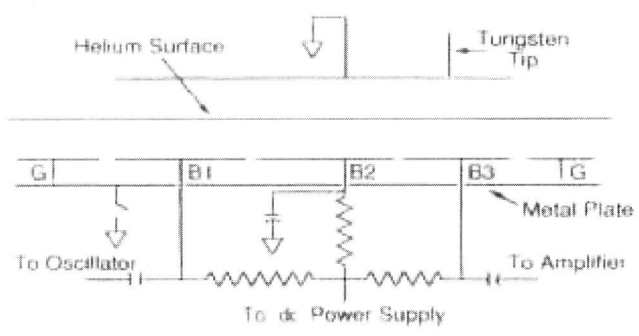

Fig. 15-21. Schematic of cell for the measurement of the motion of an electron crystal on liquid helium surface.

In one set of experiments, Dahm investigated the mobility of the electron crystal as it moves across the liquid helium surface as a function of temperature. It was found that quantized surface waves on the helium, called "ripplons", affect the flow. The higher the temperature, the more ripplons and the lower the mobility of the crystal. A schematic of the experimental cell is shown in **Fig. 15-21.** The electrodes (roughly a square inch each) are separated by a gap of 2 mm. The electron layer lies between the upper and lower electrodes. An ac signal on B_1 disturbs the electrons and that disturbance shows up later at B_3. The time lag provides a measure of the mobility. The temperature was varied from 50 to 500 mK. The data showed that the mobility dropped by a factor of five by about 300 mK, but it then increased slowly as the crystal's melting temperature was approached. Then, just at the melting point, there is a sudden drop in mobility (or conversely a bump in the lattice-ripplon scattering). Dahm, along with post-doc B. M. Guenin and grad-student, Ravi Mehrotra, analyzed this behavior near the melting point and whether it may elucidate just how the lattice melts. "Ripplon-limited mobility of a two-dimensional crystal of electrons; Experiment" *Phys. Rev. Lett.* **48** 641 1982. A later paper reported that the mobility measurements and the dependence of the power absorbed on the temperature and on the driving frequency support the theory of "dislocation-mediated melting". "Evidence in support of dislocation-mediated melting of a two-dimensional electron lattice" *Phys. Rev. Lett.* **51** 1461 1983.

Experiments of this type were continued for several years, and more detailed measurements were made. These included the determination of the effect of the "holding field", i.e. the electric field which was used to hold the electron crystal on the helium surface. This field can create a dimple in the helium surface under each electron, and this dimple can contribute to the "effective mass" of the electron. When the new measurements were taken into consideration, Dahm and grad student Mark Stan concluded that their earlier melting model is insufficient. They proposed that the observed behavior at the melting point may be associated with the existence of a new phase, called hexatic, which may exist between the 2D solid and liquid phases. As the temperature is raised, the sheet of electrons passes through this phase which is characterized by increasing "local disorder" until the random "fluid" phase is reached. "Two dimensional melting: electrons on helium". *Phys. Rev.* **B40** 8995 1989. This intermediate phase would be later observed in a very different 2D system to be described a bit later.

With grad-student Hong Wen Jiang, Dahm looked at the conduction properties of a two-dimensional array of electrons on a helium film supported by a dielectric (glass) substrate. The substrate imperfections pin the lattice and limit the flow. When an alternating voltage was applied to metal electrodes in contact with the electron layer, the electron flow was measured as a function of temperature. As the temperature is raised, there is enough thermal energy to unpin the lattice and the resistance drops very quickly at the "melting point". Again, the two-dimensional system serves as a useful analog to the more complicated 3D melting in which a similar behavior had been observed. "Conduction properties of a new two-dimensional sliding charge-density wave" *Phys. Rev. Lett.* **62** 1396 1989.

To better understand melting in two dimensions, Dahm and grad student Robert Kusner joined J. A. Mann of the chemical engineering faculty in a study of an analogous, but *macroscopic*, system. A 2-dimensional array of 1.6 micron polystyrene spheres was placed between two glass plates separated by about 2.4 microns. When subjected to a high frequency electric field, a dipole electric moment is induced on the spheres so that they interact with one another via a $1/r^3$ dipole potential. Their motion was tracked by a microscopic optical system which recorded the collapse of the lattice as the density of spheres was reduced (through controlled leakage). This effectively increased the average separation, r, thus reducing the strength of the dipole interaction. The system behaved similarly to the 2D *electron* layer discussed above, passing through an intermediate hexatic phase. "Two-stage melting of a two-dimensional crystal with dipole interactions" *Phys. Rev. Lett.* **73** 3113 1994. *Phys. Rev.* **B51** 5746 1995.

The study of the motion of electrons in two dimensions was subsequently extended to a series of experiments in which the electrons move through a very thin layer of specially doped material sandwiched between two forms of gallium arsenide semiconductor electrodes (GaAs and $Al_{0.3}Ga_{0.7}As$). Two-dimensional layers of electrons like this are referred to as an electron gas. Understanding how the electrons move in this situation is key to understanding ultra-small semi-conductor devices in which quantum effects prevail. Measurements were made of the ease with which the electrons moved (more precisely the conductivity) as a function of temperature, electron density, and applied magnetic field. The results could be interpreted in terms of a model involving two regimes of "electron hopping": short hops and long hops. "Screening of the Coulomb interaction in two-dimensional variable range hopping". *Phys. Rev.* **B56** 1161 1997. This work was done with grad student Xue Long Hu, postdoc F. W. van Keuls and Dahm's former student Hong Wen Jiang. Over the course of the following six years, Dahm was joined in a study of the behavior of electrons on a helium surface by a new member of the department, theorist Harsh Mathur, and Mathur's student, Damir Herman. We'll describe this work briefly in Chapter 17.

Electrons to store information

Dahm's work on two-dimensional layers of electrons above liquid helium may very well turn out to be the key to the next generation of computers. Information can be stored as the usual bits of 0's and 1's by manipulating the energies of the individual electrons. For several years now, Dahm has been studying the possibilities that an electron, trapped in an electric potential well above the surface of liquid He, can be placed into either of two energy levels. These two quantum states (qubits) then may represent **0** and **1** respectively. If the qubits survive for a time long enough to be manipulated similarly to the magnetic bits in an electronic computer, they may provide the basis for a super-fast quantum computer. The plan is to place individual electrons, only microns apart, onto a helium surface. Tiny electrodes below the helium hold the electrons in place laterally and voltages applied to the electrodes set the electron energy levels. The electrons will be in either the ground state for this potential (representing a **0**) or the first excited state (representing a **1**). The configuration of electronic states can be prepared by application of bursts of microwave energy and can be read out by detecting electrons in the excited state which are liberated by applying an appropriate electric field.

The thing that makes a quantum computer so extraordinarily powerful is that a whole collection of qubits can team up in a single wave function. It is possible to "prepare" a system of qubits as an admixture of different quantum states. *(The wave function of this so-called "quantum entanglement" cannot be written as a simple product of wave functions for the separate qubits, and therefore convolutes the states of all the electrons.)* When this happens, the qubits can be manipulated in parallel, rather than one at a time. The theory is rather complicated, but the power of the technique can be illustrated by an example: A search program which looks for a particular item hidden among 1000 different items requires, on the average, 1000/2 or 500 operations on a PC, but only the square root of 1000, about 32 operations, on a quantum computer. There are other equally astounding examples.

Arnold Dahm was appointed "Institute Professor of Physics" in 2000. This prestigious title, awarded in recognition of his pioneering work in cutting-edge basic research, has been held by only two other members of the department: Leslie Foldy and Robert Brown. While Dahm chose to move to emeritus status in 2001, his research on electron qubits, amply funded by a grant from the National Science Foundation continues at full speed.

Chandra

Through the 1970's and 1980's, Chandrasekhar participated in a wide range of low temperature physics experiments, including magnetostriction, barrier penetration, superconductivity in Type II and high T_c materials, and liquid helium vortices. Most of this work was done with Farrell or Dahm. He joined colleagues at the University of Illinois in a series of Raman scattering experiments related to phonon density of states and others at the Meissner Institute in Germany on some exotic superconducting materials. Between 1972 and 1976 he was the Dean of the Western Reserve College part of CWRU. After several extended stays at the Meissner (near Munich), Chandra decided to leave CWRU in 1987 and remain in Germany where he would write and reflect.

Chapter 16   Experimental Particle Physics - CWRU

In 1965, after proposals seeking Science Development Program (SDP) grants had been made to the National Science Foundation by both CIT and WRU, there was increased pressure from various quarters (including the NSF) for the merger of the two institutions. Ultimately, the NSF granted $3.5 million *each* to CIT and WRU, essentially contingent on their eventual federation. Of this, a total of $2.6 million was assigned to the two physics departments. It took several years for the combined department to reach equilibrium. There were some bumps in the road. To put it simplistically, these were rooted in the differences between the "pure physics" research, "university" atmosphere, and the BA degree at WRU and the "applied physics" research, "tech school" atmosphere, and the BS degree at CIT. The two departments had a total of 31 faculty in 1963, which grew to 53 by 1966. By 1971, four years after federation, this number had dropped back to 34, following the departure of such established players as Reines, Smith, Kisslinger, Weinberg, Zilsel, Klein, Reichert, as well as the non-reappointment of several young tenure-track assistant professors who had been hired around the time of federation. There were no new hires between 1974 and 1980. *(Thanks to Chandrasekhar for his compilation of these numbers.)*

Jenkins, Blanpied, Frisken, Sullivan: accelerator physics - counters

In the allocation of the SDP funds, it had been agreed that three new experimental, accelerator-based, particle physics groups would be formed. The groups were referred to as the "counter group" (Jenkins, Frisken), the "bubble chamber group" (Keith Robinson, Fickinger), and the "intermediate energy group" (Willard, Bevington).

Fig. 16-1. Tom Jenkins

Thomas Jenkins, whom we met in Chapter 8 when he was working on neutrinos with Fred Reines, put together the "counter group". His picture, taken around 1980, is shown in **Fig. 16-1**. Similar "counter groups" sprang up in the 1950's and 1960's at accelerator laboratories and universities across the world. Geiger counters, wire chambers, Čerenkov counters, scintillation counters, and spark chambers were the building blocks for experiments in which interactions involving specific particles traveling in specific directions were being studied. This technique was in contrast to experiments using bubble-chamber detectors. The latter allowed more of a "fishing expedition", with particles traveling in all directions. The two approaches complemented one another. In some cases, counters looking for certain signals were used to trigger a bubble chamber in "hybrid" experiments.

Jenkins was joined in 1966 by **William R. Frisken** who had been doing counter physics at Brookhaven National Laboratory. He is the fellow on the left in **Fig. 16-2**, squatting in front of an array of wire chambers. Frisken was born in 1933 in Ontario. He earned a BS and MS at Queens University in Kingston, Ontario. He completed his

Fig. 16-2. Bill Frisken (left)

doctorate at the University of Birmingham in 1960, having studied proton-proton and proton-neutron scattering at the new Birmingham synchrotron. "Isotopic Spin Dependence of Nucleon-nucleon Cross Sections between 600 and 1000 MeV" *Nuovo Cim.* **21** 581 1961. He then spent two years at the Radiation Lab of McGill University where he did some nuclear physics, specifically, looking at protons coming from long-lived levels in ^{25}Si produced at the 100 MeV proton accelerator. Frisken moved on to Brookhaven National Lab, where, for two years he worked with a Cornell counter-physics group.

Frisken was appointed associate professor in the Case department in 1966, just before the departure of Fred Reines and federation. He continued to work with his collaborators at Brookhaven on a series of accelerator-based high energy counter experiments. This work led to a series of papers on the elastic scattering by protons of pions, kaons, and antiprotons, beams of which had just become available at the Alternating Gradient Synchrotron (AGS). The measurements established the role of baryon exchange processes, i.e. collisions which are almost head-on in which the particles essentially reverse directions in the center of mass. "Backward Peaks in Elastic π p Scattering" *Phys. Rev. Lett.* **19** 460 1967. "High-Energy π^-p, K$^-$p and pbar-p Elastic Scattering" *Phys. Rev. Lett.* **21** 387 1968.

Jenkins and Frisken, along with two young post-doctoral colleagues, **Alan Strelzoff** (b. 1937; PhD Columbia 1964) and **Charles Sullivan**, (b. 1934, PhD Vanderbilt 1966) and three graduate students, constructed a large array of spark chamber detectors which they installed in the π^- beam at the Zero Gradient Synchrotron (ZGS) at Argonne National Laboratory near Chicago.

A spark chamber consists of a set of parallel metal plates, a few centimeters apart, with a high voltage between them. When a charged particle crosses the gap between the plates, a visible spark appears in the gap. Cameras are triggered by nearby scintillation counters, and the pattern of sparks in the gaps is photographed. The film is then scanned for interesting events either by trained technicians or, as in the case of the Jenkins experiment, by an electronic scanning device controlled by an IBM 1802 computer (This refrigerator-sized computer had 16K of RAM (yes, K not M) and cost about $200K.)

In a run at the Argonne National Lab ZGS accelerator, using a beam of negative pions at 5.9 GeV/c momentum, the Jenkins group measured the differential cross section for charge exchange scattering (π^- p $\rightarrow \pi^o$ n) at large center-of-mass angles. They detected both the neutron (which rescattered in their chambers) and the two photons from the decay of the π^o. For the 5.9 GeV sample, 946,000 photos were scanned for acceptable candidates. Of around 6000 such, 782 passed all cuts. The authors compared the results with π^- p *elastic* scattering. In the forward direction, both channels are dominated by meson exchange, where the target proton emits a virtual meson (a ρ meson in this case) which is struck by the incident pion. In the backward direction, one sees baryon exchange, where the target nucleon takes on most of the momentum of the incident particle. Interference among the various exchange channels produces interesting structure in the angular distributions. The charge-exchange experiment at the ZGS was later extended to lower incident pion momenta, 3.7 and 4.8 GeV/c. "Measurement of the Reaction π^- p $\to \pi^o$ n at Large Momentum Transfers" *Phys. Rev. Lett.* **26** 527 1971; "Measurement of High Momentum Transfer π^- p $\to \pi^o$ n at 5.9 GeV/c" *Phys. Lett.* **51B** 390 1974.

These experiments are typical of the period, as many interactions involving two-body to two-body channels were being described theoretically in terms of scattering by a virtual particle ("one-particle-exchange model").

Fig. 16-3. Bill Blanpied

In anticipation of the generous NSF SDP money, the Case department added **William A. Blanpied** to its particle physics program in 1966. Born in Rochester in 1933, Blanpied received his PhD from Princeton in 1959. His photo is in **Fig. 16-3**. At Princeton he had studied polarization effects in the scattering of protons at 17 MeV. Blanpied was briefly on the faculty at Yale, working with Vernon Hughes on photoproduction experiments at the Cambridge Electron Accelerator. During his first two years at Case, he worked as part of a collaboration with a counter group at Harvard, looking at muon-pairs at the Brookhaven AGS accelerator. "Observation of Muon Pairs Produced by High-energy Negative π Mesons" *Phys. Rev.* **18** 929 1967. *(It later turned out that important advances in particle physics would derive from the study of muon pairs with the Nobel prize winning discovery of charmonium in 1974.)*

In another AGS experiment, (in collaboration with Harvard and McGill), Blanpied and his student Larry Levit looked at the decays of neutral K mesons into three pions. They were especially interested in whether there were any differences between the momentum distributions of the π^+ and π^- in the decay $K_L^o \to \pi^+\pi^-\pi^o$. This was a very hot topic at the time, as the kaon had recently been shown to violate CP conservation in an earlier Brookhaven experiment. *(CP non-conservation is that subtle imbalance in Nature which could account for the fact that there seem to be more protons than antiprotons in the Universe. As a consequence, a few remain after the rest have annihilated, leaving us and something for us to study.)* In the Blanpied *et al.* measurement, the momenta of all three pions were measured and the transverse momentum distributions of the two charged pions were compared. No differences were observed. "Search for a CP-Nonconserving Asymmetry in the Decay $K_L^0 \rightarrow \pi^+\pi^-\pi^0$ " *Phys. Rev. Lett.* **21** 1650 1968.

Both Argonne and Brookhaven established boards of overseers which were responsible for the operation of the labs. Membership on these boards included representatives from affiliated universities. Since the experimental particle physicists at CWRU made use of the accelerator facilities at both of these labs, the university participated in MURA (Midwest Universities Research Association) and AUI (Associated Universities, Inc.).

Bill Frisken left CWRU in 1971, returning to Canada to take a position at York University in Toronto. He has continued in accelerator based particle physics as well as becoming active in environmental physics. His colleague, Bill Blanpied left in 1972, taking a position with the public sector programs division of the AAAS and later with the National Science Foundation, where he became Senior International Analyst. Both Frisken and Blanpied had been awarded tenure at CWRU and their departures depleted the experimental particle physics program. Their two tenure slots were not regained by that program. Tom Jenkins subsequently hooked up with Glenn Frye on the next generation of balloon (and satellite) borne detectors, as described at the end of Chapter 8.

Willard, Bevington, Baer: nuclear & intermediate energy

Harvey Willard (Fig. 16-4) was hired to set up the second SDP-supported particle physics group. After Martin Klein and Chandrasekhar stepped down as interim co-chairmen of the newly formed CWRU physics department, the 42-year-old Willard was selected as its first chairman in May of 1967. It would become his disagreeable responsibility to pare down the size of the combined department. Within four years after federation, between 15 and 20 faculty would have to leave. No new tenure-track faculty would be hired until 1980, leaving the department with a generation gap which would not be rectified until the 1990's.

Willard had completed his PhD at MIT in 1950. Using the proton beam from the MIT electrostatic accelerator, Willard studied (p,n), i.e. proton-in, neutron-out, reactions on the lithium nucleus. He continued this type of nuclear physics at the van de Graaff accelerator of Oak Ridge National Laboratory in Tennessee from 1951 until 1967. He was on the faculty of the University of Tennessee from 1963 until 1967. Among his several dozen publications from Oak Ridge, we shall mention two (because they are sort of interesting).

Fig. 16-4. Harvey Willard

In the first, the object was to check on the reversibility of time in nuclear reactions. If the physics of an interaction does not depend on the direction in which time "flows", then one would expect reversed reactions like d O $\rightarrow$ α N and α N $\rightarrow$ d O to be identical if kinematic corrections are made to compensate for the difference in masses.

Willard's experiment at Oak Ridge showed that in this reaction this was true to the one-half percent level. "Test of Time-reversal Invariance in the Reaction $^{16}O(d,\alpha)^{14}N$ and Its Inverse" *Phys. Rev. Lett.* **21** 447 1968 and *Phys. Rev.* **C3** 1065 1971.

In the second, Willard looked at some short-lived nuclear states. When a proton is shot at a ^{14}N nucleus, it can become incorporated in the nucleus which then becomes ^{15}O. The probability for this happening depends very much on the energy of the incident proton, with peaks where the total energy of the new oxygen nucleus matches one of its excited states. One can scan through a range of beam energies to find these excited states, and then deduce some of their properties (like spin) by looking at the angular distribution of the γ's which shoot out when the nucleus returns to its ground state. Willard *et al.* presented evidence for 14 such states in ^{15}O. "Level Structure in ^{15}O from the Proton Bombardment of ^{14}N" *Phys. Rev.* **179** 1047 1969. *Experiments like these, done at many energies and with many projectiles and target nuclei, formed the basis for our understanding of the atomic nucleus. They were done at reactors and accelerators by hundreds of research groups worldwide.*

Willard began his research at CWRU with an experiment which used the deuteron beam from the department's van de Graaff accelerator. Recall that Erwin Shrader (Chapter 7), who was instrumental in getting this machine up and running, had worked with a 3 MeV deuteron beam in 1963 and Rolf Scharenberg had done nuclear physics using the van de Graaff's beams in the mid 1960's. A key player in the van de Graaff program was the machine's operator, technician Larry Hinckley (Chapter 11) who had come from MIT with the machine. Bob Lescovec, the engineer introduced in Chapter 7, was a major contributor to the design and construction of detectors and electronic logic circuits. Both Shrader and Scharenberg had left the department by the time Willard started using the van de Graaff.

Nuclear Physics at the van de Graaff

Willard studied the reaction $d\ ^{16}O \rightarrow n\ ^{17}F$. In this reaction, the deuteron leaves its proton behind in the target nucleus, which thus becomes ^{17}F. The residual neutron continues on to be subsequently detected. The ^{16}O nucleus consists of neatly filled spherical shells of 8 protons and 8 neutrons, while the ^{17}F has these shells as a core and a loose proton running around the outside. The ^{17}F nuclei were produced about half the time in the ground state and half the time in the first excited state.

Recall that the van de Graaff provided a pulsed beam of deuterons, with a 1.2 ns-long bunch arriving every microsecond. Because the *time* of the collision was known, time-of-flight measurements on the neutron could give its speed and energy. In this experiment, the neutrons were detected in a "polarimeter". In the polarimeter, the neutron scatters off a helium nucleus and the recoil α particle's (i.e. the helium nucleus) direction and energy are recorded. Left-right asymmetries of the recoil α give a measure of the degree of polarization of the neutrons (i.e. whether the spin of the neutron had a preferred direction). For each run, a deuteron beam energy was selected, and the scattering angle of the neutron was fixed by placement of the polarimeter. For each event, the neutron time of flight was determined along with the energy deposited in the polarimeter by the α particle. The α was detected by one of two scintillators, one to the left and one to the right of the neutron direction, so that the average left-right asymmetry could be determined for each run.

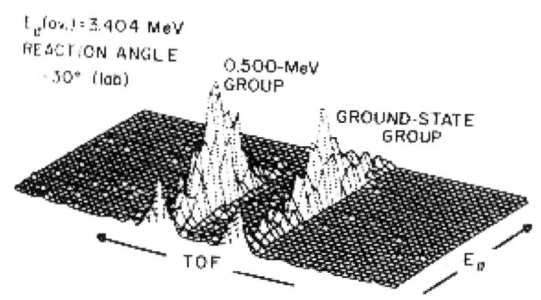

The **Fig. 16-5** shows the number of counts as a function of neutron time-of-flight versus the energy deposited by the α in the polarimeter. The two mountain ranges correspond to the two ^{17}F states, the ground and the first excited. The second plot, **Fig. 16-6,** shows the angular distribution of the neutrons for three incident deuteron energies for events in the ground state (on the left) and the excited state (on the right). The curves are the theoretical predictions of a Born approximation model, i.e. a model in which only single scattering is considered." ^{16}O(d,n) Polarizations and Cross Sections from 3 to 4 MeV" *Phys. Rev.* **C6** 1513 1972.

Fig. 16-5. α energy vs. neutron time-of-flight in (d,n) reaction on ^{16}O.

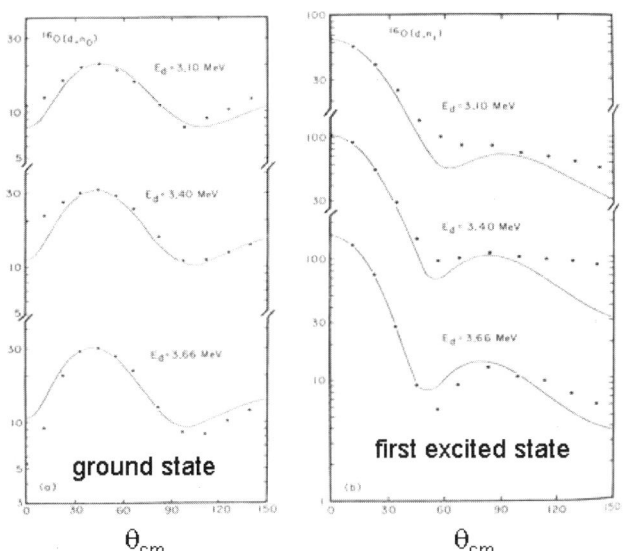

Fig. 16-6. Neutron angular distributions for two ^{17}F states in (d,n) reaction on ^{16}O.

In 1970, **Philip R. Bevington** joined Willard as the second member of the "intermediate energy" group. Born in New York City in 1933, Bevington completed his AB at Harvard and then earned his doctorate at Duke University in 1960. His photo is shown in **Fig. 16-7**. Bevington remained at Duke for an additional three years, doing van de Graaff based nuclear physics. He then spent four years as assistant professor at Stanford University, where he became an expert in the use of computers in the analysis of nuclear physics experiments. Bevington was severely injured in an automobile accident before coming to CWRU, and was confined to a wheelchair until the time of his death in 1980. In spite of his disability, he was a successful teacher and was instrumental in setting up a state-of-the-art PDP9 computer facility for use by many members of the department. One can see Phil Bevington's hand in Fig. 16-5 for the Willard van de Graaff experiment. It was surely he who produced the multi-dimensional plots with the aid of the PDP9.

Les Foldy gave me his file of Bevington memos for the PDP9. Memo #1, February 1969, describes the machine: central processor – 8192 words of 18 bits each (i.e. 8K RAM), 1.5 μsec cycle time; paper-tape reader and punch; hardware multiplier; memory expansion to 16,383 words (temporarily leased for 18 months), two DEC magnetic tape drives; X-Y oscilloscope display with Light Pen; price $46,100 on delivery with $20,000 more in 18 months. Around $200,000 in 2002 dollars, and imagine, no monitor, not even a mouse!)

In a paper published just before his arrival at CWRU, Bevington presented a user-friendly programming system, written in Fortran. "Real-time Reduction of Nuclear Physics Data" *IBM J. Res. Develop.* 1969. The system allowed the user to identify and fit structure in experimental spectra, using the latest techniques in curve-fitting and error analysis. One of Bevington's longest-lasting accomplishments is his widely used book on computerized data analysis. Keith Robinson, whom we shall meet later in this chapter, has expanded and updated this book over the past two decades. (The 3rd edition appeared in 2002; *Data Reduction and Error Analysis for the Physical Sciences*, McGraw Hill.)

Fig. 16-7. Phil Bevington

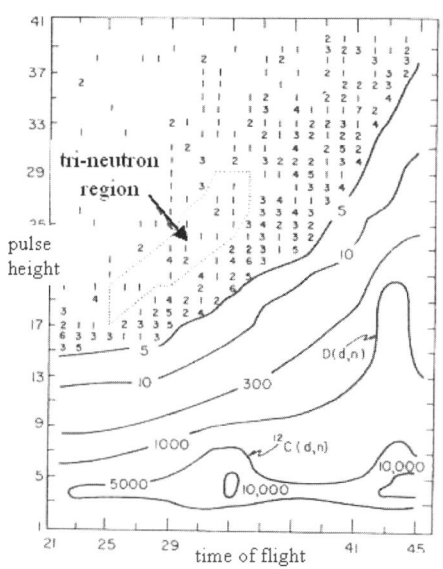

Fig. 16-8. Search for tri-neutron.

At CWRU, Bevington set up an experiment using the van de Graaff accelerator in a rather interesting search for a new nucleus which some theorists thought might exist. The three-nucleon system has two bound states: ^{3}H (pnn) and ^{3}He (ppn). These are held together by the attractive nuclear force, in spite of (in the case of ^{3}He) the repulsion between the two protons. Some calculations indicated that the ^{3}n (nnn) state might also be bound, in spite of the fact that the Pauli principle requires that one of the neutrons would have to be in the second energy shell. Bevington and graduate student Kenneth Koral used the pulsed neutron beam from the van de Graaff incident on a target of ^{7}Li. The proposed reaction was $n + {}^7\text{Li} \rightarrow {}^3n + {}^5\text{Li}$. The tri-neutron would be detected in a helium filled detector where the arrival time and recoil energy of the helium nucleus would be recorded. The **figure 16-8** shows a scatter-plot of time-of-flight (running from right to left) versus pulse height. The vast majority of hits (tens of thousands) comes from single neutrons, while an insignificant handful of events falls into the zone marked "TRINEUTRON REGION". The authors conclude that there is no evidence for the ^{3}n. "A Search for the Bound Trineutron from ^{7}Li + n Reactions" *Nucl. Phys.* **A175** 156 1971.

The versatility of the van de Graaff was again exploited when the helium isotope, ^{3}He, was accelerated and delivered in short pulses to a nuclear target. Bevington and his student S. K. Bose studied the reaction in which the incident ^{3}He nucleus left its two protons behind in a ^{20}Ne nucleus, leaving a lone neutron and a ^{22}Mg nucleus in the final state. Runs were made with ^{3}He energies ranging from 2.6 to 4.0 MeV. Measurement of the neutron time-of-flight allowed the determination of the total energy of the excited ^{22}Mg nucleus. Analysis of the energy dependence of the reaction cross-section and the angular distribution of the outgoing neutrons provided a test for theoretical models. It was found, for example, that in this reaction both "direct interaction" (when the projectile just grazes the target) and "compound nucleus" (when the projectile is absorbed by the target which later decays) processes contribute. *Nucl. Phys.* **A219** 115 1974. A similar study was made of the excited states in ^{30}S with graduate student Alberto Kogan. *Jour. of Phys. G: Nucl. Phys.* **4** 422 1975.

Bevington, along with two grad students, B.D.Anderson and Frank Cverna, worked on the development of a new type of multiwire proportional chamber. They found that they could use a plane of thin metal foil as the electron-emitting cathode, rather than a series of parallel thin wires. The wires of the anode would be enough to signal the location of the ionizing particle. They reported that such chambers could be used at higher voltages than the usual (wire-cathode) type and would result in higher efficiencies over a broad range of operating voltages. "Investigation of some properties of multiwire proportional chambers with planar cathodes" *Nucl. Inst. and Meth.* **129** 373 1975. A second paper on detector instrumentation was written with Bob Lescovec. CWRU Multiwire Proportional Counter Readout Sytem" *Nucl. Instr. and Meth.* **147** 431 1977.

Intermediate energy experiments: Los Alamos

Four years later, in 1974, a third member of the "nuclear/intermediate energy" group joined Willard and Bevington. **Helmut Baer** was born in China in 1939, the son of American missionary parents. He completed his doctorate at the University of Michigan in 1967 where he did experimental nuclear physics at the Cyclotron Lab. He came to CWRU after several years at the Lawrence Berkeley Lab. A photo of Baer is shown in **Fig. 16-9**.

Fig. 16-9. Helmut Baer

At Berkeley, Baer used the pion beam from the 184-inch cyclotron to measure radiative capture reactions, i.e. reactions in which a negative pion is absorbed by a nucleus and a photon is subsequently emitted. He and his colleagues at Lawrence published a 100 page review of this work in an article which began: "The central question to be addressed in this article is: What can be learned about nuclear structure by stopping negatively charged pions in targets of nuclei from various regions of the periodic table and examining the emitted photon spectra between 50 and 150 MeV with a resolution of ≤ 2 MeV?" The authors review their work and that of other groups on pion capture by nuclei from hydrogen to lead. One area of interest was pionic atoms, in which the pion goes into electron-like orbits around the nucleus, and where its wave function reaches deep into the nucleus. The principal goal, however, was the study of nuclear states which are excited when the pion disappears into the nucleus. "Radiative pion capture in nuclei" *Advances in Nucl. Phys.* **9** 177 1977.

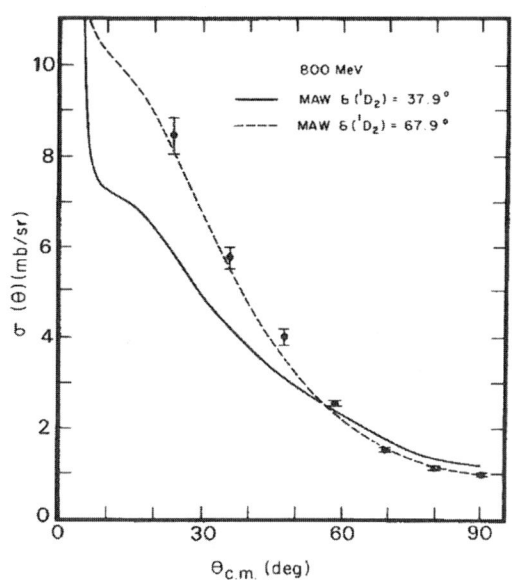

Fig. 16-10. Phase shift analysis, 800 MeV pp elastic scattering.

The first "intermediate energy" experiment mounted by the CWRU group was the study of proton-proton elastic scattering at the Los Alamos Meson Physics Facility (LAMPF). The list of authors included Willard, Bevington, and Baer. **Fig. 16-10** shows the resulting differential cross section from the 800 MeV data, along with the predictions of a phase shift analysis done by a group at Lawrence Radiation Lab in Livermore. The authors found that there was better agreement (dashed curve) with the data if one of the phase angles was changed from the Rad Lab value. "Absolute differential cross section measurements for proton-proton elastic scattering at 647 and 800 MeV" *Phys. Rev.* **C14** 1545 1976.

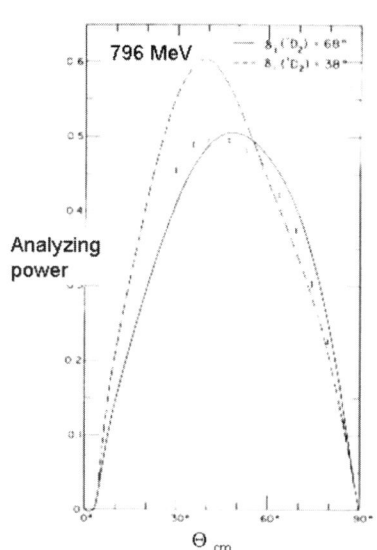

Fig. 16-11. Analyzing power in pp scattering.

This work was followed by a similar experiment, but this time the incident protons were polarized, i.e. with spins preferentially oriented. A polarized ion source had been installed at the LAMPF accelerator and the group wanted to exploit it to perform a more complete measurement of the phase shift parameters. "In this experiment a beam of protons with transverse polarization up to 0.92 was obtained from the LAMPF accelerator and focused (typically 4 mm diam) onto a CH2 target." Both protons were detected in the wire chambers built by Bevington and Lescovec so that elastic proton scatters on hydrogen could be identified. The beam polarization direction was flipped every three minutes. Data were taken at scattering angles from 30 to 90° in the center of mass (cms) and the quantity $\varepsilon = (L-R)/(L+R)$ was determined, i.e. the difference between scatters to the left and those to the right, over the total. This quantity equals the product of the polarization analyzing power of the pp scattering and the beam polarization, $\varepsilon = A_y(\theta) P_{beam}$. **Fig. 16-11** shows the resulting analyzing power as a function of the cms scattering angle. The dashed curve shows the predictions for the analyzing power as calculated from published phase shifts, and the solid curve includes the modification of one of the phase shifts described in the previous paper. The authors suggest that because the beam polarization P_{beam} is so well measured at LAMPF and most probably more difficult to measure at other accelerators, other experimenters could use these A_y results along with their own measurement of the pp asymmetry ε to deduce the polarization of their beam. "Polarization Analyzing Power $A_y(\theta)$ in pp Elastic Scattering at 643, 787, and 796 MeV" *Phys. Rev. Lett.* **41** 384 1978. *Phys. Rev.* **C23** 838 1981.

Bevington and Willard followed this with a study of 800 MeV protons scattered by *deuterons*. They once again measured polarization analyzing powers as well as differential cross sections. This was a natural next-step after the proton-proton experiment, as now there would be two nucleons in the target, and the interferences between scatters off of the proton, the neutron, or successive scatters by each, would provide meaningful tests of the models. **Fig. 16-12** shows the terribly complicated structure in the measured analyzing power. "Proton-deuteron elastic scattering at 800 MeV" *Phys. Rev.* **C21** 2535 1980.

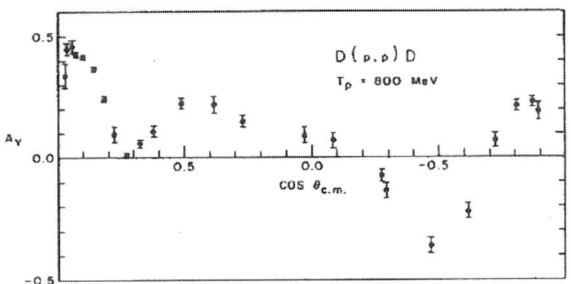

Fig. 16-12. Analyzing power in pd scattering.

The next step in this progression of experiments, from pp elastic to polarized incident protons to deuteron targets, was to look at some inelastic channels, in particular single and double pion production in pp scattering. A magnetic spectrometer based on multi-wire proportional chambers was set up at the Los Alamos facility to identify and measure the momentum of π^+ and π^- mesons. The reactions pp → pnπ^+ and pp → pp$\pi^+\pi^-$ were studied and the momentum and angular distributions of the outgoing pions compared to several current models. These channels had been studied in detail twenty years earlier in bubble chamber experiments, as will be described in the next section of this chapter. Why then would one repeat the study? Counter experiments usually measure vector momenta of hundreds of thousands of individual particles resulting from high energy collisions. Bubble-chamber experiments, on the other hand, typically measure all the particles produced in each collision, but only a few thousand events. It's a question of more statistics versus more details. "Single and double pion production from 800 MeV proton-proton collisions" *Phys. Rev.* **C23** 1698 1981.

Helmut Baer left CWRU in 1978 after fewer than five years in the department, taking a position at LAMPF. The tenure-track during this period in the CWRU physics department was a dead-end. After a productive decade of research at LAMPF, Baer tragically died of cancer in 1991 at age 52. His death has been coupled with those of at least two other LAMPF experimentalists, resulting in a wrongful death suit against the lab. A 1998 Court of Appeals decision found in favor of the defendant, the University of California, which operates the lab.

In 1980 Phil Bevington succumbed to complications from his old injuries. His computer software expertise had been vital to many of his colleagues in the department.

Harvey Willard held the position of Dean of Science from 1970 until 1976. Left without a research group, he decided in 1981 to accept a permanent position at the NSF as Section Head for Nuclear Science. Two of the SDP-funded experimental "sub-atomic" programs – the "counter group" and the "intermediate energy group", had for various reasons enjoyed somewhat truncated lifetimes.

Robinson, Kikuchi, Fickinger, Eisner: bubble chambers

The third group to be set up under the NSF grant was initiated on the WRU side of the fence with the hiring of **D. Keith Robinson.** Its history would be somewhat longer than the other two groups. If this section of the narrative seems more detailed than some of the earlier parts, the reason is obvious.

Robinson was born in 1932, raised in Nova Scotia, graduated from Dalhousie in Halifax and went to Cambridge for his doctorate. When his advisor, Dennis Wilkenson, moved to Oxford, Robinson went with him and completed his degree there in 1960, writing on the properties of hyperfragments. These states, occasionally produced in high energy collisions, are nuclei in which a nucleon has been replaced by a Λ or Σ strange baryon (or, on a different level, a down-quark by a strange-quark). Robinson studied their production and disintegrations as recorded in nuclear emulsions. (We described the emulsion technique in Chapter 7.) Robinson's picture is shown in **Fig. 16-13**.

Fig. 16-13. Keith Robinson

In 1960 Robinson took a position at Brookhaven National Laboratory in the newly organized bubble-chamber group of E.O. Salant. He was responsible for the analysis software written for the new IBM 700-series supercomputers (32K RAM, programming language: assembler or Fortran, punch card input). He developed a data summary system ("Corregram") which was subsequently used for years by bubble-chamber groups at many labs. His first experiment at Brookhaven was the study of 2 GeV kinetic energy protons striking protons in the Brookhaven 20-inch hydrogen-filled bubble chamber. The proton beam was produced by the Cosmotron synchrocyclotron.

William Fickinger (your narrator) was born in 1934, raised in New York City and Rhode Island, and graduated from Manhattan College. He spent the summer after his junior year in Salant's lab at Brookhaven, learning about nuclear emulsions. He later went to Yale for his PhD. Working under Horace Taft and Earle Fowler, he and fellow grad student Jim Sanford planned to do an experiment at the Cosmotron using positive pi mesons (π^+) incident on the 20-inch hydrogen bubble chamber. To create a beam of π^+'s the team built an electromagnetic beam separator. This device was a 20-ft long, 4-ft diameter, stainless steel, evacuated cylindrical tank containing two long parallel plate electrodes, 2-in apart, with a potential difference between them of 500,000 V, all sitting in a transverse magnetic field of 700 Gauss. What a toy to give two 24 year-old grad students! It worked just fine. The incoming beam was a mix of protons and π^+'s, all with the same momentum, so the lighter π^+'s were traveling faster than the protons. The crossed electric and magnetic fields were adjusted in strength so that the fast π^+'s would pass through undeflected, and the slower protons would be nudged slightly aside. Fickinger was responsible for the vacuum system – based on titanium ion pumps. This was the first "separated beam" at BNL. The students and their toy are shown in **Fig. 16-14**.

Fig. 16-14. Jim Sanford, Bill Fickinger and the beam separator.

(An aside for graduate students: Somehow it was my job to make the drawings for the 20-foot long stainless steel tank. By copying the notation I found on other drawings, I seem to have specified that the 4-foot diameter end-plates must be parallel to one another to something like two thousandths of an inch across the diameter. The manufacturer in Pittsburgh phoned to say that would be impossible, so I said make it half-an-inch. He found that pretty funny.)

Early work at Brookhaven

Before the Yale group had the chance to run their π^+ experiment, the "pole-face" magnets of the Cosmotron burned up, shutting down that machine for a year: the type of delay dreaded by any graduate student. Earle Fowler took Fickinger over to see Salant and somehow talked him into turning over half of the film from the 2 GeV proton run for analysis at Yale (and for a dissertation for Fickinger). Robinson was livid! Thus began the Fickinger-Robinson collaboration which was to play for forty years.

The first paper, on the production of Δ baryons by one-pion-exchange, was based on that exposure and analyzed by Horace Taft's reconstruction and fitting programs and Robinson's data summary programs. The Yale and BNL sets of data for the reaction p p $\rightarrow$ p n π^+ were merged and the paper written on the lawn of Robinson's house overlooking Long Island Sound. "One Pion Exchange in p-p Collisions at 2 BeV" *Phys. Rev. Lett.* **7** 196 1961. "p-p Interactions at 2 BeV, I. Single-pion Production" *Phys. Rev.* **125** 2082 1962. "Evidence for a Two-Pion Decay Mode of the ω Meson" *Phys. Rev. Lett.* **10** 457 1963.

Armed with state-of-the-art software and some experience in using it, Fickinger went off like Johnnie Appleseed, installing the bubble-chamber analysis software on the computers of the University of Kentucky, the Centre d'Études Nucléaire de Saclay in France, and then Vanderbilt University.

Robinson remained at Brookhaven, working in collaboration with Mark Sakitt, and with Arnold Engler's group from Carnegie-Mellon University. They produced a series of important papers on the properties of K-mesons, a very hot topic at the time. "Measurement of the K^o-K^+ Mass Difference" *Phys. Rev.* **168** 1534 1968. "Measurement of the Lifetime of the Short-Lived Neutral K Meson" *Phys. Rev.* **171** 1418 1968. "Study of the $\Delta S = \Delta Q$ Rule in the Leptonic Decays of the Neutral K Meson" *Phys. Rev.* **D1** 3031 1970. "Search for $K_S \to \pi^+ \pi^- \pi^o$" *Phys. Rev.* **D3** 1557 1971. Partnerships with Sakitt and his group at Brookhaven and Engler's group at Carnegie were to play an important role in the work of the CWRU group for the following 20 years.

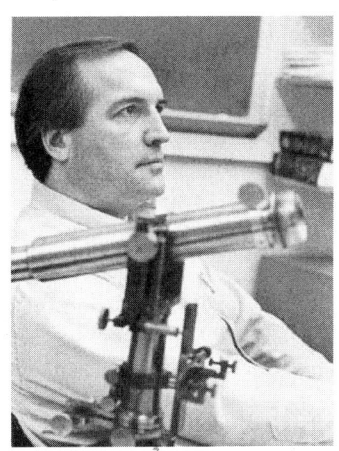

Fig. 16-15. Bill Fickinger.

In 1966, Robinson accepted a position in the Western Reserve physics department where he set up the third of the NSF-SDP programs. Projection machines to measure the bubble-chamber pictures were built by the WRU machine-shop, scanners trained, and the PDP8 data-acquisition system developed. In the fall of 1967, Fickinger joined Robinson at CWRU. He was interviewed by two chairmen (Martin Klein and Chandrasekhar) and by two deans, and was probably the first appointee to the faculty of the new CWRU department. His photo is in **Fig. 16-15.**

About bubble chambers

What is a "bubble chamber experiment"? First, a proposal is made to the program committee of the accelerator lab, asking for a certain number of pictures with a certain target liquid (propane, hydrogen, deuterium, methyl iodide), a certain beam and beam energy, giving physics arguments for the exposure. Often several university groups would join forces, and share the film, each selecting specified types of "events". The physicists, grad students and technicians would go to the accelerator laboratory for a few weeks, help set up the beam (magnets, counters, collimators), and sit in (24/7) on the run, while the laboratory physicists and technicians would tend to the chamber itself. The film, often hundreds of thousands of pictures, would be developed at the lab and taken back to the home institutions for scanning and measuring.

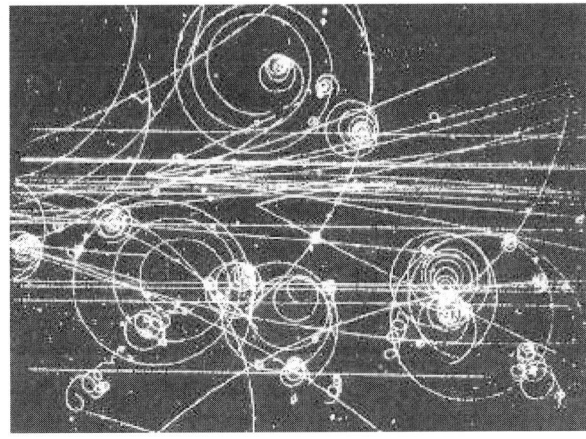

Fig. 16-16. A typical hydrogen bubble-chamber picture.

A bubble chamber is a container of liquid which is at a pressure and temperature just below the boiling point. Shortly before the beam is shot into the chamber, a piston drops, lowering the pressure on the liquid, leaving it in a superheated state. Bubbles form along the ionized path of the moving charged particles. **Fig. 16-16** shows a typical bubble chamber photograph. The bubbles are allowed to grow to optimum size for photography (for about a millisecond) and then flash lamps illuminate the chamber. Three or more cameras record the strings of bubbles as seen from different angles. The chamber is usually inside a strong electromagnet – with a 10 to 20 kilogauss field. Measurement of the curvature of a track gives the momentum of the particle which caused it. A former Case undergraduate, Donald Glaser (BS 1946) won the Nobel Prize for inventing the bubble chamber. He is shown in **Fig. 16-17** with Polykarp Kusch, another Nobelist with a Case physics degree (BS 1931), when they visited the CWRU campus.

Fig. 16-17. Donald Glaser BS '56 and Polykarp Kusch.BS '31.

A team of trained "scanners" examines the pictures at "measuring tables" equipped with movable crosshairs attached to digitizers (forerunners of today's mouse). A series of x-y coordinates are recorded along each track, in each stereo view. This is the "data acquisition phase", controlled by small computers. Robinson designed and maintained this system. The group employed 4 to 8 full-time scanners, working two shifts on four measuring machines. In **Fig. 16-18**, Robinson hovers over Gracie Broadway.

In the "data analysis phase", the x-y coordinate pairs are combined to make a three-dimensional track which is fitted with a curve, allowing for optical distortions, energy loss, and magnetic field inhomogeneities. The resulting vector momenta are fed to a second, "kinematic analysis" program which assigns trial masses to each track, and then checks to see how well relativistic energy and momentum are conserved by a given mass assignment. In this way, each event may be associated with a particular reaction, for example, p p → p n π^+. Because there are four energy-momentum conservation equations to work with, it is possible to determine the momentum of one *unobserved* neutral

Fig. 16-18. Gracie Broadway measuring bubble chamber events.

particle (in this reaction, the neutron). Fickinger was principally concerned with the analysis phase, and maintained the resulting data base of hundreds of 2400-ft magnetic tapes. Over the course of about 15 years, this procedure was applied to about five major experiments, mostly in collaboration with friends at Carnegie Mellon University.

As early as their period on the staff at Brookhaven, Robinson and Fickinger became associated with the use of deuterium as the target liquid in the bubble chamber. The idea was that one could study interactions with the neutron in events in which the proton was just a spectator to the collision. This allowed the study of some multi-particle states not accessible with a hydrogen target.

The first set of film measured and analyzed at CWRU was a continuation of the K-meson work which Robinson had begun with the Brookhaven and Carnegie-Mellon groups. Robinson was joined by three new colleagues, instructors **Tadashi Kikuchi** and **Clarence Tilger** and RA **Walter Carnahan**. The experiment was based on an exposure of the Brookhaven 30-inch deuterium filled bubble chamber to 600 MeV/c K^+. "Determination of the Sign of the $K_L^0 - K_S^0$ Mass Difference" *Phys. Rev.* **D4** 7 1971.

Bump hunting (a.k.a. building the Standard Model)

The next bubble chamber experiment tackled by the new CWRU group was an exposure of the Brookhaven 80-inch deuterium-filled chamber. The beam was K^- mesons at momentum 4.9 GeV/c. The run produced 100,000 sets of photos taken with four cameras. The object was to study interactions of the K^- with the neutron, especially those including neutral particles in the final state: e.g. $K^- n \to \Lambda \pi^-$ or $\Lambda \pi^- \pi^0$ or $K^0 \pi^- n$. The measurements were recorded on punchcards and analyzed on the CWRU Univac 1108 computer.

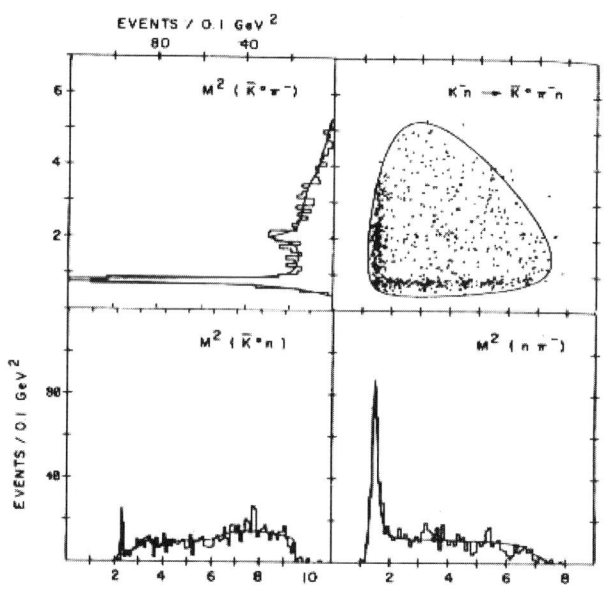

Fig. 16-19. Dalitz plot and projections for the $K^0 \pi^- n$ final state.

The physics was similar to that studied in the proton-proton experiment at Brookhaven. In the former case the properties of non-strange baryonic resonances were studied, for example, $p\,p \to \Delta^{++}(1238)\,n \to p\,\pi^+ n$. In the K^- experiment, reactions like $K^- n \to K^{*-}(890)\,n \to K^0\,\pi^-\,n$ or $K^- n \to \Sigma^{*-}(1385)\,\pi^0 \to \Lambda\,\pi^-\,\pi^0$ yielded information on strange mesons and baryons, like the $K^*(890)$ and the $\Sigma(1385)$. The figure (one of 42 in the paper) shows a Dalitz plot for the $K^0\,\pi^-\,n$ final state. The scatter plot in **Fig. 16-19** has one dot per event located at the intersection of the Kp and the np effective masses (squared). If the three outgoing particles are uncorrelated, this plot would be uniformly populated. The two highly populated regions correspond to the $\Sigma(1385)$ and $\Delta(1238)$ resonant states, as can be seen in the two tall peaks in the projections. Cross sections and angular distributions for these strongly decaying states were measured. All these data are included in the Particle Data Group

Tables which are published each year by a world-wide consortium. The PDG compilation is the raw data for the quantum chromodynamics part of the Standard Model. This experiment was done entirely by the CWRU group, including three grad students (Bernard Burdick, John Korpi, and Sergei Al Gourevitch), and was published in a single 45-page paper. "K⁻d Interactions at 4.9 GeV/c" *Nucl. Phys.* **B41** 45 1972.

Multipion resonances

In collaboration with Carnegie-Mellon, the group next tackled a 930K picture exposure to a beam of 6 GeV/c π^+ of the deuterium filled 30-inch bubble chamber at Argonne National Laboratory. This very large exposure was to occupy both the CWRU and the CMU groups' measuring capacity for at least three years. Of particular interest were those reactions in which the neutron became a proton, and the incident π^+ became a new neutral heavy meson which decayed into two or more pions. Events with more than one neutral particle in the final state cannot be analyzed (too many unobserved momentum components). The reactions studied included $\pi^+ n \to p\ \pi^+ \pi^-$; $\pi^+ n \to p\ \pi^+ \pi^- \pi^0$; $\pi^+ n \to p\ \pi^+ \pi^- \pi^+ \pi^-$; $\pi^+ n \to p\ \pi^+ \pi^- \pi^+ \pi^- \pi^0$, thus giving access to neutral multipion systems of 2, 3, 4, and 5 pions. As a result, the experiment produced many contributions to the PDG compilation. This is the stuff of which the quark picture of the Standard Model is made. The masses, widths, angular distributions, decay channels, etc. all enter into the establishment of the quark constituents of the state. The literally hundreds of baryonic and mesonic states, discovered mostly in bubble chamber experiments in the 1960's and 1970's, were shown to be combinations of three quarks or a quark and an antiquark. More impressive is the fact that **no** so-called "exotic" particle has been found, i.e. one which does not fit that picture – not, certainly, for the lack of looking. (The hunt was not abandoned, however, and three decades later, in 2003, convincing evidence was reported by laboratories in Osaka, Moscow, Virginia and Bonn for a qqqq-qbar baryon, called the Θ^+, with mass 1540 MeV.)

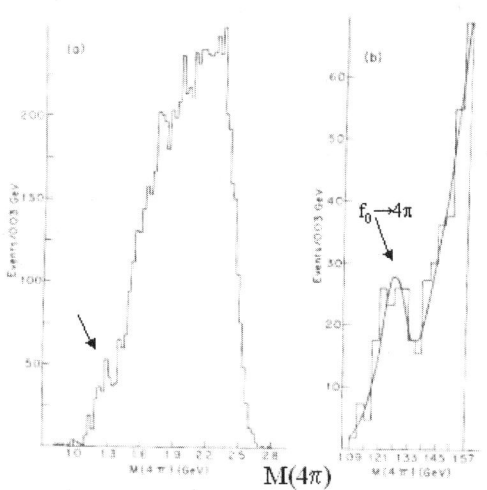

Fig. 16-20. Four-pion mass distributions

As an example, the effective mass distribution of the three-pion system $\pi^+ \pi^- \pi^0$ contained about 800 examples of the ω meson (mass 780 MeV). This large sample of events allowed the determination of the "spin-density matrix elements", that is, the factors which describe how the three-pion system breaks up. The data were compared with models based on the exchange of various spin-one mesons. "Study of the Reaction $\pi^+ n \to p\ \omega$ at 6.0 GeV/c" *Phys. Lett.* **45B** 165 1973.

The $\pi^+ \pi^- \pi^+ \pi^-$ system has an enhancement in the region of the 1272 MeV f⁰ meson which previously had been only known to decay to $\pi^+ \pi^-$. The left hand side of **Fig. 16-20** shows the four-pion effective mass for all events and the right hand figure shows how one could pull out the signal for the f° by selecting events in which the "quadri-pion" traveled along the same direction as the incident pion (i.e. in a peripheral interaction). The ratio of the four-pion to two-pion decay rates was 3.7 ± 0.7 %. The breakup of the f° into four pions was further analyzed in a search for possible intermediate steps, e.g. via 3-pion or 2-pion states. Theoretical models predicted such possibilities as a $\rho°\rho°$ intermediate state, and the data were found to agree rather well with that picture. "Observation of the Decay f° → $\pi^+ \pi^- \pi^+ \pi^-$" *Phys. Rev. Lett.* **31** 562 1973.

There was even evidence for two states which decay into five-pions, and, as has been found for many of these massive mesons, the decay proceeded in sequential steps: $A_2 \to \omega \pi^- \pi^+$ followed by $\omega \to \pi^- \pi^+ \pi°$. "Evidence for the $\omega\pi\pi$ Decay Modes of the A_2 and $\omega(1675)$ *Phys. Rev. Lett.* **32** 260 1974. The π^+n experiment yielded a variety of results in addition to the multipionic states. Because the exposure was rather large (for the time) and the statistics quite good, one could look at the details of some rare processes. During this period the group included assistant professor Charles Sullivan, who had earlier worked with Tom Jenkins. Graduate student David Matthews and research associates Frank DiBianca (Carnegie Mellon), John Malko (Ohio University), and José Diaz Bejarano (CERN) were essential to the success of the program. "Backward Production of ρ and ϕ Mesons in π n Interactions at 6 GeV/c" *Phys. Lett.* **50B** 275 1974. "Features of the π d → $\pi \pi \pi$ d Reaction at 6 GeV/c" *Phys. Rev.* **D12** 1272 1975. "$\pi\pi$ Scattering in the Energy Region 0.5 to 1.42 GeV" *Phys. Rev.* **D10** 2070 1974. "Resonant Structure in the $\pi^+ \pi^+ \pi^- \pi^-$ System between 1.5 and 1.9 GeV" *Phys. Rev.* **D23** 595 1981.

In 1974, a new member joined the group as assistant professor. **Robert Eisner** had completed his doctorate at Purdue University and had been working with a bubble chamber group at Brookhaven Lab. He pointed out the possibility of combining the CWRU data on $\rho°$ production with data from his former group concerning a related reaction in which a strange meson, the K*, is produced. The resulting papers stressed the similarities between the angular distributions for production and decay of these two spin-one mesons. "A Study of Inclusive Vector Meson Production" *Phys. Lett.* **63** 461 1976 and *Nucl. Phys.* **B119** 1 1977. Eisner and CWRU post-doc J. F. Owens built further on the collaboration with the Brookhaven group, publishing a series of papers on resonance production by pion and kaon-induced collisions.

Polarized Protons at Argonne

The next major experiment for the CWRU-Carnegie Mellon collaboration was based on two 120K photograph exposures of the enormous (12-foot diameter) Argonne hydrogen bubble chamber, one at 6 GeV/c and the other at 12 GeV/c beam momentum. Many proton-proton experiments had been done, but these were different in that the incident beam was about 50% polarized. The expectation was that several spin-dependent features of the collisions could be explored. This turned out to be true for high-statistics *counter* experiments done with Argonne's ZGS polarized beam, but the limited statistics of a bubble-chamber experiment made it difficult to extract significant information. Nevertheless, three papers were published in which a few standard deviations worth of polarization-dependent effects were coaxed out of the data. Specifically, the 6 GeV/c events of p↑ p → n Δ^{++} in which the Δ^{++} is produced backward in the center of mass (i.e. the target proton becomes the Δ), showed a polarization-induced decay asymmetry of 30 ± 10% in a particular range of momentum transfer. Such effects arise from interference between the amplitudes for various production channels. (A technical note: to avoid asymmetries caused by scanning or other experimental biases, the direction of polarization of the beam protons was reversed on each accelerator pulse, and its sign was imprinted on the film.)

The events in which the strange particles, K^o's or Λ's, were produced also showed some tantalizing effects, such as the forward-produced K^o's which presented "the rather remarkable experimental situation in which 70% of the K^o's traveled to the right and 30% to the left." These effects would have to wait for counter experiments, preferably with both polarized beam **and** target, for confirmation and exploitation. "Λ and K^o production in p↑p interactions at 6 GeV/c" *Nucl. Phys.* **B123** 361 1977; "Study of the reaction p↑p → n Δ^{++} at 6 GeV/c with polarized beam" *Phys. Rev.* **D20** 596 1979; and "Effects of beam polarization on Λ and K^o inclusive production in pp interactions at 12 GeV/c" *Phys. Rev.* **D21** 10 1980.

By 1980, it was time to get out of the bubble chamber business. Most other bubble chamber groups were switching over to counter experiments. The labor-intensive, one event at a time, photographic technology was being replaced by reaction specific all-electronic methods. The funding agencies (DOE, NSF) no longer wanted to support teams of "scanners". Fickinger and Robinson had been chasing strings of bubbles for twenty years, and the time had come to switch to electronic detection.

Counter experiments at Brookhaven

During the following decade or so, Fickinger and Robinson were part of three "counter-physics" collaborations at Brookhaven: with BNL's Mark Sakitt, with Syracuse's Ted Kalogeropoulos, and with Boston University's Lee Roberts.

The K⁺n system

The first experiment with Sakitt was a search for a previously reported resonant state in the K^+n system. This state, called the S-hyperon, has *positive* strangeness; it cannot be constructed from only three quarks, because positive strangeness is carried only by an antiquark. This then would be one of the "exotic" particles mentioned above. The experiment involved scattering a K^+ off a neutron (one which had been sitting comfortably in a deuteron for billions of years). The interaction point in the liquid deuterium target would be found by detecting the *spectator* proton. The neutron must then enter a second target, this one filled with liquid hydrogen, and

Fig. 16-21. Wire chambers built at CWRU.

scatter off a proton. This recoiling proton must then be detected, indicating whether the neutron scattered to the left or right. Any asymmetry gives the neutron polarization, whose value depends on the existence of the S-hyperon. Once again, the group would be working with liquid deuterium (i.e. neutron) targets and with polarization effects.

Twelve planes of wire chambers, one of which is pictured in **Fig. 16-21**, defined the incident beam. They were built by Larry Hinckley (the van de Graaff man) at CWRU. All the reconstruction and analysis of the data was done at CWRU. It eventually became clear that, in spite of the valiant analysis efforts of graduate student Oscar Rondon Aramayo and post-doc V. A. Sreedhar, the combination of limited spatial resolution, limited statistics, and large background would preclude a meaningful result. The moral of the story may be that one can never do too many Monte Carlo studies before attempting an experiment of this complexity.

The pbar-p system

The second counter experiment with Sakitt had three parts: antiprotons bouncing off protons, a search for a pbar-p bound state, and antiprotons bouncing off nuclei. The Low Energy Separated Beam (LESB) at the AGS was "partially separated", i.e. the antiproton to negative pion ratio was enhanced by electromagnetic separation; the antiprotons were subsequently tagged by time-of-flight and pulse height signals. The beam provided on the order of 1000 antiprotons per accelerator pulse. **Fig. 16-22** shows a drawing of the experimental layout.

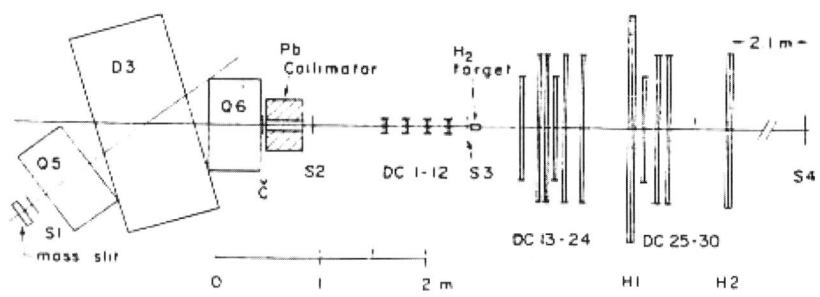

Fig. 16-22. Experimental setup for antiproton-proton scattering.

A liquid hydrogen target was placed in this beam and a measurement was made of pbar-p elastic scattering. Runs were made at eleven different incident momenta, and the angular distributions were fitted to a combination of Coulomb and nuclear terms, including interference between them. Two parameters were extracted from the data: the b which appears in the e^{-bt} part of the nuclear term and the ρ which is the ratio of the real to imaginary parts of the p-pbar scattering amplitude. Plots of ρ and b are shown in **Fig. 16-23** for two different analyses: one including the possibility of spin-flip (open circles), the other without (solid circles). The results favored the theoretical model for pbar-p scattering advanced by the "Paris Group". "Measurement of the Real-to Imaginary Ratio of the pbar-p Forward-Scattering Amplitudes" *Phys. Rev. Lett.* **54** 518 1985.

One feature of pbar-p scattering which was of great interest was whether or not there are any pbar-p bound states. Between 1974 and 1985, there were more than a dozen experiments which looked for such states, either in the elastic channel or in annihilations into pions. There were several claims of rather narrow (10-20 MeV) peaks with rather large cross sections, but there were enough "non-observations" to raise some doubts. A new measurement was needed, and the AGS beam was waiting. The same beam layout was used as just described, with a liquid hydrogen target which had Mylar walls only 15 mils thick.

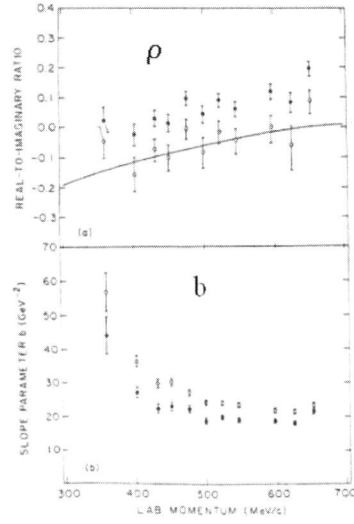

Fig. 16-23. pbar p elastic scattering.

Eleven beam momenta were chosen (387 to 682 MeV/c), corresponding to pbar-p effective masses from 1914 to 1988 MeV. Again, all the analysis was done at CWRU. There were 2000 2400-ft 1600 bits-per-inch magnetic tapes. That's 11 Gigabytes on two tons and $85K-worth of tape! The PDP 11/40 (16 kilobyte RAM, 2 Megabyte disk) was run continuously for more than one year. The results shown in **Fig. 16-24** indicate no structure between 1930 and 1980 MeV in the elastic channel (left) or in the pion-producing annihilation channel (right). The earlier claims for an "S-meson" at 1938 MeV would have made a bump many error-bars tall. (The figure has more than eleven points because the target could be divided into three slices of antiproton energy when the energy loss in the hydrogen is considered.) This experiment put an end to the controversy: no pbar-p bound state. The non-existence of the S-meson was a significant contribution to the picture of how quarks can arrange themselves: seemingly not as two quarks with two antiquarks. Graduate student Richard Marino wrote his dissertation on this experiment. "Search for the S meson in antiproton-proton interactions" *Phys. Rev.* **D34** 3332 1986.

The third experiment was a "quicky" measurement of the scattering of 514 and 633 MeV/c pbars by *nuclei*: Al, Cu and Pb. Strips of metal were mounted in place of the hydrogen target. The measured differential cross sections (**Fig. 16-25**) were fitted to an optical potential which had terms describing both elastic scattering and absorption (analogous to what happens when a beam of light passes through a cloudy crystal ball). It was useful to compare these data with those for *proton*-nucleus scattering,

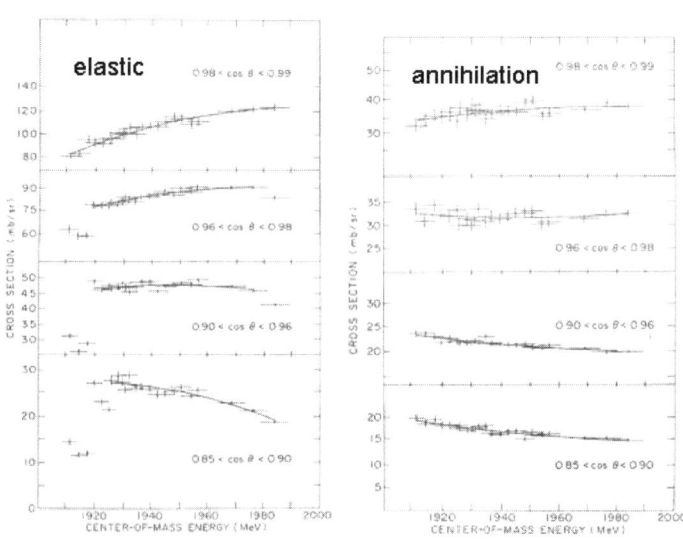

Fig. 16-24. Search for S(1938) meson. Left: elastic channel; right: annihilation channels.

since in *antiproton*-nucleus scattering, annihilation channels would make additional contributions to the absorption of the antiprotons (i.e. to the "cloudiness" of the crystal ball.) "Low energy antiproton-nucleus elastic scattering" *Phys. Rev.* **C30** 1080 1984.

Antiproton annihilations in deuterium

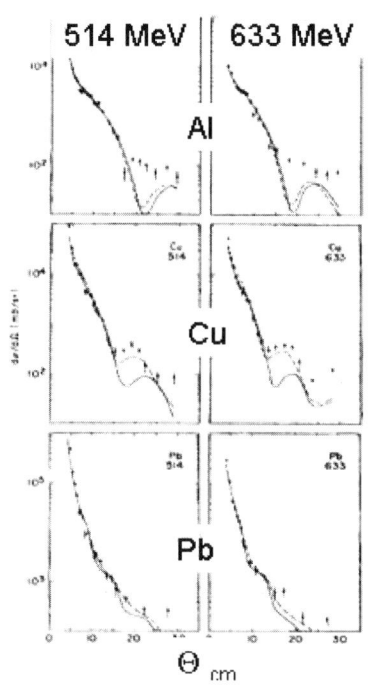

Fig. 16-25. Antiproton scattering from nuclei.

Ted Kalogeropoulos of the University of Syracuse headed a group which planned an experiment in the same beam at the AGS. Fickinger and Robinson and grad student Ramiro Debbe joined the collaboration. It was another "bump-hunting expedition", this one looking for neutral multipionic resonances. The technique was rather interesting: antiprotons were brought to rest in liquid deuterium, where they annihilated either with the proton or the neutron. Alongside the deuterium was a large bending magnet, with some scintillators and wire chamber planes before and after it. Charged particles from the annihilations passed through the magnet. The trajectory gave the momentum of the particle, and the time of flight gave its velocity, so that π^+, π^- and protons could be tagged. **Fig. 16-26** shows the setup.

Pions produced in pbar annihilations on a proton must have identical π^+ and π^- momentum distributions, but those produced in annihilations on a neutron can produce different π^+ and π^- spectra. In particular, if pbar annihilation at rest with a neutron produces a neutral mesonic object X^o along with a π^-, there should be a bump in the π^- momentum spectrum: pbar d $\rightarrow \pi^- X^o p_s$, where p_s, the spectator proton, is assumed to have negligible momentum. Here is the tricky part: to extract the signal from neutron annihilations, one subtracts the observed π^+ momentum distribution from the observed π^- distribution: making a so-called "difference spectrum". **Fig. 16-27** shows the result, where the large mountain is ascribed to an X^o with mass 1485 MeV. If this "bump" were significantly narrower, comparable to the experimental resolution, it would be much more credible as evidence for a new particle. "Evidence for a New State Produced in Antiproton Annihilations at Rest in Liquid Deuterium" *Phys. Rev. Lett.* **56** 211 1986 and *Phys. Lett.* **B180** 313 1986.

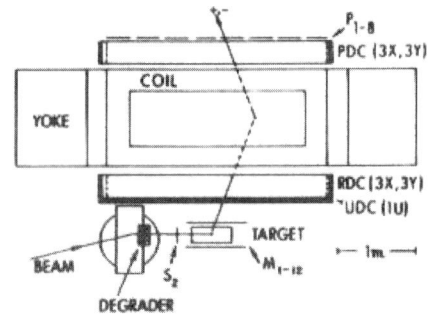

Fig. 16-26. Antiproton annihilations on deuterons.

Hyperon radiative decays

Having several years experience working with the Brookhaven and Syracuse groups at the AGS LESB, Fickinger and Robinson signed on in 1987 to a new and very different series of experiments in the same beam. This time, it was with a multi-university collaboration headed by Lee Roberts of Boston University. It involved the detection of photons from rare reactions and decays. A proposal was submitted by the CWRU group to the NSF for the support of their participation in these experiments. However, the group lacked the critical size needed to compete for funding, and the NSF declined to renew its support. Fickinger and Robinson nevertheless each spent about two months per year for three years at Brookhaven installing and operating major components of the data acquisition system. The CWRU PDP 11/45 and the Los Alamos data acquisition software were at the center of the series of runs.

Fig. 16-27. Search for new neutral meson states.

The first phase of the experiment was built around three large sodium iodide crystal detectors which were large enough and sensitive enough to detect and identify rare events. The first run involved stopping K$^-$ mesons in a liquid hydrogen target where they would be captured by a proton. Most of the time, the resulting product is a Λ or Σ hyperon along with a π meson. About once in a thousand captures, the hyperon is accompanied by a *photon* rather than a pion. Determination of the rates for these "radiative captures" is of theoretical interest in that the final states involve only the weak and electromagnetic interactions, free from the strong interaction complications associated with the pions. The key to the measurement is the ability to extract a tiny signal from the many other processes in which photons are produced, mostly from π^o or Σ^o decays.

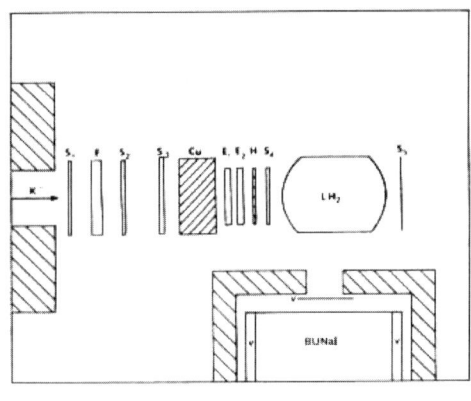

Fig. 16-28. Radiative capture experiment.

The BUNaI (Boston University sodium iodide) detector was a very large cylindrical single crystal (56 cm long, 50 cm diameter) surrounded by veto scintillators, enclosed in a lead and steel box. It was used in the second phase of the run. **Fig. 16-28** shows the setup, with the incident K$^-$ coming from the left, the copper degrader in the middle, and the liquid hydrogen target sitting opposite the window to BUNaI. **Fig. 16-29** shows the total photon spectrum at the top, where the bumps of interest at 220 and 280 MeV can hardly be seen. Below are blown-up portions

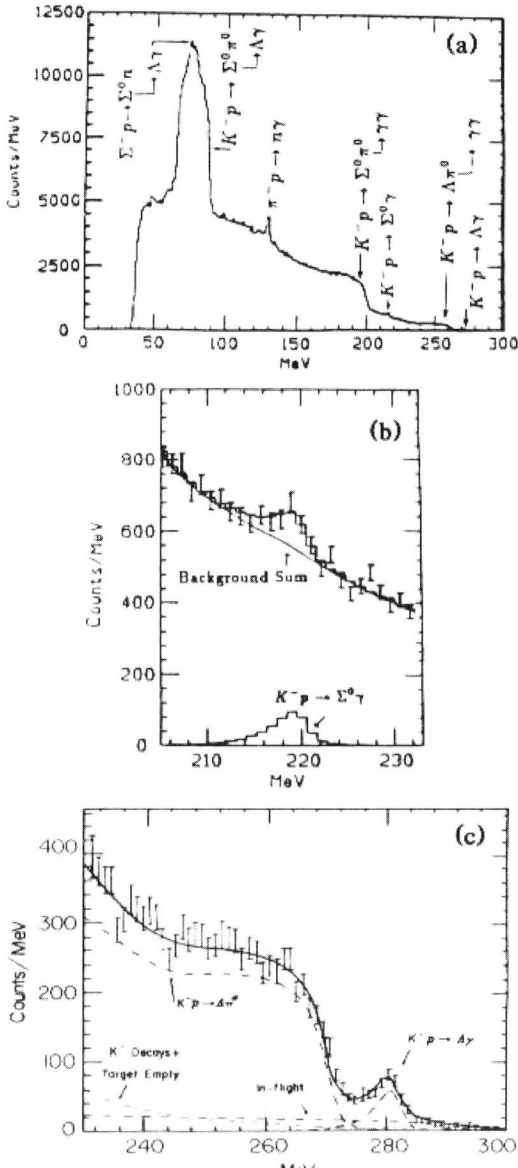

Fig. 16-29. Observation of radiative capture of K mesons. (a) total photon spectrum (b) $\Sigma\gamma$ signal (c) $\Lambda\gamma$ signal.

of the spectrum showing several hundred events in each of the two radiative capture peaks. "Radiative Kaon Capture at Rest in Hydrogen" *Phys. Rev. Lett.* **63** 1352 1989.

Stopping kaons and BUNaI were used once again, this time with liquid deuterium in the target. The goal was to detect the radiative process $K^-d \to \Lambda n\gamma$. This reaction can provide information on two interesting quantities: the radiative capture rate and the Λn scattering length. The former was determined directly from the size of a bump at the high end of the photon spectrum. The Λn scattering was of interest to theoretical groups who were studying low energy baryon-baryon scattering. The predictions of the Nijmegen Model relative to the shape of the high-end tail of the photon spectrum were found to compare favorably with the data. "Radiative Kaon Capture on Deuterium and the Λn Scattering Lengths" *Phys. Rev.* **C42** R475 1990.

The next phase of this experiment looked at the Σ^+ hyperons produced by stopping kaons: $K^-p \to \Sigma^+\pi^-$. A very small fraction of these Σ^+ decay by the emission of a photon, $\Sigma^+ \to p\gamma$. The trick in this experiment was the detection of the monoenergetic π^- to signal the creation of the Σ^+, along with the determination of the direction and energy of the photon. This latter measurement required a *segmented* photon detector, so BUNaI was retired and replaced by a 49-element rectangular array of NaI crystals: the "49'er". **Fig. 16-30** shows the setup, with the big pion detectors above and below the hydrogen target and the 49'er out in back. The resulting Σ^+ radiative decay sample of 408 events was twice the previous world total. The radiative rate was $(1.45 \pm 0.26)\,10^{-3}$ times the total decay rate. "A Measurement of the Branching Ratio for the $\Sigma \to p\gamma$ Decay" *Zeit. Phys.* **C42** 175 1989.

BUNaI and the 49'er served well as detectors of photons from stopping K's, but when the opportunity arose to borrow the "Crystal Box" from LAMPF, the group soon put it to work at Brookhaven. This detector consisted of 396 optically isolated NaI crystals. It covered about half of the total solid angle around the target, a great improvement over the earlier detectors. A sketch of the detector is shown in **Fig. 16-31**. The ability to catch 3, 4, or even 5 γ's in the Crystal Box opened up several interesting reactions. The 3 γ events included a signal for K⁻p → Λ π° followed by Λ → n γ and π° → γγ; some of the 4 γ events are from K⁻p → Σ° π° followed by Σ° → Λ γ and Λ → n γ. More than 1800 Λ radiative decays were found among the 3 and 4 γ samples, giving a branching ratio of $(1.75 \pm 0.15) 10^{-3}$. "The Weak Radiative Decay Λ → n γ and the Radiative Capture K⁻p → Σ(1385) γ" *Phys. Rev.* **D47** 799 1993.

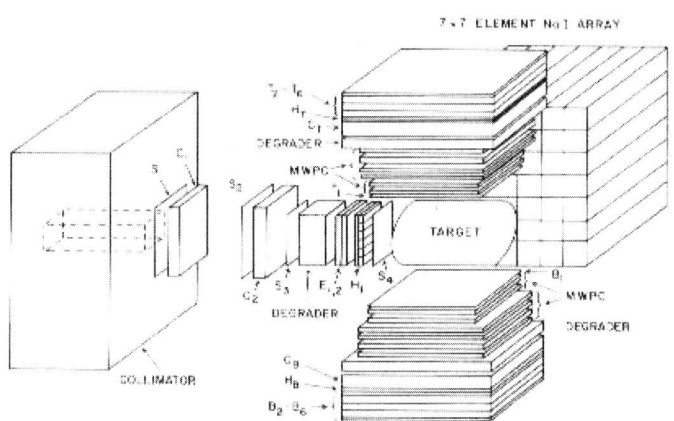

Fig. 16-30. Catching gammas from K⁻p interactions in the BU 7x7 detector.

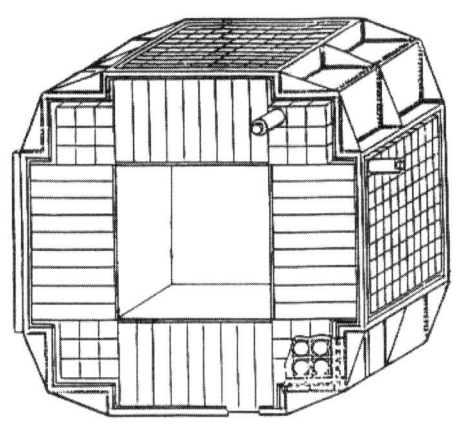

Fig. 16-31. The LAMPF Crystal Box detector with 396 NaI crystals.

This series of counter-based Brookhaven runs were the last experiments in which Fickinger and Robinson participated. Fickinger became director of undergraduate studies under chairman Lawrence Krauss who joined the department in 1993. Robinson took on the job as director of the introductory laboratories. Fickinger retired at the end of 1999 and Robinson in 2002. Experimental accelerator-based particle physics would resume at CWRU when, in the mid-1990's, theorist Cyrus Taylor put together a new team to work at Fermilab and CERN, work which continues to the present time. (Chapter 18)

Chapter 17   Condensed Matter Theory

Reitz,	**PTaylor,**	**Leff,**	**Coopersmith,**	**Silvert,**
1954-1965	1964-	1964-1971	1964-1969	1966-1969
Segall,	**Petschek,**	**Lambrecht,**	**Mathur**	
1968-2000	1983-	1996-	1995-	

Starting in the early 1940's, the experimental study of electrons in metals and their interaction with the atoms of the crystal lattice played a major role in physics research. The goal was to measure the mechanical, thermal, optical, and electrical properties of metals, insulators, and other forms of condensed matter. These measurements would then lead to the development of theoretical models based on the laws of electromagnetism and quantum mechanics. The models in turn would lead to the creation of new materials for applications in electronics, optics and many other areas. We have looked at the *experimental* work of Smith, Gordon, Eck and Schuele in Chapters 7 and 12. Here we shall sample some of the work done by our condensed matter *theorists* in explaining the experimental observations in terms of fundamental physics principles.

John Reitz

Fig. 17-1. John Reitz.

Les Foldy (Chapter 9) was the first theorist in either the Case or the WRU department. In addition to his papers on atomic and subatomic phenomena, he published work in condensed matter theory, often in areas of interest to his experimentalist colleagues. In 1954 the Case department added a dedicated condensed matter theorist, **John Richard Reitz**. **Fig. 17-1** The 31-year old assistant professor had completed his BS at Case in 1943 and was at the Harvard Underwater Sound Laboratory during the war (another "Dayton Miller acoustics" contribution to the war effort). Reitz' PhD at Chicago in 1949 was based on calculations of atomic electron wave functions. They were performed on the Electronic Numerical Integrator and Calculator (ENIAC) at the US Army's Aberdeen Proving Grounds. "The Effect of Screening on Beta-Ray Spectra and Internal Conversion" *Phys. Rev.* **77** 10 1950. Reitz then spent three years on the staff of the Los Alamos Scientific Laboratory, LASL.

Electrons in materials

One of Reitz's first works at Case was a 95-page "how-to handbook", which he described as an "attempt to consolidate the methods developed during the past fifteen years for calculating electron wave functions and electronic structures in solids." "Methods of the One-Electron Theory of Solids" *Sol. St. Phys.* **I** 1 1955.

He then turned to calculations of properties of materials being measured in Chuck Smith's lab at Case. It seems that Reitz got directly involved with the experiments as well. "Elastic Constants by the Ultrasonic Pulse Echo Method" *J. Appl. Phys.* **29** 683 1958. This was an experimental paper with Stephen Eros, one of C. Smith's students. In a paper with Smith, Reitz looks at the behavior of the shear moduli (e.g. the resistance of the material to being deformed) of dilute alloys as a function of the average number of electrons per atom, which in turn depends on the concentration of foreign atoms. Experimentally observed discontinuities in the moduli at particular values of the electron/atom ratio are successfully reproduced in the model. The model proposed that at certain concentrations, the extra electrons upset the symmetry of the crystal so that it loses some of its stiffness. "Calculation of the Elastic Shear Constants of Magnesium and Magnesium Alloys" *Phys. Rev.* **104** 1253 1956 (with experimentalist Chuck Smith).

Reitz coauthored a largely experimental paper with PhD student W. J. Tomasch in which the output voltage of thermocouples was measured as a function of the composition of the alloys being tested. They observed rather striking variations in the voltage as a function of the percentage of indium. These were ascribed to specific changes in the electronic band structure. "Thermoelectric Power of Dilute Indium-Lead and Indium-Thallium Alloys" *Phys. Rev.* **111** 757 1958. He then joined Bill Gordon and Tom Eck in publishing a paper based on their de Haas-van Alphen measurements (see Chapter 12). "Evidence for Spin-Orbit Splitting in the Band Structure of Zinc and Cadmium" *Phys. Rev. Lett.* **7** 334 1961. In "Peltier Coefficient at High Current Levels" *J. Appl. Phys.* **32** 1623 1961, Reitz discusses the optimum choice of materials for semiconductor junction devices. *(The Peltier effect concerns the transfer of heat from one place to another when current passes through a junction from one metal to another.)*

In an exercise quite different from his electrons-in-matter research, Reitz joined Foldy in tackling a problem in classical electromagnetic hydrodynamics. Starting with Maxwell's equations, they derived expressions for the force and torque acting on a low-conductivity sphere moving in a conducting fluid at right angles to an external magnetic field. "The force on a sphere moving through a conducting fluid in the presence of a magnetic field" *Jour. Flu. Mech.* **11** 133 1961.

While at Case, Reitz maintained his connections with LASL, spending extended periods at the New Mexico lab. He worked there on a variety of theoretical topics. "Thermoelectric Properties of the Plasma Diode" and "Efficiency of the Plasma Thermocouple" *Jour. Appl. Phys.* **30** 1439 1959 and **31** 723 1960. "Elastic Scattering of Slow Electrons by Cesium Atoms" *Phys. Rev.* **131** 2101 1963. "Magnetic Breakdown in Metals" *J. Phys. Chem. Solids* **25** 53 1964.

In 1964, John Reitz accepted a position with the Ford Motor Company in Dearborn MI where he remained until his retirement in 1987. With Case colleague, Fred Milford (Chapter 9), Reitz authored a widely-used undergraduate text which has gone to four editions: *Foundations of Electromagnetic Theory* (Benjamin Cummings 1993).

Phil Taylor

In 1962, a second condensed matter theorist joined the Case department. The 25-year-old London-born post-doc, **Philip L. Taylor**, had done his BSc at London and his PhD at Cambridge. His doctoral research included an analysis of how electrons are scattered by impurities as they bounce around in metals. **Fig. 17-2.**

A condensed matter glossary

*This might be a convenient place to enumerate some of properties of materials which are fair game for experimentalists to discover and theorists to explain. Basically, it's all electromagnetism and quantum mechanics. The ions and electrons have electric charge; the atoms and molecules might have magnetic moments; the molecules might have electric dipole moments (positive at one end, negative at the other), and they all talk to each other through Maxwell's equations. The quantum mechanics comes in because all the pieces are described by wave functions which specify positions, momenta and energies. You might consider the next six paragraphs as a **glossary** which you may come back to later as the terms come up.*

Fig. 17-2. Phil Taylor.

The **thermal conductivity** is a measure of the rate at which heat flows through a sample when its two ends are held at different temperatures. The **electrical conductivity** determines how much current flows when a voltage is applied to a sample. It depends on how easily the electrons move through the material. The **Hall effect** describes the appearance of a voltage at right angles to a current flowing through a conductor when a transverse external magnetic field is applied. The **thermoelectric power** measures the electric field which appears within the sample when a temperature gradient is applied across the sample. The **dielectric constant** is a measure of how much the electric field inside the sample is reduced when its molecules line up with their own dipole fields opposing the externally applied field.

Pyroelectricity is a rather subtle effect which occurs when a dielectric material is heated. In such a material, all the little electric dipoles line up with an applied electric field, so that surface charges appear on the opposite faces of the sample. If the sample is heated, the dipoles lose some of their alignment and the effective surface charge changes, resulting in a measurable potential difference between the surfaces. These materials can be used as detectors of heat, for example, in infra-red night goggles.

Ferroelectrics are materials which exhibit an electrical polarization even in the absence of an applied electric field. The direction of this polarization can be re-oriented by the application of a strong electric field. The name *ferroelectric* was chosen in analogy with *ferromagnetic* because both types of materials display hysteresis behavior on field reversal. Iron has nothing to do with ferroelectrics. The **piezoelectric effect** refers *to the appearance of a static electric field in certain materials when they are mechanically stressed.*

Now, some magnetic properties: **magnetic permeability** *measures the effective strength of the magnetic field within the sample when an external magnetic field is applied. In* **diamagnetic** *materials, the atoms have no permanent magnetic moment, but the atomic electronic orbits shift a bit so as to slightly* reduce the field. **Paramagnetic** *materials do have atoms or molecules with small permanent magnetic moments, and these line up to slightly* increase the field. **Ferromagnetic** *materials have very strong magnetic moments so that the field in the sample might be several thousand times stronger than the applied field. The term* **ferrimagnetism** *is used for materials in which the reinforcement of the field is lessened because of magnetic interactions internal to the lattice which force some of the dipoles to point the wrong way. This is beginning to sound like a biology textbook – but it's a fact that materials can respond in all these different ways to applied fields.*

The **speed of light** *in the material depends on two of the properties listed above, the dielectric constant and the magnetic permeability. The dielectric constant can depend on the frequency of the applied electric field (e.g. the color of the light), and so therefore does the speed of light. The* **index of refraction** *of the material is the ratio of the speed of light in vacuum to that in the material. The* **optical activity** *of a material is a measure of how much the direction of polarization of a beam of light is rotated as it passes through the material.*

The importance of **band structure** *in materials was described in Chapter 12. Recall that this refers to the allowed energy levels for electrons in a material, and how certain bands of energy are not allowed. The existence of these bands is extremely important in determining the electrical and optical properties of the materials. The whole of semiconductor technology depends on band structure. The experimentalists and the theorists of the department have been measuring, calculating, and predicting band structure for the past 50 years.*

In his first paper at Case, Taylor looked at the connection between the electronic band structure of metals and the temperature dependence of the voltage produced when one metal is placed in contact with another metal. This is related to the thermoelectric power studied by his predecessor, John Reitz. (You may have experienced such a voltage when you touched a silver filling in your tooth with a steel fork.) Taylor pointed out that the inclusion in the theory of interactions between the electrons and the phonons leads to a better understanding of the experimental results. "The Thermoelectric Power of Metals" *Phys. Lett.* **3** 245 1963.

One of Taylor's more important contributions to the theory of electrons in solids was a paper entitled "The Boltzmann Equation for Conduction Electrons" The Boltzmann equation describes the probability that an electron will change its momentum from one value to another. In the case where there are no fields or temperature gradients, the equation predicts that the rate at which electrons leave a certain volume in momentum space equals the rate at which they enter it. That is, not much is happening. Taylor looks at the case where there *are* electric and magnetic fields, as well as a possible temperature gradient, so that there is a lot of direction-dependent rearranging of the electrons and their vector velocities. *Proc. Roy. Soc (London)* **A275** 209 1963.

During his first dozen years at Case, Phil Taylor worked on the theory of electrons in disordered systems. Dilute alloys, in which the uniform crystal structure is disrupted by foreign atoms, and glasses, in which there is no uniform structure at all, are examples of disordered systems. The subsequent decade-and-a-half would be Taylor's "polymer period", and then in the 1990's he would tackle the theory of liquid crystals.

Disordered Systems

The properties of crystals in which small amounts of foreign atoms are introduced provide interesting tests for the theorist. Taylor published two papers showing that there should be band gaps in the allowed energies of electrons in an amorphous semiconductor which has been modified by the introduction of randomly distributed inhomogeneities. "Energy Gaps in Disordered Systems" *Proc. Phys. Soc.* **88** 753 1966 and **90** 233 1967.

In 1973, Taylor and his student James Gubernatis tackled the problem of how the magnetism of amorphous ferromagnetic materials behaves as they are warmed up. Recall that the degree of magnetization depends on the degree of alignment of the electrons' and atoms' magnetic moments, and that these tend to get jostled around as the temperature is raised. If one looks at such a system quantum mechanically, there can be certain wave-like excitations in which the magnetic moments of neighboring atoms change directions smoothly across the lattice, like a line of twirling ballerinas, each one a bit out of phase with her neighbor. These excitations are called spin-waves, and their existence affects the dependence of the magnetization on the temperature. "Spin-wave Spectrum of an Amorphous Ferromagnet" *Phys. Rev.* **B9** 3828 1974.

Thermoelectric effects in dilute alloys would occupy Taylor's interests for the better part of a decade. He points out some of the pitfalls in this endeavor in a whimsical introduction to a review paper. "The theory of thermoelectricity is to be found in a looking-glass land where all our best-loved folk-theorems are cruelly violated. Everyone knows that pseudopotentials are better than Hartree potentials in calculating scattering amplitudes, that higher-order electron-phonon interactions terms are negligibly small, and that the electronic relaxation time for phonon scattering may be approximated by a function that is symmetric about the Fermi energy. And yet .... Thermoelectricity is a subject whose siren song has lured many an unwary traveler into errors for which history has been brutally unforgiving." *Thermoelectricity in Metallic Conductors* (Plenum Publishing 1978)

Polymers

Starting in the mid-1970's, the experimental and theoretical study of polymers became a major activity in the department. Since that time there have been ongoing collaboratories with researchers in the materials science and engineering, macromolecular science and engineering, chemical engineering and chemistry departments, as well as extramural collaborations with Kent State University and the University of Akron. The earliest polymer work in the region dating back to the nineteenth century developed in the "Rubber City", Akron.

Polymers comprise a special class of condensed matter, consisting of chains of repetitive groups of atoms called monomers. Some polymers occur naturally as in tars and fibers and some, like nylon and vinyls, are produced from petrochemicals. Thousands of different structures display a wide variety of mechanical, thermal, electrical and other properties. Phil Taylor became interested in understanding these materials and their properties in terms of the underlying physics.

Taylor, and his collaborator Anton Hopfinger of the CWRU Department of Macromolecular Science, published a series of papers on the problem of crystallization and melting of polymers. By computing the energy of interaction between each monomer and its nearest and next-to-nearest neighbors along the chain, and it nearest neighbors in adjacent chains, they were able to determine the temperature at which "melting is approached through increasing anharmonicity as phonon interactions become dominant and changes in lattice cell dimensions occur." "Theory of melting in simple polymers. I." *J. Chem. Phys.* **67** 353 1977.

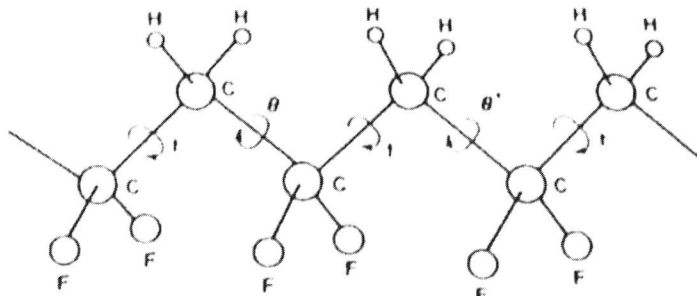

Fig. 17-3. A typical polymer structure.

In the introduction to a 1979 paper, Taylor and his collaborators write, "We believe this to be the first reported work in which such a rich variety of stable and metastable phases of a polymer has resulted from a calculation in which there are no adjustable parameters, and of which the starting point is the calculation of potential energy by means of conformational analysis." **Fig. 17-3** shows the molecular structure of the "PVF2" polymer. The object of the calculation was to look at the energy of the system as a function of the angles between successive monomers and between adjacent rows of polymers, for varying applied stresses and temperatures. Minima in these energies successfully predict different experimentally observed phases for PVF2. "Theory of Structural Phase Transitions in Poly(Vinylidene Fluoride)" *Phys. Rev. Lett.* **43** 456 1979. (in collaboration with Hopfinger)

When an electric field is applied to certain polymers, there results a degree of alignment of subunits called crystallites. This process of reorientation is called "poling". Taylor became interested in the process and the way in which a wave of realignment moves through the material. A computer simulation of the motion of a kink moving along the polymer was used to determine the speed of propagation, and this compared reasonably well with experimental observations. "Kink Propagation as a Model for Poling in PVF2" *Phys. Rev.* **B21** 3700 1980.

The **Hall effect** has been known since 1879, but it was not until about 100 years later that it was discovered that, at low temperatures, when the magnetic field across a two-dimensional (i.e. very thin) sample is raised, the Hall voltage rises not smoothly, but in jerks, like the steps of a staircase. Taylor and his student Olle Heinonen produced a theoretical model of the "quantum Hall effect" based on a thermodynamical equilibrium approach, which correctly described the experimental behavior. In this model, the electrons in the bulk material travel in circles with quantized energies, contributing nothing to the current. At the edges of the conductor, however, the electron orbits are interrupted by the surface, and a net current flows, with negligible dissipation. The following year, a second paper addressed the observation that at a certain critical value of the current density, the resistance to the flow of current rises precipitously. Noting that the breakdown occurred when the drift velocity of the electrons reached the speed of sound in the material, they decided to investigate the role of phonons (i.e., those quanta of vibrational energy in the lattice) in the onset of resistance. It appears that the interaction with phonons disturbed the neatly filled energy levels of the electrons when the electron density gets too high. "Conductance Plateaus in the Quantized Hall Effect" *Phys. Rev.* **B28**, 6119 1983 and "Electron-Phonon Interactions and the Breakdown of the Dissipationless Quantum Hall Effect" *Phys. Rev.* **B30**, 3016 1984.

Experiment had shown that certain membranes made from polymers exhibited markedly different transport properties for positively and negatively charged ionic species. Taylor and colleagues applied their modeling techniques to this phenomenon of "permselectivity". As in all their calculations, energy is the key. This time it is the electrostatic energy between the ions and the membrane and the elastic energy of the polymer. The goal is an understanding of the preferential direction of ionic flow, based on the properties of the surface presented to the bath by the membrane. The application of effects such as these may provide better materials for such technologies as fuel-cells. "Simple Model for Clustering and Ionic Transport in Ionomer Membranes" *Macromolecules* **17**, 1704 1984.

A particular one-page paper stands out as unusual. It was written by Taylor and colleagues, Les Foldy and Rolfe Petschek, and concerns the application of some of these polymer materials to the detection of high energy particles. They point out that incident particles can heat the polymer, destroying its polarization, resulting in a pyroelectric voltage which can be detected. Whether this suggestion was acted upon by particle physics experimentalists is not clear. "Pyroelectric Materials as Electronic Pulse Detectors of Ultraheavy Nuclei" *Phys. Rev. Lett.* **54** 1089 1985.

Everyone knows that when a liquid is cooled it eventually changes phase and becomes a solid. Taylor and his post-doc Vladimir Pines investigated the way in which the transition spreads through the liquid. You may have admired the patterns of ice which grow across a cold window pane, or the fern-like patterns of salts coming out of solution in a petrie dish. At the simplest level, the key parameters are the latent heat and the rate of heat flow. The speed of the phase boundary depends on how much heat must be removed and how fast can it be moved away. "Stability and Instability in Crystal Growth-I. Symmetric Solutions of the Stefan Problem" *Phys. Rev.* **B32**, 5362 1985.

Most magnetic materials involve metals with unpaired electrons whose magnetic moments can become aligned. Taylor and postdoc Kashi Nath looked into the possibility of building a "ferromagnet" out of carbon, hydrogen, oxygen, and nitrogen – the stuff we are mostly made of. They point out that one way to get the magnetic moments to line up would be to hang the atoms on polymer chains, most probably arranged in parallel helices. They calculated the magnetization and critical temperature for such arrangements for various numbers of unpaired spins per helix loop, and find that "macroscopic ferromagnetism might occur in a bulk sample of these materials." "Electronic Structure and Magnetic Properties of some Possible Organic Ferromagnetic Polymers" *Molecular Crystals and Liquid Crystals* **205** 87 1991.

The interaction between ions in a polymer depends to some extent on the screening of their fields by electrons. "…many-electron theory can be used to derive an approximate expression for the effective Coulomb interaction between pairs of charges in an insulating organic solid". "Dielectric Screening of Coulomb Interactions in Polymers" *Macromolecules*, **25** 1694 1992.

Liquid Crystals

In the early 1990's, Taylor turned his attention to the study of polymer liquid crystals. This work would complement the experimental work on liquid crystals by Chuck Rosenblatt who joined the department in 1987 (Chapter 18). Liquid crystals consist of molecules consisting of a few hundred atoms which are typically longer than they are thick and which tend to line up roughly parallel to one another. There are various ways in which the molecules might align themselves, for example mostly parallel, but with their centers randomly distributed in the crystal (called **nematic** - Greek for thread – or like uncooked spaghetti in the box), or mostly parallel, with centers lying in layered planes normal to the molecules' axes (called **smectic** - Greek for soap – more like uncooked lasagna in the box). The degree of orientation tends to get washed out as the temperature is raised. Liquid crystals are bi-refringent, that is, their index of refraction depends on the angle between the plane of polarization of the light and the "director" or principal axis of the sample.

It was discovered that these liquid crystal molecules could be attached in an orderly fashion along the length of a polymer. In this way, they could still line up to control the index of refraction, but they are anchored in place and no longer float around in liquid form. It was a natural move for Taylor to apply his polymer expertise to such materials. The theorist's interest would be to determine the degree of the liquid crystal alignment along with the thermal, mechanical, electromagnetic, and optical properties. With the widespread application of liquid crystals to PC monitors, TV screens, and other displays, for example, it is important to know how completely and how quickly one can change the optical properties of the liquid crystal with the application of external electric fields.

Another thing which might be asked is at what temperature will a solid crystal change into a liquid crystal. Again one must look at the energy associated with the interactions among the molecules and the various degrees of freedom within the molecules. In one paper Taylor determines how the transition temperature decreases as the molecules are elongated by the introduction of monomers which act as spacers. "Theory of the Solid-Nematic Transition in Thermotropic Main-Chain Liquid-Crystalline Polymers" *Phys. Rev.* **A44** 821 1991.

Between 1991 and 2003 Taylor, along with members of his group, published an additional eighty papers on the theoretical modeling of polymers and polymer liquid crystals. Many of these papers were co-authored by his colleague, Chuck Rosenblatt, who leads the department's experimental liquid crystal program.

Phil Taylor authored a widely-used advanced text in 1970: *A Quantum Approach to the Solid State* (Prentice Hall, 1970, 321 pp). Thirty two years later, he and his former student, Olle Heinonen (then at the University of Central Florida), collaborated on a updated version which reflects the enormous advances made in the intervening years: *A Quantum Approach to Condensed Matter Physics* (Cambridge University Press, 2002, 424 pp). He remains active in teaching and research. Taylor has always been a champion of the environment and has regularly taught a popular course in "Environmental Physics". His research has been funded principally by the National Science Foundation, which has made possible the support of a long line of graduate students and post-doctoral collaborators. Taylor has been research advisor to about twenty doctoral students, many of whom returned to campus in 1997 on the occasion of a Festschrift honoring him on his 60th birthday.

Harvey Leff, Michael Coopersmith and William Silvert

Between 1964 and 1966, three new young theorists were hired, two at Case and one at Reserve. Each of them, like Phil Taylor, did "condensed matter theory". New faculty were needed in each department to handle increasing teaching loads. One attraction of hiring theorists is that they require no initial set-up expenditures for labs and equipment. Ultimately, each of these three new theorists would remain six years or less.

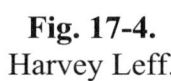

Fig. 17-4.
Harvey Leff.

The 27 year-old **Harvey Leff** had completed his doctorate under Max Dresden at the University of Iowa in 1963. **Fig. 17-4.** His principal research interest was statistical mechanics. During the six years he spent at Case and CWRU, Leff published a dozen papers on a variety of problems in statistical mechanics, papers which have been described as significant attempts to explain the complicated mathematics associated with that area of physics theory. In particular, he sought ways to link the macroscopic variables used in thermodynamics with the descriptions used in statistical mechanics. Among his publications at Case are the following representative papers: "Class of Ensembles in the Statistical Theory of Energy-Level Spectra" *J. Math. Phys.* **5** 763 1964. "Statistical Thermodynamics of Incompletely Specified Systems" *Jour. of Chem. Phys.* **41** 596 1964. "Asymptotic Densities in Statistical Ensembles" *Phys. Rev.* **136** A355 1964. "Difference-Equation Solutions for the Linear Ising Model and Nearest-Neighbor Fluid" *Amer. Jour. Phys.* **36** 591 1968.

Leff's special interest in teaching is illustrated by an article he wrote for the *American Journal of Physics*, a favorite journal of teachers of university physics. The abstract closes with the remark, "The canonical structure, which is demonstrated, is pedagogically useful as (1) a memorization device, (2) an example of the utility of Legendre transformations in statistical mechanics, and (3) an aid in understanding the relationships of the various ensembles to one another." "On the Connections between Thermodynamics and Statistical Mechanics" *Amer. Jour. Phys.* **37** 65 1969.

Leff was promoted to associate professor in 1969 but, as was the case for several others among his colleagues, his 1970 bid for tenure was not successful. Leff was a popular and successful teacher, with 100 of his students signing a petition in his behalf. By the end of the semester, he had accepted a position as chair of the physical sciences department at Chicago State College. In 1979, he joined the Energy Information Systems Program at Oak Ridge National Laboratory.

The second 1964 addition to the department was **Michael H. Coopersmith**. Fig. 17-5. The 28-year-old was born in Brooklyn, did his BA at Swarthmore, and his PhD in 1962 at Cornell where he worked with Robert Brout on the theory of phase transitions. His dissertation was on the "Statistical Mechanical Theory of Condensation". *Phys. Rev.* **130** 2539 1963. He was a National Science Foundation Fellow at the École Normale Supérieure and then spent two years as a postdoc at the University of Chicago.

Fig. 17-5.
Michael Coopersmith.

One of the problems Coopersmith tackled, first at Chicago and then at Case, was the behavior of free electrons dancing around in a cold helium vapor. This is quite different from electrons in a crystal lattice, where the forces are long range. Here the helium atoms are hard spheres, and the electrons essentially bounce off them elastically. Experimentally, it had been observed that the mobility of the electrons drops by a factor of a thousand when the pressure of the vapor is raised above a certain critical value. Coopersmith reported that his "bubble model", which looks at the scattering of electrons within a cluster of stationary, hard-core helium atom scatterers, provides an adequate description of the situation. "Multiple Scattering and Many-Body Theory: Free Energy of Electrons in Helium" *Phys Rev.* **139** A1359 1965. "Relaxation-Time Approximation for the Mobility of Electrons in Helium" *Phys. Rev.* **161** 168 1967 (with NASA-Lewis collaborator, Harold Neustadter).

Coopersmith and Leff teamed up to produce two even more mathematical statistical mechanics papers on the distribution of hypothetical hard-core particles in a one-dimensional fluid. "Translational Invariance Properties of a Finite One-dimensional Hard-core Fluid" *J. of Math. Phys.* **8** 306 and 434 1967.

In a series of five single-author papers, Coopersmith looked at the behavior of magnetic or fluid systems undergoing phase transitions near a critical point. The work was summed up in the last paper, whose abstract states that "we have arrived at a picture of a phase transition which is in accord with all currently observed behavior in the neighborhood of a critical point." "Analytic Free Energy: A Basis for Scaling Laws *Phys. Rev.* **172** 230 1968.

Coopersmith resigned from the CWRU department in the summer of 1969. He took a position on the physics faculty at the University of Virginia where he is currently a member of the emeritus faculty.

Fig. 17-6.
William Silvert.

The third among the new condensed matter theorists was **William L. Silvert**. **Fig. 17-6.** Silvert did his PhD at Brown under future Nobelist Leon Cooper. Cooper is the "C" of the BCS theory which is described briefly in Chapter 15. Silvert's research was on the theory of superconductivity. He published a series of papers on surface superconductivity and on non-homogeneous superconductors as a post-doc at Michigan State. In 1966 he was recruited by Chairman Chandrasekhar to a position as assistant professor in the WRU department. Silvert was given a year's leave of absence so that he could participate in a US-USSR Academies of Sciences exchange program. He spent the 66-67 academic year at the Institute of Theoretical Physics in Moscow. His paper on non-homogeneous superconductors was published soon after his move to Cleveland. The problem which Silvert addressed was the modification of the electron-electron interaction when two different superconducting metals are placed in contact. "Solution of the linearized energy-gap equation in nonhomogeneous superconductors" *Zh. Eksp. Teor. Fiz.* **53** 1693 1967; *Sov. Phys. JETP* **26** 971 1968. Silvert soon moved on to faculty positions at the University of Kansas and at Dalhousie in Halifax.

Segall: band structure in solids

In Chapter 12 we wrote about the experimental work done by Gordon and Eck on the behavior of electrons in a metallic lattice and what one could learn from measuring the magnetization of the material as an applied magnetic field is varied. The looping orbits of the electrons in the material provide information on its band-structure. We discussed there how determination of the Fermi surface and related band-structure fixes the allowed energy levels for electrons which find themselves in the periodic potential of a crystal lattice. It is the *disallowed* energy levels, in the so-called "energy gap", which give the material its interesting properties.

It was clear that the newly merged CWRU department would benefit from the addition of an expert on the theory of electrons in materials. Benjamin Segall joined the department in 1968 and would be productive in this area for the next thirty years. He completed his BS at Brooklyn College in 1948 and his PhD at the University of Illinois Urbana in 1951. **Fig. 17-7.** His dissertation research was in particle theory, under Geoffrey F. Chew. It involved the calculation of the cross section for the photodisintegration of the deuteron (γ d $\rightarrow$p n) in the 50 to 100 MeV range. After postdoctoral positions at Illinois and at the Institute for Theoretical Physics in Denmark, Segall decided in 1956 to leave the academic world and join the research department of General Electric Laboratory in Schenectady. He would remain there for thirteen years. Not only did he leave academe, but he also left behind his work in particle and nuclear physics, switching abruptly to condensed matter theory. He would be calculating electronic band structure within the year.

Fig. 17-7. Ben Segall.

Segall's early work at GE was the development of a technique to calculate band structure in a crystal lattice with more than one atom per unit cell. At the end of the 1950's, the first digital computers became available for research laboratories, and Segall had access to state-of-the-art computer power at GE. (typically 16K RAM, punch card or paper tape input, $50-l00K each!)

For a theorist, band structure and Fermi surface calculations begin with an assumed crystal structure and a form for the periodic potential in which the electrons find themselves, and proceed iteratively to find self-consistent wave functions. Most of the technical expertise is in the computational methods used in seeking the solution. Segall worked on the electron structure of such pure metals as copper and aluminum, and on a variety of doped semi-conductors. It gets interesting when impurities are included. The results can be compared with experimental measurements of electron mobility, resistivity, magnetoresistance, the Hall coefficient, and optical properties. Segall published about two dozen papers while at GE, most of them in the *Physical Review.* A significant number were written in collaboration with experimenters, and addressed in-house measurements.

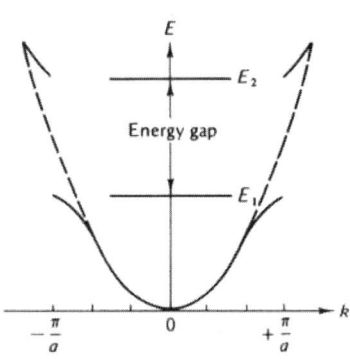

Fig. 17-8. Energy vs. wavenumber in a periodic potential.

*An aside on the quantum mechanics "particle-in-a-box" problem and band structure. A particle in a one-dimensional infinite well of width **a** must have a wavelength so that an integral number of half wavelengths just fits in the well, $a=n\lambda/2$; just like a standing wave on a piano string. The electron's momentum is $p= h/\lambda$, its wave number is $k=2\pi/\lambda$, and its kinetic energy is*

$E = p^2/2m$. Therefore, only certain values of λ and k and p and E are allowed. A plot of E vs k for a single particle is a parabola. For a periodic potential such as one would find in a one-dimensional crystal, the allowed values of E spread out into bands of energy. This happens because the E vs k curve takes abrupt jumps at the single particle "square-well" k values. As a result, there are big gaps in the allowed energies. That means, for example, that no electrons have energy between E1 and E2 in **Fig 17-8**. It's the size of this band-gap which determines many important properties of the material. Now, all the theorist has to do is find the allowed and forbidden energy values for a three-dimensional periodic potential and predict the material's electrical and optical properties.

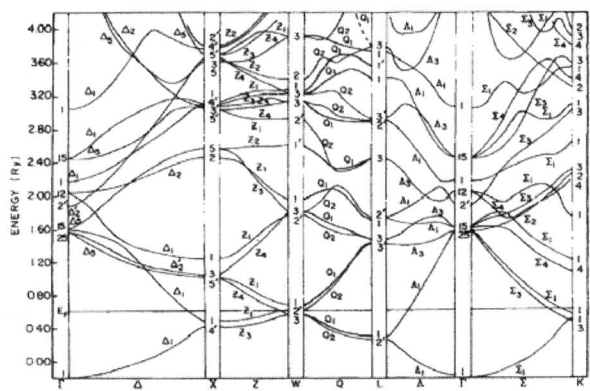

Fig. 17-9. Computed electronic energy levels for a typical crystal..

A portion of the abstract from a 1974 paper gives an idea of the technique: "A new empirical energy-band parametrization scheme based on the Green's-function method has been developed and was applied to Cu and Ag. The scheme utilizes the logarithmic derivatives associated with an *ab initio* muffin-tin potential $V^{(0)}(r)$.. The scheme can be understood in terms of the addition to $V^{(0)}$ of energy and angular momentum dependent square-well potentials, the depths of which are adjusted to yield the correct (empirical) energy bands." The "muffin tin potential" sets the potential to some constant level in a sphere around each atom, and zero elsewhere. This potential is then "tuned-up" to match the experimentally observed band-structure. This was one of a series of papers co-authored with post-doc An-Ban Chen. *Phys. Rev.* **B12** 600 1975.

As an example of the electronic energy levels which Segall and his group have calculated, we show in **Fig. 17-9** a rather amazing plot from a paper on the band structure of aluminum. It is a far cry from the little broken parabola in Fig. 17-8. The vertical stripes correspond to different directions in the crystal lattice. One can see portions of many parabolae. Perhaps a bit more understandable is the upper curve in **Fig. 17-10** which shows the calculated x-ray spectrum which follows from the electron energies. (The x-rays are emitted when an electron drops from one level to another, so this spectrum corresponds to

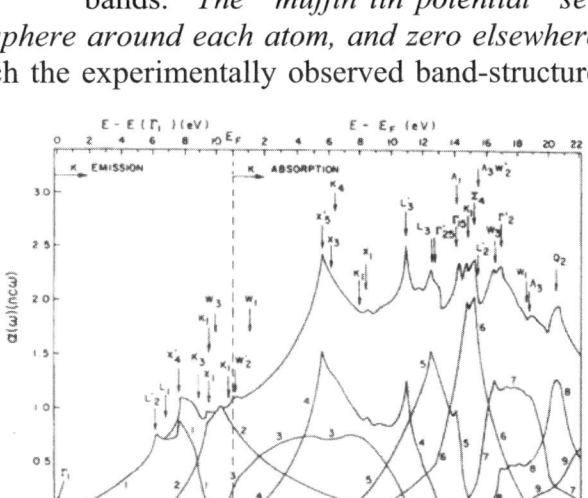

Fig. 17-10. Corresponding computed x-ray spectra for a typical crystal..

energy differences.) Measurements of the x-rays provide a test of the theoretical calculations. This work was done with grad-student Frank Szmulowicz. "K x-ray absorption in aluminum" *Phys.Rev.* **B21** 5628 1980.

Segall was interested in the properties of a great variety of materials, some of which had been measured experimentally, and some of which were hypothetical materials which had not yet been fabricated. In addition to establishing mechanical properties like lattice structure and elasticity, calculations based on first principles, *e.g.* atomic numbers, can predict the dielectric, optical and electrical transport properties crucial to technology. As in the polymer and liquid crystal studies done by Phil Taylor and Rolfe Petschek, the ultimate goal is to design structures with interesting technological potential.

Segall tackled a very different type of calculation when he joined his colleague, the versatile theorist Les Foldy, in a series of papers on the lattice dynamics of certain alkali halide "mirror pairs". These involved pairs of elements such that one is a positive alkali ion, like potassium, and the other is a negative halide ion, like chlorine, which sits at the opposite end of the line in the periodic table. Another such pair is rubidium and bromine. It was found that materials incorporating such pairs have remarkably similar optical properties, as in the comparison between $RbCl_xBr_{1-x}$ and BrK_xRb_{1-x}. "Anion-cation mirror symmetry in alkali halide ion dynamics II" *Phys. Rev.* **B29** 2293 1984.

Band-structure calculations can be done not only for bulk materials consisting of a single lattice, but for *heterostructures* (two solids in close contact) or *superlattices* (inter-leaved layers of two solids). Of special interest to semiconductor technology are interfaces between two different solids, for example, the contact between a semiconductor and a metal. It is important to understand how the electron energy levels behave in the region where the two lattices meet. Another possibility is the study of the interface between two solids of the same material, but with different crystal alignments.

Segall and his colleague **Walter Lambrecht** did extensive calculations on wide band-gap semiconductors such as silicon carbide and aluminum and gallium nitride. These materials have band-gaps well into the ultraviolet *(that is, you need an ultraviolet photon to promote an electron across the gap)* and have robust mechanical and thermal properties, with possible applications to high power electronic applications. Another feature of the nitrides is that they can be used in light emitting diodes (LED's) which produce green and blue light.

To keep this chapter in somewhat chronological order, we'll pick up on Ben Segall's research a bit later when we describe his partnership with Walter Lambrecht. But first, we back up to the 1980's and the arrival of Rolfe Petschek.

The 1980's – New Faculty, New Directions

If one looks at the "time-line" chart in the introductory chapter, it is easy to follow the arrivals and departures of physics faculty members. In the late 1960's, chairman Willard had the unenviable job of slimming down the oversized department. Then in the

1970's, as the newly merged university worked its way through some difficult political and financial challenges, neither chairman Ken Kowalski nor his successor Don Schuele succeeded in adding new faculty. The "intermediate energy" group (Willard, Bevington and Baer) was disbanded and not replaced, and the "high energy" and "cosmic ray" experimental groups were unable to add essential junior members. The situation began to turn around in the early 1980's, with the hiring of Gary Chottiner (Chapter 12) in 1980 and of Rolfe Petschek in 1983. The pendulum was swinging from nuclear/particle/cosmic ray research back to condensed matter research. Thus began two decades, not so much of expansion, but of rejuvenation (literally), as new young faculty were brought in to replace those who decided to retire.

Rolfe G. Petschek

Condensed matter theorist, **Rolfe Petschek**, earned his S.B. degree at MIT and his Ph.D. (1980) under B. I. Halperin at Harvard. Rolfe was born in Los Alamos, New Mexico, where his father worked as a theoretical physicist. **Fig. 17-11.** His doctoral research concerned certain aspects of dynamics near continuous phase transitions, including the rate at which magnetically aligned protons in a crystal of ammonium chloride "relax", i.e. lose their alignment. Each proton in this crystal lattice sits between a nitrogen atom and a chlorine atom. Measurement of the relaxation times as a function of temperature provides information on the interactions between the protons and their neighboring ions, and ultimately on the effect of defects on the dynamics of the phase transition. "Proton-spin-resonance relaxation times near the ordering transition in NH_4Cl" *Phys. Rev.* **B19** 166 1979.

Fig. 17-11. Rolfe Petschek.

Polymers

Petschek became a graduate research assistant at Los Alamos and later had postdoctoral positions at UC Santa Barbara and UC San Diego. He worked with the eminent physical chemist John C. Wheeler and French physicist Pierre Pfeuty on modeling the conformations of interacting polymers. "Bicriticality in the Polymerization of Chains and Rings (*Phys. Rev. Lett.* 50, 1633 1983)

Petschek joined the CWRU department in 1983. During the following four years he continued the work on polymer growth and phase transitions. In a series of solo papers, he presented mathematical models for the way in which certain polymers undergo phase transitions. Then, along with Wheeler and Pfeuty, he investigated the applicability of a model originally proposed for magnetic systems to analogous behavior in polymers.

A key parameter in the description of these materials is T_c, the *critical temperature*. At such a critical point, the material goes through a phase transition and its properties change abruptly. Petschek described an interesting transition in polymers. When elemental sulfur is heated, its properties change very rapidly at around 160 C. At this temperature the 8-atom-rings of low temperature sulfur rearrange themselves into polymer chains which have higher entropy than the rings. Such transitions in systems that form polymers have deep connections to magnetic critical phenomena. Theoretical models for polymers attempt to determine how properties change as the critical point is approached, from above or from below. In a paper with Wheeler and Pfeuty, Petschek looked at the critical points of polymers which could contain both rings and long-chain components. The structure and volume fraction of long-chain polymers suggested *bicriticality*, i.e. behaving as one would expect at a point in the phase diagram at which two lines of critical points merge. From the abstract: *"The bicritical nature of the critical point is the result of a competition between a transition to form long-chain polymers and a transition to form an infinite-rings condensate. As a result the fraction of monomers incorporated in chains and in rings varies with temperature according to a different power law than that for the total fraction of polymerized material."* "Equilibrium polymerization of chains and rings: A bicritical phenomenon" *Phys. Rev.* **A34** 2391 1986.

Liquid Crystals

Petschek soon became interested in the theoretical analysis of liquid crystals (LCs). In 1987, he and grad student Kimberly Wiefling looked in particular at **ferroelectric** fluids. Recall that ferroelectric materials can have a permanent electric polarization. Here is their abstract: "While there is no fundamental reason that fluids should not be ferroelectric, the only known ferroelectric fluids are chiral smectic-C liquid crystals. In this paper we argue that a variety of nonchiral smectic phases composed of properly designed oligomeric or polymeric molecules will also be ferroelectric. These ferroelectric fluids, and the solids formed by quenching them, should be of fundamental and practical interest. The nature of these phases, the design of the relevant molecules, and the experimental techniques for identifying the phases are discussed qualitatively."

We introduced some LC jargon earlier in this chapter, e.g. nematic (aligned molecules), smectic (aligned, and in layers). Here is a bit more: If the constituent molecules are **chiral**, i.e. if they look like right-handed or left-handed corkscrews, the LC is called **cholesteric**. Molecules that form liquid crystals can be chemically bonded to each other in various ways. A single chemical unit which forms a liquid crystal is a liquid crystal **monomer**. If a few such monomers are bonded to each other, this is called an **oligomer** (from the Greek "few parts") – if many are bonded together this is called a **polymer**. The molecules proposed by Petschek and Wiefling were fraternal twins – two rather different liquid crystal molecules tied to each other.

An aside on the design of ferroelectric liquid crystals: the goal is to have monomers which are electric dipoles (positive on one end, negative on the other) and which can be stuck onto a polymer chain so that their dipole moments line up parallel with one another. The resulting material has enhanced electrical effects, including non-linear optical effects, just what you want in electrically switched liquid crystals. Some chiral smectics and bowl-shaped or disk-shaped LCs show this desired behavior.

The authors stress the potential applications of ferroelectric fluids to liquid crystal technology. From their concluding remarks: "These phases would have practical applications. Solid phases, made by rapid quenching of these ferroelectric liquids, may have desirable mechanical and electrical properties such as plasticity and a well-defined, hard-to-change polarization. For such applications, it may be that long, polymeric molecules are superior. It may be possible to use these fluids in displays or other devices which require rapid change of a liquid crystal's properties." Recall that LCs can rotate the direction of polarization of light, so that when they are sandwiched between two polarizers, they can act as electrically activated switches which can turn the transmitted light on and off in microseconds. "Novel Ferroelectric Fluids" *Phys. Rev. Lett.* **59** 343 1987.

Petschek and Dennis Perchak (initially at CWRU, then at Eastman Kodak Research Laboratories) expanded on this work, collaborating on an extensive program of computer simulations of ferroelectric LCs. "Computer simulations of quasi-lattice models for novel ferroelectric liquid crystals" *Phys. Rev.* **A43** 6756 1991.

Petschek, along with new collaborators at the Liquid Crystal Institute at nearby Kent State University, continued the work on ferroelectric LCs. From the abstract of a 1988 paper: "We examine the possibility that molecules with permanent dipole moments can form a ferroelectric nematic phase, the least ordered conceivable ferroelectric phase. We show that reasonable electric dipole interactions between disk-shaped molecules may lead to such a phase, and calculate the phase diagram using mean-field theory." "Ferroelectric Nematic Liquid Crystals: Realizability and Molecular Constraints" *Phys. Rev. Lett.* **60**. 2303 (1988). These interactions with Kent State, both with its physics department and with the Liquid Crystal Institute, grew over the years, resulting in a large science and technology center called ALCOM (Advanced Liquid Crystalline Optical Materials). With about twenty investigators from CWRU, Kent and the University of Akron ALCOM engaged in important LC research for over a decade.

Post-doc Eugene Terentjev, now at Cambridge, worked with Petschek as part of the ALCOM center. His research was focused on various aspects of liquid crystal molecules which have been linked together. He continued Petschek's study of ferroelectric liquid crystals, suggesting that liquid crystal mesogens which are connected to each other in a "head-to-tail" fashion may form ferroelectric phases.

Grad student George Hinshaw joined Petschek in studying critical phenomena in systems which they found can be described by a theory related to one which has had great success in particle physics. Surprisingly, this "Yang-Mills" theory had practical applications for liquid crystals. In their abstract: "Possible consequences for designing molecules for chiral smectic-C phases with increased polarizations are discussed." " Transitions and modulated phases in chiral tilted smectic liquid crystals" *Phys. Rev.* **A39** 5914 1989.

Supporting the experimentalists

In 1989, Petschek began to work extensively with Chuck Rosenblatt and his experimental liquid crystal group. Their goal is to understand, design, fabricate, and test materials with the most desirable properties, such as transparency, stability, and rapid response time. This work will be briefly described in the "after-1990" Chapter 18. Petschek has played an important role as the "house theorist" for the LC group, collaborating on over a dozen papers in as many years. As an example, an early paper describes an experiment in which a specially designed chiral nematic LC was shown to display an "**electroclinic**" effect, i.e. the angle of tilt of its molecules rotates in an applied electric field. (The samples were typically around 25 microns thick and the rotation was very small: on the order of a tenth of a milliradian for a field gradient of one V/micron.) They studied how this effect changes as the temperature is raised, taking the sample through the smectic-to-nematic transition. "Nematic electroclinic effect" *Phys. Rev.* **A41** 1997 1990. The figure (**Fig. 17-12**) shows the rapid drop in dθ/dE as the phase transition is approached. The paper concludes with a proposed model for this unexpected behavior.

Between 1990 and the present (2005), Petschek has continued to work with Rosenblatt and the many students who are measuring liquid crystal properties, jointly publishing more than two-dozen papers. By the mid-1990's, Ken Singer had established his experimental program in optical materials, and Petschek soon became a regular collaborator with that group as well. By this time, the Rosenblatt and Singer groups would provide doctoral research projects for about half of the department's grad students, and Petschek has played an essential role in contributing theoretical support. More-

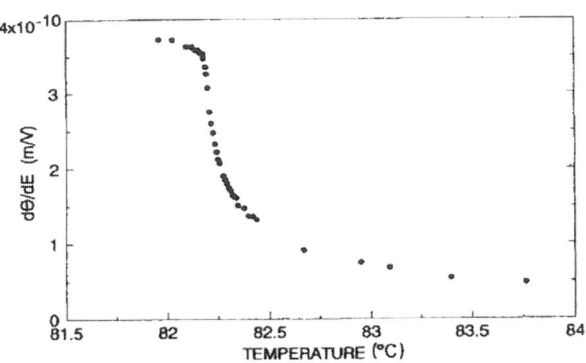

Fig. 17-12. Computed electroclinic effect vs. temperature, through phase transition.

over, he has continued to produce theoretical work on polymer liquid crystals with graduate students Hinshaw, Richard Sones, and Jonathan Stott and post-doc Terentjev. His statement on the department website summarizes this ongoing work. *"We use analytic and numerical tools of statistical mechanics and quantum mechanics, symmetry analysis, and extensive discussions with near-by experimentalists in order to discover new principles which control the statistical or optical properties of the wide variety of actual or potential structures. ... Our work is simultaneously both applied and basic, with impact on electro-optic and optical technologies, and with impact on the tools with which physicists study the world."*

Fig. 17-13. Walter Lambrecht.

Lambrecht, Segall: electronic structure group

In 1987, **Walter R. L. Lambrecht** joined Ben Segall as a senior research associate. Walter had completed his PhD at the University of Ghent in Belgium in 1980. Subsequently he was a visiting professor at the Universidad Nacional del Sur in Bahia

Blanca, Argentina. At the Max-Planck Institute for Solid State Research in Stuttgart, Walter calculated electron energies in diamond-like structures. At CWRU, Walter has worked with Ben Segall on band-structure calculations for over a decade. Their first paper was on heterojunctions, addressing the question of how the energy bands line up at an interface between two semiconductors. They presented a list of their calculated "offsets", comparing them with measured values for junctions of a series of materials, e.g. silicon in contact with GaP, germanium with AlAs, etc. "Theory of Semiconductor Heterojunction Valence-band Offsets; from Supercell Band-Structure Calculations toward a Simple Model" *Phys. Rev. Lett.* **61,** 1764 1988.

A different type of system was studied by Segall, Lambrecht and visiting Mexican physicist, Carlos Amador. They looked at the phase diagram of an alloy in which there is a large "size mismatch", in this case nickel-platinum. How do the melting point and the lattice structure change as the percentage of nickel is changed? "Strain Effects on the Phase Diagram of Ni-Pt alloys" *Phys. Rev.* **B47** 15276 1993.

Lambrecht and Segall have regularly interacted closely with experimentalists, both in the analysis of experimental data and by suggesting interesting materials to be explored. One such experimenter is John Angus, professor of chemical engineering, who is CWRU's "diamond man". (We met him in Chapter 12 as a collaborator of Gary Chottiner.) For many years, Angus has studied the properties, fabrication and applications of diamond crystals and diamond *thin films.* The diamond structure, a cubic array of carbon atoms, is particularly simple. There is a possibility that diamond can be used as the basis of a new semiconductor technology, with advantages over the silicon-based devices of today. One theoretical question concerned the choice of substrate to be used in growing diamond films. Segall and Lambrecht investigated the feasibility of using boron nitride, whose properties seemed to be an ideal match for diamond. "Electronic structure of diamond/BN interfaces and superlattices" *Phys.Rev.* **B40** 9909 1989. They were later joined by Angus and members of his group in a study reported in *Nature.* "Theory of Diamond Nucleation by Hydrogenation of the Edges of Graphitic Precursors," *Nature* **364**, 607 1993.

During their fourteen year collaboration, Lambrecht, Segall, and their Electronic Structure Group have published more than fifty papers on electronic structure of materials which are likely to have applications to semi-conductor technology. The emphasis throughout has been on the optical responses such as X-ray, ultraviolet and visible reflectivity and absorption, all deduced from the band-structure. The group has a special interest in wide-band-gap semiconductors such as diamond, silicon carbide and gallium nitride.

The underlying computational technique, the "density functional method", manages the great complexity of electrons interacting with the crystal lattice in the presence of applied fields. The success of the approach has grown over the years as more powerful computers have become available. It is possible to predict essentially all the properties of a materiel (e.g. structural, vibrational, mechanical, optical, magnetic, electronic transport) from "first principles" such as the atomic numbers and the crystal structure.

The group's website summarizes: "By calculating the optical response functions from the underlying electronic band structure, and analyzing them, we assist in the interpretation of various types of optical spectroscopy: X-ray, ultraviolet and visible reflectivity and absorption." A recent line of work has been on the integration of *magnetic* materials with semiconductors. These may eventually provide the basis for *spintronics,* a novel form of electronics which uses the spin degree of freedom.

Ben Segall became professor emeritus in 2000. He continues to interact with the department on a regular basis and, with his wife Annette, to promote the goals of a progressive liberal society. Walter Lambrecht, who was promoted to faculty status in 1996, along with his group of post-docs and grad students, continues a vigorous program of materials theory, providing critical support to experimentalist colleagues in the physics department and materials researchers in the college of engineering.

Harsh Mathur

In 1995, two years after Lawrence Krauss (Chapter 18) arrived as chairman, **Harsh Mathur** was appointed to the faculty. Mathur, having done a bachelor of technology degree at the Indian Institute of Technology - Kanpur, had recently completed his PhD at Yale. He came to CWRU after a postdoctoral appointment at Bell Labs and a semester as a general member of the Institute for Theoretical Physics in Santa Barbara.

Fig. 17-14.
Harsh Mathur.

Mathur had already established himself as an extraordinary young theorist, while still a graduate student, with his work on subtle quantum effects of electron spin on electrical conduction in metals at low temperature. This is an area of great interest now, following the birth and explosive growth of the field of *spintronics*. Another significant paper that Mathur wrote while he was at Yale revisited the classic Foldy-Wouthuysen analysis of the Dirac equation (described in Chapter 9 and in Appendix G) and reinterpreted it in terms of a newly identified quantum effect called Berry's phase. "Thomas Precession, Spin-Orbit Interaction, and Berry's Phase," *Phys. Rev. Lett.* **67**, 3325 (1991).

We'll briefly discuss here some representative topics which Harsh has studied while at CWRU. *Briefly*, because my intended soft cut-off was 1990, and *here*, because his work is in condensed matter theory. In Chapter 18, we will offer similarly brief descriptions of the work of the twelve other department members who arrived after 1986.

Mathur's main interest is in quantum effects in condensed matter, particularly effects due to interactions between particles and between particles and impurities. A large part of his work has been concerned with the mobility of electrons in *dirty* conductors. At low temperatures, quantum interference can cause electrons to become *localized* as they scatter from impurities. This leads to a measurable increase in the electrical resistance as the temperature is lowered. Mathur explains: "An important problem in 'localization physics', one that is still only partially solved, is to understand *dephasing*. Dephasing refers to processes that suppress quantum effects as the temperature is raised. A broad fundamental problem of physics (sometimes called the decoherence problem) is to understand the processes that allow the classical world to emerge from the underlying laws of quantum mechanics."

Electrons deposited on the surface of a pool of liquid helium form a flat conducting sheet that is particularly well suited for the study of such localization phenomena. Electrons on helium have been the principal interest of experimentalist Arnie Dahm and his group for many years (as we described in Chapter 15). Thus it was natural for Mathur and his student Damir Herman to collaborate with Dahm and his student Ismail Karakurt in the study of dephasing in this system. The CWRU group found that the electrons are localized by their collisions with helium gas atoms in the vapor above the liquid. The dephasing is caused by the slow thermal motion of the vapor atoms. They developed a theory of such dephasing that proved to agree with their measurements. "Localization scaling relation in two dimensions: Comparison with experiment" *Phys. Rev.* **B56**, 13263 1997. "Damping of quantum interference of electrons on helium" *Phys. Rev.* **B68**, 33402 2003.

Silicon MOSFETs (metal-oxide semiconductor field-effect transistors) are another arena in which localization physics can be studied. They are the building blocks of modern electronics, used, for example, to control high speed switching in integrated circuits. In a MOSFET, a flat conducting sheet of electrons is to be found between a layer of silicon and a layer of silicon oxide. The roughness of the semiconductor-oxide interface determines the electrical properties of the device, including how long it will last. Thus, characterizing and controlling the roughness is a matter of great technological importance and a potential roadblock to further device miniaturization. Mathur and his collaborator Harold Baranger (Duke University) have shown that at low temperature the scattering of electrons from the rough interface can cause localization and that electrical resistance measurements in a magnetic field can be used to measure the roughness of the interface non-destructively. "Random Berry phase magnetoresistance as a probe of interface roughness in Si MOSFET's" *Phys. Rev.* **B64** 235325 2001.

Mathur has extended his study of electrons confined to small regions by tackling the theory of *quantum dots*. He explains: "the ability to fabricate quantum dots is a major development in semiconductor physics. A semiconductor quantum dot is a pool of electrons confined to a small region, often less than 100 nm in size. The number of electrons in the dot can be controlled with exquisite precision and varies from a few electrons in small dots to a few hundred in larger micron sized dots."

Aside from the possibility that they will play a role in some future nanotechnology, quantum dots can also be of great fundamental interest. Mathur and his collaborators, Ganpathy Murthy (University of Kentucky) and Ramamurti Shankar (physics chairman at Yale) have predicted that, in large quantum dots with low electron density, the electrons will form a *new state of matter* different from the ordinary Fermi liquid state in which electrons are normally found in semiconductors and metals. Mathur points out that this new state has many remarkable properties - for example the electrons are predicted to carry a persistent current even in the absence of any applied electric fields, even though the new electronic state is not superconducting. "Diamagnetic persistent currents and spontaneous time-reversal symmetry breaking in mesoscopic structures" *Phys. Rev.* **B69** 41301 2004.

Condensed matter theory continues to be an important component of the CWRU physics department and it attracts a significant proportion of its graduate students. They appreciate the opportunities to interact with a variety of physics and materials science experimental groups. Many will go on to careers in which they will apply their calculational and instrumentational talents to a rapidly expanding technology.

Chapter 18   New People, New Physics

This book's subtitle implies that we will cut off the narrative at 1990. But I'm sure you would like to know something about the physics research done at CWRU this past decade and a half. In fact, in several earlier chapters, I have described some of the more recent exciting work done by my colleagues. For example, I explained how Bob Brown has exploited his expertise in electromagnetism and quantum mechanics to become a world player in medical imaging; and how Dave Farrell has advanced from basic research on superconductivity to medical applications of biomagnetism; and how Arnie Dahm has capitalized on his skill in manipulating individual electrons to take a first step into the world of quantum computing.

If you were to visit the department in 2005, you would find twelve (about half) of its faculty, along with a dozen research associates and about two dozen graduate students conducting exciting, world-class programs in three research areas:

- optical materials and nanotechnology,
- particle/astrophysics/cosmology theory,
- astrophysics and cosmology experiment.

In this chapter, I present brief descriptions of the research interests of these twelve physicists and their groups.

Optical materials: light on matter
Rosenblatt   Singer   Kash   Shan
1987   1990   1994   2002

Charles Rosenblatt (PhD Harvard 1978) arrived in 1987 to establish a program in liquid crystal research. **Kenneth Singer** (PhD Pennsylvania 1981) came three years later to begin a new program in optical materials. **Kathy Kash** (PhD MIT 1982) was the first new member of the faculty after the arrival of Lawrence Krauss as chairman. More significantly, she was the first woman full-time faculty member in the 165 year history of the department. In 1994, Kash set up a laboratory for research in mesoscopic systems. The next addition to the department is its second woman physicist, **Jie Shan** (PhD Columbia 2001), who works in ultra-fast spectroscopy. This group of four experimentalists, working in related frontline areas of applied physics, is complemented by the four condensed matter theorists: Phil Taylor, Rolf Petschek, Walter Lambrecht, and Harsh Mathur.

Rosenblatt – liquid crystals

In Chapter 12, we described some of the experimental work done by Gordon and Schuele on the properties of liquid crystals (LC's). Then, in Chapter 17, we mentioned Petschek's theoretical investigation of ferroelectric LC's. Chuck Rosenblatt's group, which currently includes one post-doc and four grad students, has established a significant liquid crystal research program. The first among several topics they are investigating is the study of phase transitions from one liquid crystalline state to another. An example is the transition between the nematic (molecules aligned in a preferential direction) and the isotropic (random alignment). A second area if interest is the study of LC's which have ferroelectric behavior. A third topic concerns the interface of the LC and its "container". Since liquid crystals are usually confined between two flat surfaces, their interaction with the substrate plays an important role in their behavior. Rosenblatt and his group have explored the possibility of influencing the way in which the molecules align themselves at the interface. They have, for example, investigated coating the substrate with a polymer layer, and then to mechanically score alignment grooves in the polymer with the stylus of an atomic force microscope. From the group website: "... we scribe patterns onto a polymer-coated substrate as small as 10 nanometers in length, approximately one ten-thousandth the width of a human hair. The liquid crystal molecules are forced to align parallel to the scribing direction, allowing us to study phase transitions and elastic behavior on very tiny length scales."

The ultimate goal of these studies is to develop LC's which are fast, stable, cheap and applicable in many areas of optics, including such things as the control of optical signals in communications and computer technology.

Fig. 18-1.
Chuck Rosenblatt.

Singer – nonlinear optics

Ken Singer and his group (currently 5 grad students and one post-doc) are interested in "non-linear optics". Ordinarily, when light passes through matter, like a glass lens, it gets partly absorbed and is slowed down by interactions with the electrons in the medium. The frequency of the light remains unchanged. This is *linear* optics. If the light is very intense (as light from a laser), its electric fields push the electrons around, resulting in an oscillating polarization of the medium. The oscillating electrons produce their own radiation. In this case, the transmitted beam can have small components with twice, thrice, or other multiples of the frequency of the incident beam. This is called "harmonic generation." (A hundred years earlier, Dayton Miller was studying sonic harmonics generated in musical flutes.) There are other interesting and potentially useful non-linear effects, such as dependence of the speed of transmission on the frequency, or the shifting of the phase of the light beam. The optical materials which behave in this way usually consist of organic molecules which can be linked together to form polymers. The polymers in turn spontaneously organize themselves into crystals and liquid crystals, sometimes in a single molecular layer.

Fig. 18-2.
Ken Singer.

From the group website: "Members of our research group measure optical, electronic, and structural properties of organic electronic and photonic materials with the aim of gaining an understanding of the physical origin of optical and electronic properties, as well as the potential of particular materials for optoelectronic devices." "We also study how these materials might be used in devices, such as solar cells, displays, optical switches, image processors, and other electronic and optical devices."

Kash – reduced dimensionality, crystal growth

Kathy Kash is doing research in nanotechnology, specifically in the preparation of semiconductor devices for the study of linear and nonlinear optical processes. In a "semiconductor quantum well", electrons and holes are restricted to move in planes between alternating layers of low-band-gap and higher-band-gap semiconductors. When particles are constrained to very small regions of space, in this case to a two-dimensional sheet, their motions are determined by quantum mechanics. Kash has studied the quantum confinement of excitons, i.e. electron-hole pairs created by a laser. Excitons can be further restricted to one-dimensional "quantum wires" or even to "quantum dots". From the group's website: "The goal of the effort is to study linear and nonlinear optical processes as they are modified by quantum confinement, and to study phase transitions of collections of particles in these systems of reduced dimensionality."

Fig. 18-3.
Kathy Kash.

Most recently, Kash has been collaborating with colleagues in Chemical Engineering and Macromolecular Science. One such project involves the growth of large single crystals of gallium nitride at near-atmospheric pressure. Currently, this can be done only at extremely high pressures, and success of the new approach being investigated by Kash and her colleagues would have a very important impact on industrial applications.

A second research area concerns the electrochemical deposition of semiconductors. As Kash describes it, the aim is to "make devices 'in a bottle', very cheaply, and using schemes that could be made environmentally friendly". This work can involve nanometer sized "molecular templates" as patterns for very small wires and dots, whose remarkable quantum properties can be exploited. Currently, three grad students and one post-doc are working under the direction of Professor Kash.

Jie Shan – femtosecond pulses

Jie Shan is the fourth in this group of experimentalists who are studying the interaction of light with matter on the microscopic scale. Jie's specialty is "time domain spectroscopy" in which ultrashort electrical pulses are generated using femtosecond optical pulses to study charge transport properties of materials. The new radiation has frequencies in the terahertz or far-infrared region, with typical wavelengths a few hundred micrometers and photon energies a few milli-electronvolts. A 100 femtosecond pulse is typically only tens of micrometers long as it flies through the sample. Photons in this meV range match the energies of many fundamental excitations in solids and molecules, including phonons, low-frequency vibrational modes, rotations, and certain collective electronic excitations. One can use these pulses to investigate electrical charge transport in insulators and in nano-scale structures, such as the ones studied by Kathy Kash.

Fig. 18-4. Jie Shan.

From Jie's website: "Currently we are focusing primarily on applications of ultrafast spectroscopy .... to the study of various condensed-phase systems including conventional materials and strongly correlated systems. Although both equilibrium and nonequilibrium properties are investigated, we emphasize pump/probe techniques to examine the dynamical properties of materials, such as carrier and spin dynamics and energy relaxation. At present, we are particularly interested in probing the effects of reduced dimensionality associated with interfaces and nanostructures."

Essential to the research of each of these four condensed matter experimental groups is the department's **"nanoscale facility"** which is equipped with state-of-the-art instruments, such as an atomic force microscope and a near-field scanning optical microscope. The four faculty members are aided by a total of fifteen graduate students and three post-docs. Each group uses specially tailored light pulses to elucidate the quantum behavior of technologically interesting systems and materials: Rosenblatt studies mono-layers of innovative liquid crystals; Singer uses laser pulses to produce nonlinear effects in optical materials; Kash looks at the propagation of electron-hole pairs in two-dimensional quantum wells; and Shan uses extremely short bursts of radiation to examine dynamical properties of materials. Each group does "basic physics" research, using the newest techniques, to find out how materials behave under unusual or extreme conditions. Much of what they learn will reappear as "applied physics" as new technologies are perfected.

Particle/astrophysics/cosmology theory

C. Taylor	Krauss	Starkman	Vachaspati
1988	1993	1995	1996

Cyrus Taylor - factotum

Cyrus C. Taylor joined the department in 1988. He had completed his PhD at MIT in 1984, where he worked on field theory. He was the first particle theorist to be hired at CWRU since Bob Brown's arrival 18 years earlier. Taylor began his research at CWRU, working with graduate student Evalyn Gates, on a study of the quantization of phase spaces associated with various field theories. Later, he joined Bob Brown and research associate Shlomo Shvartsman in the development of the theoretical treatment of the quantum electrodynamics of heavy fermions. As we saw in Chapter 13, Brown and Shvartsman soon afterward moved from particle theory to medical imaging. At the same time, Taylor, too, made a significant move away from theory, being reincarnated as a particle experimentalist.

In 1993 Cyrus joined J. D. Bjorken of the Stanford Linear Accelerator Center (SLAC) in proposing an experiment to be run at the Fermilab Tevatron. The experiment, called MiniMax, was originally conceived as preliminary to a more complex run at the super-conducting super collider (SSC) in Texas. Following the defunding and demise of the SSC, the group, along with new collaborators, has submitted a proposal to undertake a similar experiment (christened FELIX) at the large hadron collider (LHC) currently being built at CERN in Switzerland.

Fig. 18-5
Cyrus Taylor

The MiniMax run at Fermilab took place in 1996 and the results were soon published in a Physical Review paper with authors from seven institutions. Included among the collaborators were three CWRU undergrads, two grad students, and professors Tom Jenkins and, a second converted theorist, Ken Kowalski. The experimental setup consisted of a large array of detectors placed alongside the intersection of the colliding protons and antiprotons. The total energy of the collisions was 1.8 TeV. Charged mesons and photons produced at very small angles relative to the beamline were detected. The goal was to determine the ratio of the number of neutral pions (which decay to two gammas) to the number of charged pions produced in the very high energy collisions. The tiny region in which the proton and antiproton annihilate into a burst of mesons has an extraordinarily high energy density, higher than anything short of a supernova. The theoretical motivation for the measurement was the proposal that, in such a situation, the symmetry of the charged and neutral mesons might be different from that observed in less frantic neighborhoods. The theorists' name for such behavior is "disordered chiral condensate". (Cyrus had, in fact, worked on the theory of chiral condensates as early as his grad school days at MIT.) Earlier experimental clues for DCC had been seen in atypical cosmic ray events. MiniMax produced several events which suggested a signal for DCC, certainly enough to support plans to continue the search in Geneva.

Theorist and experimentalist Taylor put on a third hat as inaugurator and director of the department's Physics Entrepreneurship Program (PEP). This is a 2-year master's degree program which, as described on its website, "provides studies in technology innovation and state-of-the-art physics, practical business instruction, and real-world entrepreneurial experience to individuals with a bachelors, masters, or PhD in a physics-related field". This program, which in its first four years has won several national awards, has been imitated in several other Case departments. In the summer of 2005, Cyrus Taylor succeeded Lawrence Krauss as chairman of the physics department.

Krauss – astrophysicist, communicator

Lawrence Maxwell Krauss joined the department as its new chairman in 1993. He had completed his PhD at MIT in 1982 where he worked on theoretical gravitation and big-bang physics. During the following three years at Harvard and six on the faculty at Yale, Krauss wrote prolifically on topics at the intersection of particle physics and cosmology. Several of his papers, including a few with his sidekick (now Nobelist) Frank Wilczek, described ways to detect and identify various dark matter candidates, papers of great value to the several groups who were building detectors to do just that.

In 1992, a distinguished external visiting committee was invited by the Dean of Arts and Sciences to evaluate the department and its potential They strongly recommended that, on Bill Gordon's retirement, a new chairperson should be sought from outside the university. Krauss, who was at the time on the Yale physics faculty, was already recognized as a leader in particle/cosmology theory and as the author of widely read popular books on modern physics. With the impending retirement of about half of the CWRU physics faculty, Krauss recognized the opportunity to build a new and competitive department. Central to his plan was to establish a nationally significant astrophysics effort. Consequently, during the following decade, two additional astro/cosmology theorists and four young astrophysics experimenters were recruited to the department.

At CWRU, Krauss has worked with research associates, Peter Kernan, Mark Trodden and Craig Copi, and with grad student Hong Liu, in studies of the formation of nuclei in the big bang, of the strange quantum properties of black holes, and of the weakly interacting massive particles (WIMP's) which have been proposed to account for the missing mass in the universe. This theoretical work, some of which directly addressed observational techniques, has been especially appropriate given the presence in the department of the four experimental groups to be described later in this chapter.

Fig. 18-6
Lawrence Krauss.

In 1995 Krauss and Michael Turner of the University of Chicago wrote a prescient paper on the "return of the cosmological constant", an idea which dominates current cosmology. This refers to the repulsive force originally proposed and then rejected by Einstein in an attempt to keep the universe from collapsing. More recently new data from observations of very distant supernovae indicate that the rate of expansion of the universe is actually increasing and the "cosmological constant" is at the center of the expanding universe discussion.

Krauss has been instrumental in the creation of CERCA, the Center for Education and Research in Cosmology and Astrophysics. With the support of the Kavli Foundation, the eight faculty members described in this chapter, along with members of the CWRU department of astronomy and colleagues at the Cleveland Museum of Natural History, have formed this center to provide "an interdisciplinary framework for interactions between faculty, postdoctoral researchers, graduate and undergraduate students, educators and the public. CERCA is particularly active in supporting the world class research activities of scientists at CWRU and making new connections to bring the excitement of this research to the public at many levels." (from the CERCA website)

Taking science to the public has always been a major component of Krauss' interests. While, at the same time chairing the department, producing significant research, and lecturing at all levels to audiences the world over, he has authored seven widely read books published in many languages. Krauss is known internationally as a champion of scientific integrity, speaking out against the abuse, misuse, and misunderstanding of science.

Starkman and Vachaspati – the particle-cosmology link

Within a year after Krauss' arrival, two young and well-published "particle-astrophysics-cosmology" theorists joined the department: Glenn Starkman and Tanmay Vachaspati.

After completing his BS at Toronto and his doctorate at Stanford, **Starkman** held post-doctoral positions at Princeton and Toronto. His work during these years concerned Big Bang nucleosynthesis, the role of neutrinos in the early universe and various candidates for dark matter. At Case, working with a string of post-docs, Glenn has studied topics concerning the early universe and how it got to where it is. One paper, "Does Chaotic Mixing Facilitate $\Omega<1$ Inflation?", takes the prize for the most succinct abstract: "Yes, if the Universe has compact topology."

Fig. 18-7
Tanmay Vachaspati.

Vachaspati earned his doctorate in 1985 at Tufts, working with Alexander Vilenkin on the role of cosmic strings in the early universe. During the following decade, Tanmay divided his time among positions at the University of Delaware, Cambridge University, and Tufts. Vachaspati is especially interested in "cosmological defects". Consider that if the early universe, just after the big bang, were absolutely homogeneous, it would still be so. Some local variations, or defects, had to be there from the start in order to result in the universe we see today. Tanmay has looked at the potential role of such candidates as magnetic monopoles, cosmic strings and domain walls as responsible for the observed structure. At the opposite extreme, he is studying the impact of these "cosmic defects" on the properties of fundamental particles, for example in a paper titled: "An Attempt to Construct the Standard Model with Monopoles,"

Glenn and Tanmay have collaborated on a variety of projects, including a paper on galactic cosmic strings as sources of primary antiprotons. In 1999, they joined the mounting excitement in speculations about the accelerating universe in a paper with Mark Trodden on "the fate of the universe". Tanmay took this a bit further in work with former grad student, Levon Pogosian, in speculations about the "dark energy" which seems to be responsible for the cosmological constant. They look at the observed large scale structure in the universe and the results of cosmic microwave background (CMB) surveys and deduce a picture in which we (along with all our brethren in our particular universe) find ourselves at a fork in the road, a coin-toss perhaps, pulling us possibly back into a big crunch.

Starkman, along with post-docs Copi, Trodden, Dejan Stojkovic, and Dragan Huterer, has pursued the nature of dark matter and the universe's acceleration in a variety of ways, looking, for example, at the possible role of large extra dimensions or of possibly observable effects on gravitons. The abstract of a recent paper summarizes the state of the art in a colorful way: "The nature of the fuel that drives today's cosmic acceleration is an open and tantalizing mystery. We entertain the suggestion that the acceleration is not the manifestation of yet another new ingredient in the cosmic gas tank, but rather a signal of our first real lack of understanding of gravitational physics."

Two other topics on Starkman's list concern the search for circles in the sky and the observation that some of the cosmic structure in the CMB may not be cosmic after all. The first was a search for identical patterns of microwave structure on opposite sides of the sky. These might be expected if the universe has a topology such that there exists more than one route which the radiation can take to reach us from distant sources. No such matched patterns were observed, thus ruling out the "possibility that we live in a universe with topology scale smaller than 24 Gigaparsecs." Room enough, I suppose.

Fig. 18-8
Glenn Starkman.

The second topic is also related to the CMB data. The usual way to summarize the observed structure in the CMB is to plot the "power spectrum". This is a measure of the temperature correlations between all pairs of points on the microwave sky as a function of the angular separation between them. Much of the current discussion on what happened after the big bang is based on the series of bumps which appear in the power spectrum – where preferred angular separations are evident. Starkman and company looked closely at these bumps and determined that the "large angle correlations" were different on opposite sides of the ecliptic plane (the plane in which the planets move around the sun). Why should cosmic processes bother about our modest little solar system? Michelson and Morley, and the persistent Miller, sought an effect of the motion around the sun, Starkman and company's lopsided patterns in the large-angle correlations in the CMB seem to signal one.

Experimental Astrophysics and Cosmology
Akerib   Covault   Ruhl   Shutt
1996   2001   2002   2005

Complementing the *theoretical* astrophysics/cosmology program is an exciting multifaceted research program pursued by a quartet of young experimenters, **Dan Akerib** (PhD 1991 Princeton), **Corbin Covault** (PhD 1991 Harvard), **John Ruhl** (PhD 1993 Princeton), and **Tom Shutt** (PhD 1993 UC Berkeley). These fellows and their groups are participants in a wide range of multi-national astrophysics collaborations, each one a search for particles or radiations from space, each one using different state-of-the-art detection systems. Their research may be described as a continuation of the pioneering neutrino work done in the 1960's by Crouch and Reines, and the cosmic ray work done in the 1970's by Frye and Jenkins, as described in Chapter 8.

Akerib - wimps

Fig. 18-9
Dan Akerib.

Daniel Akerib joined the department in 1996. He and his group of post-docs and grad students are part of the "Cryogenic Dark Matter Search" (CDMS) collaboration. The collaboration currently includes physicists from twelve institutions. Observations made during the past two decades indicate that the protons, neutrons and electrons which make up the stars, planets and sundry dust in the universe account for only a fraction of its mass. Better than 90% of the mass is unaccounted for. Whatever the missing mass is, it must interact very weakly with ordinary matter and is therefore very difficult to observe. Neutrinos, even though they are plentiful, have been ruled out as candidates for the missing mass. One possibility is that there exists a sea of "weakly interacting massive particles" (WIMPs) which are responsible for the observed gravitational effects and which have cooled down to non-relativistic energies. The CDMS and other experimental collaborations have developed instruments sensitive enough to detect the rare and tiny signals produced when WIMPs interact with nuclei.

As Dan explains on the department website: "The goal of my research is to try and detect WIMPs directly, through their elastic scattering from atomic nuclei in a terrestrial detector. If they were in fact produced in the early universe, WIMPs would have coalesced to form the dark matter halo of our Galaxy at detectable level. The experimental challenge is formidable. Because WIMPs are slow and weakly interacting, they lead to small energy transfers and very low rates."

Because the counting rates are so low, the detectors are placed in a deep mine to reduce the background from cosmic rays. Recall that Reines, almost forty years ago, reduced the cosmic ray background in his neutrino work by setting up in a deep gold mine in South Africa (Chapter 8). The Akerib group is responsible for the development of solid-state detectors which operate at temperatures of milli-Kelvins and which can detect both the burst of charge and the tiny rise in temperature caused by a WIMP hitting a nucleus. Because most of the residual background is from electrons or photons, the team has developed a detector which can pick out the rare WIMP signals. The ratio of the amount of ionization to the amount of heat for WIMP's is expected to differ from that for the background events.

The most recent run of the Cold Dark Matter Search, at the 2000 foot deep Soudan iron mine in Minnesota in the summer of 2004, has set a new upper limit on WIMP interactions. The germanium and silicon detectors (two kilograms total) were operated at 50 millikelvin. Assuming a WIMP flux consistent with their observed galactic gravitational effects, the group has been able to set an upper limit on the cross section for WIMP interactions with matter. This limit is significantly lower than that set by any other search – and, as the authors state: the result "constrains predictions of supersymmetric models". In other words, the non-observation of a signal at this level is starting to make the theorists scratch their heads.

Covault – cosmic gammas

Corbin Covault came to CWRU in 2001 after several years on the faculty of the University of Chicago. He has set up a group interested in the detection of ultra-high energy radiation and cosmic ray particles. One of the group's collaborations, STACEE (Solar Tower Atmospheric Čerenkov Effect Experiment), involves the search for extremely high energy bursts of gamma rays. A large array of mirrors (totaling 7000 square meters), spread out in the desert of New Mexico, catches the Čerenkov radiation produced at night by secondary particles moving through the atmosphere. These particles are produced in collisions in the atmosphere of incoming gamma rays in the 50 to 250 GeV energy range.

Fig. 18-10
Corbin Covault

From the STACEE website: "It is believed that observations of sources" in this range "will provide important evidence concerning the acceleration mechanisms of the most energetic objects in the Universe, including rotating neutron stars (pulsars), remnants of exploded stars (supernovae), gamma-ray bursts, and distant, but intense, active galactic nuclei (quasars)."

A second collaborative project, the Pierre Auger Observatory, involves the use of 1600 sets of Čerenkov and atmospheric fluorescence detectors spread over a 3000 km² area in southern Argentina. The goal is to detect electron showers produced by incoming cosmic rays at the highest-so-far end of the energy scale, say up to 10^{20} eV. The relative arrival times of the numerous secondary particles at each of the wide-spread detectors allow one to determine the direction of the incoming primary. The CWRU team has been responsible for the design and testing of the global positioning system (GPS) based components which are the key to the directional sensitivity of this exciting, multinational project.

Ruhl - CMB

John Ruhl arrived in 2002 to establish the third new experimental astrophysics group, this one a team in search of fine structure in the cosmic microwave background. This 2.7 Kelvin radiation, the residual "cooled down" photons left over from the Big Bang, lies at the lowest end of the energy scale, a few electron volts. The goal is to measure variations as small as a few micro-Kelvins in the CMB temperature and the degree of its polarization as one scans across the sky. As we described above in the paragraph on Glenn Starkman's work, details of the "power spectrum" provide clues to the formation of early structure in the universe.

Fig. 18-11 John Ruhl.

The key to clear viewing of the microwave sky is to get your detectors "high and dry", i.e. above the atmosphere and away from water vapor. Recall that in the 1960's, Glenn Frye and his group flew their spark chambers on high-altitude balloons over Texas and Australia, in search of cosmic gamma rays (Chapter 8). A half century later, the Boomerang collaboration (Balloon Observations Of Millimetric Extragalactic Radiation ANd Geophysics) has launched high-altitude balloon flights which circle the south pole. Suspended from the balloon are a 1.3 meter diameter telescope and an array of low-temperature bolometers ("heat" detectors). In a 2005 paper, the Boomerang team reports on the most recent flights, presenting the most detailed power spectrum to date. The paper concludes: "We characterize a series of features in the power spectrum, which extend to multipoles $\ell > 1000$, consistent with those expected from acoustic oscillations in the primordial plasma in the context of standard cosmologies". This means that, down to regions of the sky as small as a third of a degree, the observed pattern of tiny temperature variations is well explained by an expansion in the early universe including a few bounces.

John Ruhl's team is also involved in a ground-based experiment set up at the South Pole: the Arcminute Cosmology Bolometric Array Receiver (ACBAR). The expectation is that this detector, installed on the 2 meter diameter Viper telescope, has 2.5 times better angular resolution than Boomerang. And, neither least nor last, the group is also involved in the construction of the 10-meter diameter "South Pole Telescope". The emphasis here will be the study of galactic clusters and the role of "dark energy" in their formation.

Shutt – wimps too

Tom Shutt moved his research program from Princeton to Case in early 2005. Like Dan Akerib, Shutt's principal interest is the detection of WIMP dark matter. Each group is in the race to be the first to nail down a clear signal of interactions caused by these evasive particles. Akerib and the CDMS collaboration have set upper limits on WIMP flux by using kilogram-sized cryogenic detectors. Shutt and his team are members of a large collaboration called XENON which will chase down WIMP's with a ton of liquid xenon. The detector will be placed in the Laboratori Nazionali del Gran Sasso, an already existing large complex adjacent to an autostrada tunnel deep under the Appenines in central Italy. The plan is to begin with ten 100 kg modules, each with a very sophisticated signal read-out system which simultaneously picks up the electrons, photons and recoil xenon nucleus. The expected event rate, given the presumed density of WIMP's in the Milky Way, is ten events per year. The Case group has built and successfully tested a 35 kg test module. The WIMP Race at Case is underway.

Fig. 18-12
Tom Shutt.

What next

It will be fun to watch the research program in the CASE physics department as it moves ahead. What research will occupy its members in 2050, or maybe even 2015, is beyond anyone's guess. Physics is at the heart of our understanding of all natural phenomena and it will continue to be at the center of the university. The materials science programs, with innumerable links to applications, will certainly continue to flourish. With orders-of-magnitude improvements in the speed of computation, atom by atom, and even electron by electron, calculations will be possible. New materials will be conceived and analyzed by computational physicists before being turned over to the people who will build them. The same is true for areas like medical imaging which is driven by computing power. It is almost certain that physicists will become ever more involved in medical science: not just for diagnostics, but for treatment. The decoding of the various signals from space by advanced computational analysis will continue to lead to a better understanding, not only of how the universe was formed, but of the very nature of its building blocks. In particle physics and field theory, the computer's ability to find the needle in the haystack or to carry out calculations to ever higher orders will be key to new physics. *WRC, WRU, CSAS, CIT, CWRU, CASE* physics will be there, and with continued leadership and support, it will thrive.

Appendix A Bachelor Degrees in Physics from WRU, CSAS, CIT, CWRU

Year	Surname	Given
1911	Kemble	Edwin C.
1912	Venne	John C.
1917	Steiner	Oscar H.
1917	Valasek	Joseph
1921	Firestone	Floyd A.
1923	Torreson	Oscar W.
1924	Pritchard	Howard A.
1925	Glathart	Justin L.
1925	Morse	Philip M.
1925	Smith	Theodore Hunter
1927	Domizi	Dante
1928	Hartline	Ralph E.
1928	Herman	Clarence A.
1929	Davies	Gomer L.
1929	Gebhardt	Robert E.
1929	Lichtblau	Stephen
1930	Brennan	Martin J.
1930	Buxton	Chester L.
1930	Cerny	Elmer J.
1930	Herzegh	Frank
1930	Smith	Donald D.
1930	Smith	Harry W.
1930	Van Voorhis	Stanley N.
1930	Wilson	Roger R.
1931	Budd	Chester B.
1931	Eichelberger	John F.
1931	Kusch	Polykarp
1931	Lamb	Robert E.
1931	Randolph	Russell H.
1932	Eisler	Sanford M.
1932	Focke	Alfred B.
1932	Holl	Fred M.
1932	Jacobus	Harland E.
1932	Krakora	Frank W.
1932	Rudd	Milo O.
1933	Beckwith	Glen J.
1933	Boyer	Raymond F.
1933	Harrington	Robert A.
1933	Tarasov	Leo P.
1934	McRae	Homer C.
1935	Brooke	A. Wayne
1935	Donaldson	John B.
1935	Goffman	Casper
1935	Hinzmann	Paul R.
1935	Johnson	James B.
1935	Prettyman	Irven B.
1935	Rense	William A.
1935	Smith	Simmons S.
1936	Amos	Richard E.
1936	Kraft	Frank R.
1936	Ott	Howard F.
1936	Roddy	Francis J.
1937	Coltman	John W.
1937	Ellsworth	Louis D.
1937	Smith	Clare E.
1938	Rose	Gene F.
1938	Wolkov	David
1939	Follett	Thomas L.
1939	Franks	Clifford V.
1939	Hickox	Walter A.
1939	Zaffarano	Daniel J.
1940	Adams	Gail D.
1940	Fitzwilliam	James W.
1940	Heskett	Harry E.
1940	Wilson	John A.
1940	Winkel	Edwin F.
1941	Bachman	Clarence G.
1941	Beutel	Phillip R.
1941	Dorris	Robert C.
1941	Foldy	Leslie L.
1941	Fox	John M.
1941	Horvath	Robert W.
1941	Swain	Robert W.
1942	Davis	William R.
1942	Greenwood	Ivan A.
1942	Hitchcock	Burt
1942	Hudimac	Albert A.
1942	Humiston	Homer A.
1942	Little	John L.
1942	Stokes	Richard H.
1942	Strough	Robert I.
1943	Baietti	Albert L.
1943	Blachman	Nelson M.
1943	Bletcher	Arthur L.
1943	Francis	John E.
1943	Graves	Jacob D.
1943	Hopler	John W.
1943	Leiss	William J.
1943	Lewis	John W.
1943	Reitz	John R.
1943	Schwarz	Walter A.
1943	Thackeray	Ross S.
1944	Bowman	John C.
1944	Cohen	Bernard L.
1944	Jurman	Harry R.
1944	Landon	Harry H.
1944	Leary	Arthur P.
1944	Mates	James A.
1944	Nusbaum	William C.
1944	Oravec	Ralph J.
1944	Raymond	Joseph L.
1944	Schneerer	William F.
1944	Spielman	Milton R.
1944	Voelker	William H.
1945	Baker	Saul P.
1945	Crowl	Laura D.
1945	Edwards	A. James
1945	Hart	John C.
1945	Kneip	George D.
1945	Rohrer	George W.
1945	Rutemiller	Herbert C.
1945	Springer	George
1945	Stanish	Ray J.
1945	Vanderwist	Donald C.
1946	Glaser	Donald A.
1946	Hamlin	John W.
1946	Ruth	Ralph P.
1946	Webster	Katherine W.
1947	Garwin	Richard L.
1947	Green	Thomas A.

Year	Last Name	First Name		Year	Last Name	First Name	
1947	Hoffman	Richard	W.	1950	Layzer	Arthur	J.
1947	Johnson	Robert	G.	1950	Lewis	Arthur	M.
1947	Ness	Arthur	J.	1950	Lisy	Bert	J.
1947	Slater	Raymond	J.	1950	Miller	Allan	
1948	Bazeley	Arthur	J.	1950	Morris	Peter	R.
1948	Christiansen	Robert	J.	1950	Peterjohn	Robert	F.
1948	Conklin	John	R.	1950	Powell	Robert	C.
1948	Goodman	Esther		1950	Revelt	Jean	J.
1948	Hill	Edwin	R.	1950	Riehl	Warren	L.
1948	Koda	N. John		1950	Schuerger	Thomas	R.
1948	Mergler	Harry	W.	1950	Tasch	Andrew	
1948	Raske	Arthur		1950	Walker	James	E.
1948	Robey	Donald	H.	1950	Watling	Robert	E.
1948	Rondeau	Herbert	F.	1950	Wieder	Irwin	
1948	Rowland	Theodore	J.	1951	Bauman	R. Craig	
1948	Smith	Robert	J.	1951	Beggs	William	C.
1949	Baldwin	L. David		1951	Blair	L. Russell	
1949	Bebout	Donald	E.	1951	Bow	Bark	Hall
1949	Berick	Joseph	G.	1951	Brainard	Ralph	C.
1949	Berlincourt	Ted	G.	1951	Burket	Robert	E.
1949	Berlincourt	Don		1951	Buskirk	Fred	R.
1949	Clark	Richard	E.	1951	Buynak	George	R.
1949	Cope	Randolph	H.	1951	Caris	John	C.
1949	Daye	Charles	J.	1951	Cygnaski	Richard	J.
1949	Francis	Richard	T.	1951	Friedenthal	Kenneth	J.
1949	Handelman	Stanley	M.	1951	Haynam	George	E.
1949	Herzog	Bertram		1951	Kasner	William	H.
1949	Koeblitz	William	E.	1951	Kloss	John	W.
1949	Lantz	Edward		1951	Mataich	Peter	F.
1949	Leiss	James	E.	1951	Norseth	Howard	G.
1949	Likly	Frank	G.	1951	Rogel	Albert	P.
1949	Milford	Frederick	J.	1951	Rouse	Carl	A.
1949	Naegele	Eugene	L.	1951	Scott	John	C.
1949	Neighbours	John	R.	1951	Wilson	Charles	R.
1949	Piper	Ervin	L.	1952	Fitzgerald	John	H.
1949	Reitz	Robert	A.	1952	Fitzwilliam	James	W.
1949	Schmidt	William	C.	1952	Friedman	Lawrence	
1949	Shields	Robert	G.	1952	Friedman	Richard	M.
1949	Thompson	Kingsley	P.	1952	Girardeau	Marvin	D.
1949	Walker	Douglas	H.	1952	Harmon	Edward	L.
1950	Adams	Robert	B.	1952	Hon	John	F.
1950	Anders	Frederic John		1952	Norris	Russell	S.
1950	Backer	Edwin	A.	1952	Pollak	Victor	L.
1950	Blachman	Arthur	G.	1952	Sciamanda	John	A.
1950	Braschwitz	Harold	J.	1952	Shirer	Donald	L.
1950	Bratten	Frederick	W.	1952	Stafford	Kenneth	E.
1950	Brett	Allen	E.	1952	Tomasch	Walter	J.
1950	Chowanetz	Robert	E.	1953	Field	Herbert	C.
1950	Cichocki	Theodore	S.	1953	Fisher	Henry	N.
1950	Coppock	Richard	A.	1953	Hill	Dale	E.
1950	Daniels	Raymond Dewitt		1953	Kepes	Joseph	J.
1950	Fenn	Craig	S.	1953	Kissel	Donald	E.
1950	Galey	William	F.	1953	Lawrence	Gerald	C.
1950	Garvey	John	J.	1953	Misek	Albert	E.
1950	Griffin	Byron	E.	1953	Pline	Richard	A.
1950	Groth	Lloyd	H.	1953	Reichert	Jonathan	F.
1950	Habermann	Carl	L.	1954	Bonsack	Walter	K.
1950	Hellwig	Paul	W.	1954	Corll	James	A.
1950	Hendrie	Joseph	M.	1954	Dickenson	Richard	H.
1950	Johnson	Robert	E.	1954	Finnerty	John	J.
1950	Lawrence	John	C.	1954	Fotland	Richard	A.
1950	Layer	Edwin	H.	1954	Garwin	Edward	L.

Appendix A Bachelor's Degrees Awarded

Year	Last	First	MI		Year	Last	First	MI
1954	Jones	Donald	H.		1958	Kramer	David	A.
1954	Kaufman	Ronald	R.		1958	Lewis	Murrell	E.
1954	Magee	Annette			1958	LoPorto	Leonard	S.
1954	Maneri	Carl	C.		1958	Luce	David	A.
1954	Shaw	Gordon	L.		1958	McFarland	Robert	L.
1954	Valencic	Frank			1958	Mentall	James	E.
1954	Venables	John	D.		1958	Olsson	George	R.
1954	Wolf	James	S.		1958	Rohde	Paul	J.
1954	Zeleznik	Frank	J.		1958	Scheppner	Edward	E.
1955	Conway	Melvin	E.		1958	Wiley	Charles	L.
1955	Gundel	Carl	H.		1959	Asik	Joseph	R.
1955	Kagan	Morton	R.		1959	Burns	Rowland	E.
1955	Palladino	Richard	W.		1959	Carver	J. Richard	
1955	Rasmussen	Charles	P.		1959	Conti	Carl	J.
1955	Stearns	Carl	A.		1959	Dudek	James	S.
1955	Vignos	James	H.		1959	Ferrante	John	J.
1956	Abramovitz	Albert	J.		1959	Gill	David	J.
1956	Baum	John	J.		1959	Giltinan	David	A.
1956	Blickstein	Beryl	D.		1959	Gundzik	Michael	G.
1956	Brown	Dennis	W.		1959	Gurr	Henry	S.
1956	Carlson	Roland	W.		1959	Jirberg	Russell	J.
1956	Chatterton	Neil	E.		1959	Kautz	Harold	E.
1956	de Castro	Aurora	F.		1959	Labuda	Edward	F.
1956	Estock	Paul	J.		1959	Lennon	Conrad	J.
1956	Ferguson	John	H.		1959	Manista	Eugene	J.
1956	Griggs	James	L.		1959	Markworth	Alan	J.
1956	Haybron	Ronald	M.		1959	McFadden	Robert	G.
1956	Hoffman	Lanny	L.		1959	Nezrick	Frank	A.
1956	Hooks	Lawrence	E.		1959	Payne	Carl	R.
1956	Kampos	Samuel			1959	Plummer	Robert	D.
1956	Kern	Edward	L.		1959	Priebe	Ray	F.
1956	Michael	James	A.		1959	Root	F. Elwood	
1956	Miller	John	C.		1959	Sayles	Charles	W.
1956	Niles	William	J.		1959	Spero	Samuel	W.
1956	Plummer	Arnold	M.		1959	Tomsic	Richard	T.
1956	Plummer	Robert	D.		1959	Zender	Karl	F.
1956	Reitz	Leonard	M.		1960	Baran	James	A.
1956	Rettig	Walter	H.		1960	Bartels	Richard	A.
1956	Salahlekekonen	Imunahad			1960	Bell	Seymour	
1956	Sanduleak	Nicholas			1960	Buccilli	Peter	R.
1956	Wohl	Millard	L.		1960	Comella	Thomas	M.
1957	Arndt	Richard	A.		1960	Czerniejewski	Francis	R.
1957	Butler	James	H.		1960	Dent	William	A.
1957	Holzschuh	Phillip	A.		1960	Dykes	Robert	R.
1957	Kilner	Joseph	R.		1960	Gans	Daniel	J.
1957	Lieabler	Kenneth	A.		1960	Green	James	R.
1957	Liebenauer	Paul	H.		1960	Hobbs	Robert	W.
1957	Marshall	Thomas	C.		1960	Jeffers	Larry	A.
1957	Meyers	Richard	G.		1960	Kaplafka	James	P.
1957	Nezbeda	Charles	W.		1960	Klementis	Kenneth	A.
1957	Plummer	John	P.		1960	Koneval	Donald	J.
1957	Salkeld	Edwin	M.		1960	Lawrence	Glen	S.
1957	Strickland	Paul	R.		1960	Lindstrom	Walter	W.
1957	Trimmer	Donald	S.		1960	Marmer	Gary	J.
1957	Walker	Arthur	B.C.		1960	McKeever	John	W.
1957	Wolf	Michael	L.		1960	Randall	Roger	M.
1958	Andeen	Carl	G.		1960	Safko	John	L.
1958	Eto	David			1960	Scearce	Bob	D.
1958	Galloway	Louie	A.		1960	Seidlitz	Bertram	E.
1958	Garber	Donald	I.		1960	Socash	Richard	R.
1958	Gebauer	David	C.		1960	Sugiuchi	Howard	
1958	Gustke	Eric	F.		1960	Van Horn	Hugh	M.

Year	Last	First		Year	Last	First	
1960	Witalis	R.E.F.		1962	Rader	Jon	A.
1960	Wolfe	Paul	J.	1962	Richards	Bernard	L.
1960	Young	Warren	M.	1962	Rosenberg	Howard	M.
1961	Blood	Frank	A.	1962	Rosenberg	Ronald	L.
1961	Bogan	Larry	D.	1962	Scott	Roderic	M.
1961	Breckinridge	James	B.	1962	Sheridan	Terrence	E.
1961	Breckling	Jane	M.	1962	Skelton	Harold	L.
1961	Chevako	Robert	J.	1962	Stieglitz	Robert	G.
1961	Cliffel	Earl	M.	1962	Szekely	Joseph	G.
1961	Deo	Priyatama		1962	Wallach	David	L.
1961	Elder	Timothy	W.	1962	Wells	James	W.
1961	Foote	Francis	C.	1963	Axline	John	T.
1961	Fritz	Oswald	G.	1963	Bebko	David	A.
1961	Gary	S. Peter		1963	Bousek	Ronald	R.
1961	Grossenbacher	Roger	W.	1963	Boyer	Charles	F.
1961	Hewes	Ralph	A.	1963	Burwasser	David	R.
1961	Hoy	Dennis	L.	1963	Cermak	Vince	
1961	Kish	James	A.	1963	Fell	Barry	M.
1961	Kish	James	A.	1963	Johnson	Porter	W.
1961	Kropfli	Robert	A.	1963	King	William	C.
1961	Levitt	Morris	R.	1963	Koral	Kenneth	F.
1961	Mantenieks	Maris	A.	1963	Le	Chinh Dinh	
1961	Marusek	John	J.	1963	Lindow	James	T.
1961	Meier	Anthony	W.	1963	Mantsch	Paul	M.
1961	Meier	John	J.	1963	Millard	Kenneth	Y.
1961	Miller	Sheldon	S.	1963	Morgan	David	R.
1961	Minium	Wayne	S.	1963	Oran	William	A.
1961	Phillips	Warren	E.	1963	Pierce	John	G.
1961	Ramins	Peter		1963	Sampson	Thomas	E.
1961	Slabinski	Victor	J.	1963	Sank	Eli	
1961	Swartz	Karl	D.	1963	Secura	Rajaishvar	
1961	Zych	Allen	D.	1963	Sega	Gary	A.
1961	Zych	Dale	A.	1963	Seiler	David	G.
1962	Anderson	Charles	A.	1963	Skala	Dennis	P.
1962	Billinghurst	Roy	A.	1963	Smith	Gary	R.
1962	Boys	Donald	W.	1963	Spiegelberg	William	D.
1962	Brenner	Dale	M.	1963	Strnisa	Fred	V.
1962	Cap	Daniel	M.	1963	Tessin	Jerry	G.
1962	Czika	Joseph		1963	Weber	Donald	E.
1962	Delly	William	A.	1963	Weiss	Paul	F.
1962	Dierdorf	John	R.	1964	Banks	Bruce	A.
1962	Everett	Paul	M.	1964	Barnes	Gary	T.
1962	Friar	James	L.	1964	Berman	Barry	L.
1962	Hansen	Gilbert	J.	1964	Braley	Richard	C.
1962	Hetzel	Frederick		1964	Burke	Victor	B.
1962	Hite	Gerald	E.	1964	Channin	Donald	J.
1962	Houff	Harry	P.	1964	Chevalier	James	L.
1962	James	Peter	N.	1964	Ciszek	Theodore	F.
1962	Jedlicka	Richard	A.	1964	Corwin	William	C.
1962	Jones	Roy	C.	1964	Crawford	Larry	J.
1962	Jones	Sanford	G.	1964	Daugavietis	Raymond	
1962	Kaercher	John	M.	1964	Eckert	Richard	R.
1962	Krinsky	Barney		1964	English	Larry	W.
1962	Krus	David	J.	1964	Gennert	David	W.
1962	Kurfess	James	D.	1964	Gigante	Joseph	R.
1962	Lane	Stephen	S.	1964	Gross	Robert	I.
1962	Mahoney	Richard	W.	1964	Hartzler	Arden	J.
1962	Mehlhorn	Rolf	J.	1964	Harwood	Kenneth	
1962	Milliken	John	C.	1964	Heckman	Roland	V.
1962	Olson	Jon	L.	1964	Heestand	Glenn	M.
1962	Ott	James	H.	1964	Hohberger	Clive	P.
1962	Pierret	Robert	F.	1964	Jordan	Michael	

Appendix A Bachelor's Degrees Awarded

Year	Last Name	First Name
1964	Levit	Lawrence B.
1964	Lewis	Trevor J.
1964	Lohanick	Alan W.
1964	Mackey	Jack E.
1964	Martonchik	John Vincent
1964	May	James M.
1964	McDonald	William W.
1964	Mook	Delo E.
1964	Moyer	LeRoy D.
1964	Murday	James S.
1964	Packan	Nicolas H.
1964	Patrick	Richard T.
1964	Pavco	John A.
1964	Peterson	Michael V.
1964	Rosenthal	Stephen M.
1964	Rottmayer	Robert E.
1964	Schilling	Hartmut
1964	Singer	Irwin L.
1964	Stabenau	Walter F.
1964	Sukel	Gerald J.
1964	Swyt	Carol K.
1964	Szabo	Szilard I.
1964	Thomas	Carol K.
1964	Thompson	Donald R.
1964	Veirs	Val R.
1964	Vesely	William E.
1964	Vrancik	James E.
1964	Wagner	David L.
1964	Wagner	Jerome
1964	Wallace	Russell D.
1964	Zappala	Robert R.
1965	Armbrust	Wayne T.
1965	Aronson	Barry A.
1965	Bane	John C.
1965	Baxter	Calvin E.
1965	Brockett	William S.
1965	Cellarosi	Mario J.
1965	Chechile	Richard A.
1965	Clark	Christopher D.
1965	Crower	James M.
1965	Daniel	Maurice
1965	Debesis	John R.
1965	Dennis	Wiley S.
1965	Dixon	Albert R.
1965	Doherty	James E.
1965	Donlon	William T.
1965	Doyle	Thomas B.
1965	Forman	Fred L.
1965	Fugate	Robert Q.
1965	Funk	Thomas R.
1965	Gayle	David M.
1965	Gilchrist	Michael S.
1965	Gorsha	Gary L.
1965	Green	Andrew S.
1965	Grega	Michael G.
1965	Hamel	Victor J.
1965	Hendrickson	Gary W.
1965	Hutyera	Andrew R.
1965	Jun	Eva T.
1965	Kaput	Terry M.
1965	Kramer	Paul J.
1965	Leib	Rena H.
1965	Meleg	Alexander J.
1965	Moore	James A.
1965	Nichols	Matthew A.
1965	Northrop	Joh
1965	Randle	Prather
1965	Roberts	Ronald
1965	Ruth	Robert J.
1965	Steele	William K.
1965	Wiefel	Shannon J.
1965	Yauch	Janet L.
1965	Yauch	Michael S.
1965	Zack	Dennis M.
1965	Zuppero	Anthony C.
1966	Ammirato	Frank V.
1966	Bartucci	John F.
1966	Boerio	Francis J.
1966	Boord	Warren T.
1966	Camerer	Frederick I.
1966	Chandler-Horowitz	Deane
1966	Curtis	Richard A.
1966	Doljack	Frank A.
1966	Garfinkel	Charles L.
1966	Goddard	Terrence P.
1966	Gray	Roger E.
1966	Hendrickson	John R.
1966	Hicks	Charles L.
1966	Hlusak	Donald G.
1966	Hoekstra	Dirk M.
1966	Johnson	James D.
1966	Kalasky	Edward D.
1966	Loe	Richard S.
1966	Loomis	John S.
1966	Lowdermilk	Warren H.
1966	Martin	Joshua W.
1966	McDowell	Robert R.
1966	Reiss	Michael Levi
1966	Roelant	Charles L.
1966	Sekaer	Christina M.
1966	Sepeta-Wissmann	Phyllis A.
1966	Stephens	Frederick C.
1966	Sulcs	Juris
1966	Taggart	Keith A.
1966	Waluch	Victor
1966	Wissmann	Phyllis A.
1966	Young	Kenneth P.
1967	Anderson	John T.
1967	Baird	Thomas J.
1967	Battes	Lee T.
1967	Cencula	Michael A.
1967	Channell	David F.
1967	Dayton	David B.
1967	Dei	Donald E.
1967	Dong	Clifford
1967	Fox	David A.
1967	Geisler	Fred H.
1967	Ghozeil	Isaac
1967	Gordon	Harold W.
1967	Grabowski	Ann K.
1967	Holy	John A.
1967	Iammarino	Joseph
1967	Kownacki	Edward J.
1967	Kraut	Valgene L.

Year	Last Name	First Name		Year	Last Name	First Name	
1967	Leadenham	Douglas	J.	1969	Kazek	Gregory	J.
1967	Mackay	Ian		1969	King	Elizabeth	H.
1967	Mynderse	Eric	H.	1969	Krempasky	Jerome	J.
1967	Nebel	William	P.	1969	Krouse	John	K.
1967	Norton	Nancy	M.	1969	Lerch	John	A.
1967	Parker	Sam	H.	1969	Malin	John	R.
1967	Rosenberg	John	M.	1969	Mannon	C. Leroy	
1967	Royer	Thomas	C.	1969	Marshall	Dana	M.
1967	Schultz	Roger	D.	1969	Martin	Peter	M.
1967	Sewall	Roy	F.	1969	Mathews	John	D.
1967	Singleton	Chloe	J.	1969	McCune	Robert	C.
1967	Szabo	Marianne	B.	1969	McCutchan	Joel	L.
1967	Wagner	Richard	E.	1969	Meier	Michael	J.
1968	Alles	Harold	G.	1969	Rehberg	John	T.
1968	Brown	Virgil	E.	1969	Shanfield	Zef	
1968	Carlson	John	D.	1969	Sharpe	Richard	T.
1968	Clendening	Charles	W.	1969	Sobol	Arnold	D.
1968	Cole	Gary	M.	1969	Spring Dickey	Janet	H.
1968	Cooper	John	E.	1969	Telesco	Charles	M.
1968	Delaune	Carl	I.	1969	Troyanowski	Charles	L.
1968	Elmore	David		1970	Adams	Randolph	A.
1968	Gambardella	Pascal	J.	1970	Allan	Richard	E.
1968	Genova	James	J.	1970	Arps	David	F.
1968	Greenberger	Bernard	A.	1970	Bloom	Phillip	J.
1968	Halley	David	C.	1970	Comstock	William	J.
1968	Hooley	David	L.	1970	Davis	Richard	C.
1968	Horton	Richard	F.	1970	Deckman	Harry	W.
1968	Horvath	John	J.	1970	Field	Robert	A.
1968	James	Kenneth	A.	1970	Finucane	James	J.
1968	Kastelic	John	R.	1970	Frohlich	Thomas	W.
1968	Kern	Wayne	K.	1970	Fyda	Walter	R.
1968	Kolena	John	A.	1970	Gettings	Michael	B.
1968	Langner	John	W.	1970	Greenstein	Martin	
1968	Lubin	Michael	L.	1970	Hancock	Dennis	M.
1968	Lubinsky	Anthony	R.	1970	Holm	Raymond	G.
1968	Masek	Richard	C.	1970	Karberg	Wayne	E.
1968	Mitchell	Paul	A.	1970	Kinstler	John	R.
1968	Newman	Charles	M.	1970	Kummer	David	W.
1968	Pahlow	Herbert	W.	1970	Kutz	James	W.
1968	Pelnar	Thomas	J.	1970	Kyle	Gary	S.
1968	Pinkney	Christopher	G.	1970	Lagin	Alan	R.
1968	Rowland	Edward	J.	1970	Lanese	Gustino	J.
1968	Sejnowski	Terrence	J.	1970	Lobl	Elena	S.
1968	Slane	John	A.	1970	Lock	James	A.
1968	Slusser	Sandra	J..	1970	Losh	David	L.
1968	Webber	Ronald	M.	1970	Mallinak	Edward	S.
1968	Wenocur	Brian	D.	1970	Mastrom	Michael	A.
1968	Yip	Vincent	F.S.	1970	Mostrom	Michael	A.
1969	Blonski	Janina	B.	1970	Orzech	Mary Ann	T.
1969	Bocianowski	Michael	W.	1970	Prevey	Paul	S.
1969	Brill	Michael Henry		1970	Richardson	Llanda	M.
1969	Collins	Byron	R.	1970	Rickman	James	D.
1969	Craig	Joseph	N.	1970	Schapira	Morey	R.
1969	Cung	Vu Khac		1970	Slavichak	Steven	E.
1969	Derkacs	Thomas		1970	Smith	Don	H.
1969	Dykstra	Dewey	I.	1970	Sojka	Richard	J.
1969	Gerlach	Robert	W.	1970	Steiner	Terry	O.
1969	Goetz	Daniel	W.	1970	Strok	Jack	M.
1969	Hagerling	Carl	W.	1970	Swift	James	H.
1969	Herrgesell	Carl	V.	1970	Wymer	Larry	J.
1969	Janecek	Edward	D.	1970	Young	Roger	A.
1969	Joseph	Roger	A.	1971	Bastian	Bruce	L.

Appendix A Bachelor's Degrees Awarded

Year	Last Name	First Name
1971	Biel	Joseph R.
1971	Bomberowitz	Robert J.
1971	Brown	Stanley C.
1971	Coblitz	David B.
1971	Cohen	Eric C.
1971	Drotning	William D.
1971	Epperson	Merrill A.
1971	Gann	Robert C.
1971	Granrath	Douglas J.
1971	Gross	Thomas P.
1971	Isett	Lawrence C.
1971	Klauber	Gary M.
1971	Laimins	Laimonis A.
1971	Lim	Evan Y.
1971	Matthews	David L.
1971	Mercer	Herald D.
1971	Mudrak	Samuel
1971	Rako	John G.
1971	Skow	Richard H.
1971	Soroka	Michael D.
1971	Stecewycz	Joseph
1971	Szmulowicz	Frank
1971	Szuch	Gene M.
1971	Valenzeno	Dennis
1971	Wilson	J. Roger
1971	Zadzilka	Dale R.
1972	Biba	Kenneth J.
1972	Boyle	Frederick P.
1972	Carlino	Thomas J.
1972	Cuddeback	John K.
1972	Dreger	Mark A.
1972	Farmer	Jeffrey K.
1972	Gasner	John T.
1972	Goldfinger	Richard C.
1972	Jung	Paul E.
1972	Kamin	Herman
1972	Kaumans	John F.
1972	Lewanski	Andrew J.
1972	Lieberman	Abraham
1972	McGowan	Paul T.
1972	Monnier	Richard A.
1972	Olshansky	Sanford C.
1972	Puhala	Michael J.
1972	Roberts	Neal P.
1972	Rullo	Norman A.
1972	Stair	Robert D.
1972	Stark	Gregory L.
1972	Stewart	John J.
1972	Strenio	Donald J.
1972	Tuckerman	Thomas A.
1972	Verderber	Lowell M.
1972	Vernot	David E.
1972	White	Ronald D.
1972	Wojcik	Robert C.
1972	Womack	Kenneth H.
1973	Abrams	Joel L.
1973	Bair	Virginia
1973	Bilbro	Griff L.
1973	Black	Bruce W.
1973	Bowser	William M.
1973	Friedenberg	Robert A.
1973	Gasner	Donald R.
1973	Grande	Sebastian
1973	Heagy	Stuart M.
1973	Kelin	David L.
1973	Klein	David L.
1973	Kubat	Dwight A.
1973	Lizak	Paul T.
1973	Mar	James
1973	Monet	David G.
1973	Pauli	Myron R.
1973	Perricelli	Vincent Jerome
1973	Powell	Byron R.
1973	Prior	Richard W.
1973	Sedlak	Joseph E.
1973	Sejnowski	Mark F.
1973	Smalc	Martin D.
1973	Stevenson	James W.
1973	Van der Mude	Antony
1973	Weiss	Douglas L.
1973	Yezzi	Michael J.
1974	Bernard	James E.
1974	Buta	Ronald J.
1974	Carr	David M.
1974	Crouse	Neal A.
1974	Cverna	Frank H.
1974	Deissler	Robert J.
1974	Ergun	David L.
1974	Gottschalk	Stephen C.
1974	Jennings	Wayne
1974	Knox	Charles A.
1974	Kohan	Thomas D.
1974	Kuenzli	Wayne H.
1974	Matthew	Michael W.
1974	Nolan	Anne M.
1974	Phillips	John Thomas
1974	Queenan	Robert M.
1974	Sheppard	James R.
1974	Sklan	Mark L.
1974	Strauch	Thomas B.
1974	Voss	Donald E.
1974	Wolpert	Robert C.
1975	Ameling	William
1975	Eckstein	David L.
1975	Ferrigno	Stephen J.
1975	Hinshaw	George A.
1975	Kahler	Richard L.
1975	Kirkland	Earl J.
1975	Lucas	Athena M.
1975	Nees	Thomas R.
1975	Neville	James G.
1975	Norgren	Richard M.
1975	Oprea	John Francis
1975	Peterson	Harry Mitchell
1975	Platt	Christine E.
1975	Salem	David J.
1975	Shaw	Brandon H.
1975	Slaminka	Edward Eugene
1975	Szedenits	Eugene
1975	Tompkins	Walter Hal
1976	Barthelmy	Scott D.
1976	Bryan	Randy J.
1976	Forden	Geoffrey E.
1976	Hoffmann	Mark T.

Year	Last	First		Year	Last	First	
1976	Kacenjar	Steve	T.	1980	Vanderwall	Robert	L.
1976	Keller	Ronald	J.	1980	VanDoren	Clayton	L.
1976	Moody	Trent	N.	1981	Bennett	David	P.
1976	Navrotski	Gary	K.	1981	Boytim	Bradley Albert	
1976	Richards	Bruce	G.	1981	Braymer	Joseph	E.
1976	Salontay	James	J.	1981	Carande	Richard	E.
1976	Slavin	James	A.	1981	Clary	Robert	D.
1976	Speriosu	Virgil	S.	1981	Crawford	Gregory	A.
1976	Stern	David	A.	1981	Crecca	Michael	A.
1976	Stewart	Kenneth	P.	1981	Elkmann	Paul Joseph	
1976	Weiss	Alan	A.	1981	Hebboul	Saad	E.
1977	Adams	Anne	E.	1981	Koch	Steven	W.
1977	Davis	Anne	A.	1981	Kritzer	Margaret	R.
1977	Greenberg	Paul		1981	Mack	Bryan	D.
1977	Gropp	William	D.	1981	Moore	Tracy	A.
1977	Hirt	Andrew	M.	1981	Plants	Donald	G.
1977	Talvacchio	John	J.	1981	Podany	Mark	E.
1977	Tartaglia	Michael	A.	1981	Stewart	John	S.
1977	Thomas	Donald	A.	1981	Tatah	Abdelkrim	
1977	Woodward	Richard	Paul	1981	Warner	Jeffrey David	
1977	Wright	Dennis	H.	1982	Beck	Kevin	A.
1978	Alexander	Mark	A.	1982	Hardy	Richard	H.
1978	Boenke	Mark	A.	1982	Heinonen	Olle	G.
1978	Durkin	John	T.	1982	Hetrick	James	E.
1978	Happoldt	Greig	P.	1982	Jones	Brian	K.
1978	Mann	Elizabeth	K.	1982	Kellner	Thomas	M.
1978	Schumacher	Reinhard	A.	1982	Kopanski	Joseph	J.
1978	Shure	Mark	A.	1982	Meara	Robert	J.
1979	Ailes	James	R.	1982	Mego	Thomas	J.
1979	Barile	Michael	P.	1982	Messenger	William	G.
1979	Cox	James	R.	1982	Naffah	George	
1979	Gillis	Keith	A.	1982	Risser	Steven	M.
1979	Greenberg	Jeffrey	A.	1982	Ruden	Edward	L.
1979	Guerrero - Egar	Anne	H.	1982	Worrell	Gregory	A.
1979	Hartley	John	G.	1982	Yeager	John	D.
1979	Johnson	Kenneth	L.	1983	Cavano	Loran	J.
1979	Johnson	James	D.	1983	Ciolek	John	T.
1979	Laird	Carol	J.	1983	Frankel	Jesse	E.
1979	Nolder	Craig	A.	1983	Glinsky	Michael	E.
1979	Palmer	Lawrence	A.	1983	Irwin	George	M.
1979	Prosser	Alan	G.	1983	Jaszczak	John	A.
1979	Taylor	Ronald	C.	1983	Jiang	Hongwen	
1979	Tupitza	Carol	J.	1983	Monzel	James	A.
1979	Wingate	Gaylord	V.	1983	Mooney	Douglas	D.
1979	Wood	Stephen	A.	1983	Naculich	Stephen	G.
1980	Bliss	Richard	D.	1983	Palunas	Povilas	V.
1980	Brasaemle	Karla	A.	1983	Pratt	Jonathan	P.
1980	Collard	Michael	J.	1983	Recko	Timothy	P.
1980	Dewhurst	Henry	Samuel	1983	Suszcynsky	David	M.
1980	Ditz	Michael	J.	1983	Washburn	Karl	B.
1980	Fisch	David	E.	1983	Withers	Richard	M.
1980	Hess	Daryl	W.	1984	Bernath	Gregory	N.
1980	Hinkle	Nancy	L.	1984	Chow	David	H.
1980	Hoekje	Peter	L.	1984	Cognion	Rita	L.
1980	Jaffe	David	E.	1984	Crumbaker	Todd	E.
1980	Klisz	Andrew		1984	DeLaney	David	B.
1980	Koontz	Nancy	L.	1984	Fleming	George	Wang
1980	Langelo	Victor	J.	1984	Fortner	Jeffrey	
1980	Larson	Edmund	M.	1984	Fujita	Isao	
1980	Manning	Robert	M.	1984	Hilterman	Robert	J.
1980	McConnell	Mark	L.	1984	Kleppe	Gary	T.
1980	Morris	Wayne	B.	1984	Lukin	Jonathan	A.

Appendix A Bachelor's Degrees Awarded

Year	Last Name	First Name	
1984	Marchetti	Vincent	J.
1984	McBride	James	J.
1984	O'Connor	Michael	J.
1984	Widmer	Mark	T.
1985	Biery	Kurt	A.
1985	Chaffee	Kevin	P.
1985	Dunham	Bruce	M.
1985	Harkless	Curt	
1985	Hess	Carl	H.
1985	Jayne	Douglas	T.
1985	Kopan	Mark	L.
1985	Lundberg	Wayne	R.
1985	Marx	William	J.
1985	Page	Timothy	D.
1985	Profusz	Steven	N.
1985	Radivoyevitch	Thomas	
1985	Schlabach	Philip	
1985	Schulman	Martin	A.
1985	Wakefield	Mark	S.
1985	Wolf	Robert	A.
1986	Goldstein	Tammy	
1986	Horvath	John	F.
1986	Karlsson	Magnus	
1986	McCarthy	Daniel	R.
1986	Morilak	Daniel	
1986	Piascik	Jeanne	M.
1986	Reed	Joseph	A.
1986	Ryba	Martin	F.
1986	Schienman	John	Edward
1986	Terlizzi	Kathleen	A.
1986	Tornkvist	Nils	Ola
1986	Vaughan	Timothy	E.
1987	Brown	Kyle	A.
1987	Chen	Anthony	Li-Chung
1987	Clark	William	R.
1987	DiCarlo	David	A.
1987	Fray	John	G.
1987	Heath, Jr.	James	Edward
1987	Heidger	Susan	
1987	Heindl	William	A.
1987	Hotes	Scott	A.
1987	Keener	Christopher	D.
1987	McKee	W.	Shawn
1987	Notte	John	A.
1987	Rootham	Krys	M.
1987	Slotta	James	D.
1987	Steagall	Robert	N.
1987	Teomi	Oren	
1987	Wimer	Joyce	E.
1988	Bleher	Siegfried	H.
1988	Boyne	Daniel	M.
1988	Danner	Guy	M.
1988	Hintz	Eric	G.
1988	Karpathakis	Michael	
1988	Lutgen	Steven	J.
1988	Moeller	Dieter	
1988	Paonessa	Thomas	P.
1988	Park	Ioana	V.
1988	Rice	Robert	A.
1988	Richards	Bruce	M
1988	Robertson	William	G.
1988	Scheick	Xania	N.
1988	Simpson	Mark	
1988	Underwood	Thomas	C.
1988	Valeriu	Ioana	
1988	Wickert	Steven	A.
1988	Woods	David	M.
1988	Zakariya	Eimad	
1988	Zolotarevsky	Julius	V.
1989	Blasko	Christopher	
1989	Bonham	Scott	W.
1989	Brick	David	H.
1989	Burlage	David	Stanley
1989	Campbell	H.	Brent
1989	Canfield	Gregory	H.
1989	Castiglione	James	A.
1989	DeFazio	Richard	A.
1989	Di Filippo	Frank	P.
1989	Foster	John Derek	
1989	Kaziner	Alexander	
1989	Kim	Young-Hwan	
1989	Klinich	George	
1989	Olhoeft	Jeffrey	M.
1989	Petrick	Rose	M.
1989	Reineks	Edmunds	Z.
1989	Shoemaker	Neil	S.
1989	Sornsin	Elizabeth	A.
1989	Starr	Gordon	C.
1990	Boerner	Eric	D.
1990	Cheiky-Zelina	Margaret	A.
1990	Glover	David	A.
1990	Oyster	Jay	C.
1990	Parmelee	Christopher	L.
1990	Rella	Christopher	A.
1990	Shovlin	Joseph	D.
1991	Ambigapathy	Rajesh	
1991	Degen	Michael	M.
1991	DeMaria	Carla	D.
1991	Gajewski	Donald	Anthony
1991	Gillahan	Robert	Mark
1991	Hjort	Hans	Henrik
1991	Kagel	Geoffrey	Allen
1991	Klepfer	Robert	O.
1991	Lewellen	John	Wesley
1991	Lieber Bruce	Sheryl	
1991	McLean	Michael	Allen
1991	McMillan	Kara	B.
1991	Morse	Andrew	
1991	Morse	Carroll	A.
1991	Raines	Eric	M.
1991	Siembor	Richard	C.
1991	Stocker	Dean	
1992	Al-Faks	Boulos	
1992	Beck, Jr.	Rex	G.
1992	Dureiko	Richard	D.
1992	Farukhi	Zaid	H.
1992	Foster	Josh	P.
1992	Herr	Quentin	P.
1992	Hosack	Michael	Galen
1992	Jogan	Stephen	M.
1992	Kennedy	D. Monroe	
1992	Lehner	Andrew	B
1992	Meltzer	Joel	D.
1992	Nacaskul	Poomjai	

Year	Last	First	Middle	Year	Last	First	Middle
1992	Nordgren	Charles	Erik	1997	Marshall	Cheshana	
1992	Stadelmaier	Brian	E.	1997	McGinnis	Sean	Patrick
1993	Andre	Kathleen	M.	1997	Schluchter	William	Clay
1993	Apanius	Christopher		1997	Schmidt	James	Sterling
1993	Booth	Michael	Francis	1997	Shack	Elizabeth	Ann
1993	Carter	Gregory	William	1997	Snider	Neal	Everette
1993	Cronin	Daniel		1997	Weyand	Jeremy	Ross
1993	Daire	Adam	Christian	1998	Fetters	Alan	Ford
1993	Harris	Todd	Louis	1998	Gurarie	Eliezer	
1993	Jayne	Jillanne	Mary	1998	MacGillivray	Edward	Stuart
1993	Keating	Brian	G.	1998	Monkiewicz	Jacqueline	Ann
1993	Scotland	Thomas	E.	1998	Olsen	Todd	Carter
1993	Smith	Steven	W.	1998	Rubin	Neil	Alan
1993	Warren	Christopher	Paul	1998	Schober	Andrew	Michael
1994	Belcher	Thomas	F.	1998	Smith	Alexander	Kent
1994	Dash	Denver	Halbert	1998	Thompson	Daniel	Stephen
1994	Jacono	Frank	J.	1998	Wozniak	Cheryl	
1994	Ngui	Sue-Mee	Shin	1999	Aubertine	Daniel	Bouorne
1994	Rapine	Richard	Ronald	1999	Jalics	Emily	Sofia
1995	Ahmed	Ali	Hassan	1999	Lufkin	Graeme	Walter
1995	Bachman	Richard	Joseph	1999	Phillips	Adam	Brian
1995	Becker	James	S.	1999	Ruedlinger	Benjamin	Franklin
1995	Dimmock	William	J.	1999	Taft	Rachel	
1995	Fisher	Christopher	William	1999	Tobias	David	Andrew
1995	Fuerst	Russell	A.	2000	Boss Bonilla	Michael	Anthony
1995	Holt	Christopher	Thomas	2000	Hartman	Mark	Aaron
1995	Iovane	John	Vincent	2000	Hedrick	Elizabeth	Marie
1995	Kangas	Erik		2000	Hibbitts	Lorenzo	
1995	Kaplar	Robert	J.	2000	Kolthammer	Jeffrey	Allen
1995	Kaspar	Jason	John	2000	Kraig	Robert	Eugene
1995	Knepley	Matthew	G.	2000	Linton	Eric	Thomas
1995	Lewis	Matthew	Allen	2000	Olson	David	
1995	Pierre	Darren	M.	2000	Truch	Matthew	David
1995	Saito	Susumu		2001	Blackmore	Brian	Lee
1995	Siegel	Donald		2001	Chunko	John	Daniel
1996	Chiaverini	John	A.	2001	Hanneke	David	Andrew
1996	Coleman	David		2001	Katz-Hyman	Moshe	Y.
1996	Elliott	Charles	B.	2001	Kogan	Oleg	
1996	Fridman	Moses	A	2001	Lewandowska	Marta	Karolina
1996	Graf	Benjamin	J.	2001	Mathieson	Genevieve	
1996	Harey	Andrew	A.	2001	Miller	Brandon	Eugene
1996	Johnston-Halperin	Ezekial		2001	Peshek	Timothy	John
1996	Kelly	Corinne		2001	Sica	Christopher	Thomas
1996	Kinemuchi	Karen		2001	Smith	Megan	Leah
1996	Kosc	Tanya	Z.	2001	Stickrath	Andrew	Blaine
1996	McDonald	Patrick	V	2001	Wagner	Nicolas	Lynn
1996	Nelson	Stephen		2002	Charlton	Kimberly	Anne
1996	Nielsen	Aaron	P.	2002	Davis	Chad	Allen
1996	Renner	Ryan		2002	Hyland	Peter	Owen
1996	Venkat	Radha		2002	Kaib	Nathan	Andrew
1996	Watkins	Neil		2002	Keenan	Cameron	Bradley
1997	Amon	David	Patrick	2002	Khalil	Joseph	Belden
1997	Arai	Takuya		2002	Large	Evan	David
1997	Banning	Matthew	Willliam	2002	Manalaysay	Aaron	G.
1997	Brooks	Travis	Christopher	2002	Radachy	Jason	David
1997	Eagan	Timothy	Patrick	2002	Salem	Michael	Phillip
1997	Field	Ethan	Bartlett	2002	Sherwin	William	Geoffrey
1997	Foster	Christopher	Howard	2002	Steinberg	Jeffrey	David
1997	Graff	John	W.	2002	Yoder	Jacob	Luther
1997	Hornish	Michael	Jay	2003	Adalia	Aaron	Ashok
1997	Knudson	Adam	Allen	2003	Alabiso	Audry	Marie
1997	Linenweber	Martin	Robert	2003	Bing	Thomas	Joseph

Appendix A Bachelor's Degrees Awarded

Year	Last Name	First Name	Middle Name
2003	Boehm	Joshua	Adam
2003	Hous	Robert	K
2003	Huss	Andrew	David
2003	Janezic	Timothy	Joseph
2003	Kenny	Justin	Ward
2003	Kubit	Brian	Edward
2003	Mantey	Kevin	Andrew
2003	McBride	Cameron	Keith
2003	Mehandru	Sonali	
2003	Reali	Dominic	Anthony
2003	Rodgers	James	Allen III
2003	Rodney	Steven	Alexander
2003	Schmidt	Clinton	T.
2003	Stroiney	Steven	Richard
2003	Wheeler	Jonathan	Allen
2004	Berkowitz	Zachary	Michael
2004	Borchers	Nick	
2004	Bush	Stephanie	Josephine
2004	Cuson	Jeremy	Moore
2004	Davidson	Benjamin	Shoots
2004	Finnerty	Matthew	John
2004	Johnson	Benjamin	David
2004	Khamis	Samuel	Mahmoud
2004	Kubera	Brian	
2004	Kundtz	Nathan	Brion
2004	Miller	David	Torbet
2004	Minar	Michael	Alan
2004	Morgan	Justin	Allen
2004	Morley	Adam	
2004	Mullet	Sarah	Catherine
2004	Nielsen	David	Christian
2004	Richards	Justice	Tyson
2004	Ritchie	Peter	J
2004	Rousos	Michael	James
2004	Waldstein	Micah	J.
2004	Wilbanks	Matt	Cheston
2004	Winkler	Mark	Thomas
2004	Woods	Scott	Patrick
2005	Bach	Paul	
2005	Bachler	Brandon	
2005	Benish	Rachel	
2005	Helle	Michael	
2005	Hui	Chiumun (Michelle)	
2005	Lehrian	Sarah	
2005	Light	Adam	
2005	Quattrone	Marisa	
2005	Salovich	Nicholas	
2005	Shekhar	Ravi	
2005	Smith	Matthew	
2005	Thrall	Justin	
2005	Treat	Alyx	

Appendix B   Graduate Degrees in Physics

Recipients of Masters Degrees in Physics

Before 1949, the only physics graduate degrees were at the Masters level. Listed here are the dates and names of the recipients, along with the thesis title and advisor, if available. After the federation of Case and Western Reserve in 1967, most of the Masters degrees have been earned as a result of passing a set of required courses and a qualifying exam, a written thesis no longer being required. Only a handful of thesis titles appear after that date. In general, terminal Masters students are no longer admitted to the physics graduate program. Most of the students after 1967 who are listed here went on to complete the Doctorate. Some left the University after earning their Masters degrees. Therefore, the list is cut off at 1990.

1895   Woodward, Harry Wilmot
1901   Skeels, Arthur A.
1901   Springsteen, Harry W.
1902   Jones, Franklin Turner
1904   Peabody, Carroll Adelbert
1913   Dreisback, Robert Rickert
1914   Randall, Dorus Powers
1915   Hoff, Clayton Malvern
1917   Hoover, Haided Hazel
1917   Hower, Harry Sloan
1918   Schad, Lloyd W.
1920   Hodgman, Charles D.
1921   Smith, Leland Ray
1922   MacLeon, Archibald Garrard
1926   Martin, John Richard
1928   Goss, Norman Philip
1928   Grebe, John J.
1928   Ott, Lawrence Henry
       A Study of the Production of Single Crystals of Iron
1930   Jones, Matthew Turner
1931   Offner, Abe
       A Critical Survey of Ether Drift Experiments   Mountcastle, H.W.
1931   Wallace, Clarence
1932   Balas, Michael
       Dielectric Constants of Some Fluoride Gases by a Radio Frequency Method
       Mountcastle, H.W.
1932   Barton, Roger
1932   Danstedt, Rudolph T.
       Dielectric Constants of Some Fluoride Gases by a Radio Frequency Method
       Mountcastle, H.W.
1932   Taylor, Philander B.
       A Study of the Volume Changes upon Setting of Some Silver-Tin Amalgams
       Mountcastle, H.W.
1932   Wanamaker, John H.
       Some Studies of Dilatations of Quartz Crystals at Low Potentials
       Mountcastle, H.W.
1933   Buxton, Chester L.
1933   Carpenter, Otis R.
1933   Lamb, Robert E.
1933   Robner, Leopold
1933   Shankland, Robert S.
1934   White, Herbert E.

Appendix B Graduate Degrees in Physics

		The Laws of Radiation: A Critical Survey	Mountcastle, H.W.
1935	Boyer, Raymond R.		
1935	Prettyman, Irven B.		
1936	Anderson, Norma		
1936	Eckstein, Richard W.		
1936	Ford, James W.		
1936	LaGanke, Robert S.		
1936	Shear, Sidney K.	Design of an Infra-Red Spectrometer	Mountcastle, H.W.
1936	Thompson, Bruce		
1936	Wenger, Emerson	The Extinction Coefficient of Smoke Haze	Mountcastle, H.W.
1937	Donaldson, John B.		
1938	Spremulli, Paul Francis	New Terms of the First Spark Spectrum of Manganese	Curtis, C. W.
1938	Tindal, Charles H.		
1939	Hine, Jerome Brooks	Construction of a Supersonic Oscillator	Curtis, C. W.
1939	Stearns, Hoyt A.		
1939	Zapf, Kenyon L.		
1940	Krumhansl, James A.		
1940	McDaniel, Boyce Dawkins		
1940	Taylor, James Earl	Ultrasonic Emulsification and Coagulation Effects	Curtis, C. W.
1941	Byers, Robert Cummins	An Investigation of Striations in a Kundt's Tube	Curtis, C. W.
1941	Davis, Gordon Albert		
1941	Parks, J. R.		
1941	Yanko, John A.	Dielectric Constants of Some Isomers of Hexyl Alcohol by a Radio Frequency Method	Curtis, C. W.
1942	Blythe, Richard H.		
1942	Fletcher, Charles H.		
1942	Gregg, Earle C. Jr.	Physical Phenomena in Boundary Lubrication	Crittenden, E. C. Jr.
1942	Klein, Gilbert E.		
1942	Rice, Philip James		
1942	Strough, Robert Irving	The Design and Construction of a High Intensity Electron Gun for use in Electron Diffraction	Crittenden, E. C. Jr.
1943	Rogers, Edward Stanley	An Electron Diffraction Camera for Metal Surface Studies	Crittenden, E. C. Jr.
1943	Sharkey, Andrew Gans		
1944	Elashowick, Murray		
1944	Ellis, Murray		
1946	Kerr, George P. Jr.		
1947	Arrowsmith, Charles		
1947	Carstensen, Edwin Lorenz	Self Reciprocity Calibration of Electroacoustic Transducers	
1947	Krohmer, Jack S.		
1947	Levine, Marc S.	The Electrokinetics of Certain Immune Reactions	McCarthy, J. T.
1947	MacIntyre, William James		
1947	Macklin, Eugene S.		
1947	Miller, Harry Bernard	A Method of Artificial Reverberation	Shankland, R. S.

Year	Name	Title	Advisor
1947	Schenk, Harry Millard		
1948	Arthur, N. Robert	Measurement of Secondary Emission	
1948	Benedict, Theodore S.	X-Ray Study of the Anomalous Structure of Cobalt	Smith, C. S.
1948	Demuth, Herbert M.		
1948	Fawcett, Sherwood L.	A Method for the Removal of the Electron Beam from the Betatron	Gregg, E. C.
1948	Halteman, Eber Kingdon	A Double Lens Electron Diffraction Camera	Crittenden, E. C. Jr.
1948	Hart, John Charles		
1948	Hudimac, Albert Aloysius	B-H Meter for Thin Films	Crittenden, E. C. Jr.
1948	Kneip, George D. Jr.	Quanitative Analysis of Iron and Cobalt Mixtures by Fluorescent X-Ray Spectra	Smith, C. S.
1948	Landon, Harry Hill Jr.	The Thermal Expansion of the Monoclinic Crystal Li2SO4H2O	Smith, C. S.
1948	Mendelson, Alexander	The Effect of Centrifugal Force on the Flutter of a Uniform Cantilever Beam	Shankland, R. S.
1948	Rosenblum, Earl Sobel	A Double-Foccussing Nuclear Spectrometer	Shrader, E. F.
1948	Strough, Robert Irving	An Improved Magnetizing System for a Dynamic Magnetic Analyser	Crittenden, E. C. Jr.
1948	Weaver, Harry Edward Jr.	The Design and Construction of a Wilson Cloud Chamber	Crittenden, E. C. Jr.
1949	Cook, Wilbur Schuyler	Straubel-Cut Quartz Crystal by Means of the Phenomenon of Streaming	Shankland, R. S.
1949	Harkless, Earl Thomas	Scintillation Counter Characteristics	
1949	Hoffman, Richard W.		Crittenden, E. C. Jr.
1949	Mayhew, Ray Winfield	An Electron Multiplier as a Particle Detector	Shrader, E. F.
1949	Orvis, Alan LeRoy	A Study of the Evaporation of Iron from Tungsten	Smith, C. S.
1949	Pavlovic, Arthur Stephen	Secondary Emission of Metalic Surfaces and Films	Shrader, E. F.
1949	Peterson, Richard Holmes		
1949	Samuel, Edmund William	Streaming in Liquids Resulting from Quartz Crystal Vibrations	Shankland, R. S.
1949	Van Horn, David D.		
1949	Zentner, Robert J.	Ultrasonic Light Diffraction Effects	Shankland, R. S.
1950	Chanstain, Joel William Jr.	The Calibration and Operational Performance of a 30 MeV Betatron	Gregg, E. C.
1950	Dixon, Jack R.		
1950	Fairweather, Stephen H.	Photomultiplier Tube as a Scintillation Detector of Alpha, Gammas, Neutrons	Shrader, E. F.
1950	Fleischer, Harold		Gregg, E. C.
1950	Fleischer, Melvin D.		
1950	Goddard, Murray C. II	A Beta Ray Spectrometer	Shrader, E. F.
1950	Hale, Donald W.		

Year	Name	Title	Advisor
1950	Hallam, Arthur F.	A Photographic Pyrophotometer	
1950	Johnson, Robert Donald	A Circuit for Short Ultrasonic Pulses	Shankland, R. S.
1950	Male, Donald	The Lattice Parameter-Composition Curve of Cobalt-Tungsten	Smith, C. S.
1950	Polmanteer, Keith Earl	Lattice Parameter V S. Composition Curves for Binary Metallic Solid Solutions	Smith, C. S.
1950	Truby, Frank Keeler	High Temperature Electronic Recording X-Ray Diffraction Equipment	
1950	Warner, Raymond M. Jr.	Design Considerations for a Pair Spectrometer Using Scintillation Counters	Shrader, E. F.
1951	Benedict, Robert Neil	Direct Detection of Photoneutrons Produced by the Bremsstrahlung Spectrum of the Betatron	Crouch, M. F.
1951	Bing, George Franklin	The Pulse Height Distribution from Scintillation Crystals	Foldy, L. L.
1951	Bond, Angus		Foldy, L. L.
1951	Bowman, John Charles	A Secondary Electron Multiplier Design	Shrader, E. F.
1951	Delio, John Gene	An Analysis of the Dynamic Characteristics of a Typical Gas-Turbine Engine	Beth, Richard
1951	Leary, Arthur P.		
1951	Neighbours, John R.	The Elastic Constant of Nickel	Smith, C. S.
1951	Rudlin, Leonard	On the Determination of Temperatures of Flames by Means of K-Band Microwaves	Beth, Richard
1951	Trask, Richard Keith	The Lattice Parameter-Composition of Nickel-Tungsten	Smith, C. S.
1952	Anders, Frederic John Jr.	Crystal Imperfections in Thin Films in Metals	Crittenden, E. C. Jr.
1952	Broadbent, Kent D.	A Constant Energy Expander for a Field-Biased Betatron	Gregg, E. C.
1952	Burns, John Wallace	The Elastic Constants of Cu-Si Alloy	Smith, C. S.
1952	Dandois, M.		
1952	Groselle, John Benjamin	Temperature Variation of Intensity of Magnetization in Thin Films	Crittenden E. C. Jr.
1952	Krohn, Victor E. Jr.		
1952	Layer, Edwin H. Jr.		
1952	Ogrinc, Raymmond S.	Shrinkage of Proton Tracks in Water Loaded Emulsions	Shrader, E. F.
1952	Robson, John W.	Dosimetry Studies with a Scintillation Crystal	Gregg, E. C.
1952	Stearns, Robert L.	A Statistical Analysis of Interferometer Data	Shrader, E. F.
1952	Tucker, Benson Leland	A Pulsed Scintillation Counter	Gregg, E. C.
1953	Adams, Robert B.		

1953	Bacon, Roger An Electron Ejection System for the Case Betatron	Gregg, E. C.
1953	Creager, Charles Bicknell Effects of Annealing on Electron Diffraction Line Broadening in Nickel	
1953	Daniels, Raymond D.	
1953	Diller, Dwain E.	
1953	Garr, Carl R. A Coincidence Study of the $Be^9(\alpha,n)C^{12*}$ Reaction	Crouch, M. F.
1953	Long, Thomas R. The Elastic Constants of Aluminum	Smith, C. S.
1953	Schmidt, William Variation of the Adiabatic Elastic Constants of Aluminum with Temperature	Smith, C. S.
1954	Cunningham, Donald E.	
1954	Fessler, Theodore E. Magnetic Properties of Thin films at Low Temperatures	Hoffman, R.W.
1954	Foreman, Charles Edward	
1954	French, Park Neutron Characteristic Times	Crouch, M. F.
1954	Harmon, Edward L. A New Electron Accelerator	Gregg, E. C.
1955	Buskirk, Fred	
1955	Chrien, Robert Edward A Table of Covariant Quantities for a Two Particle System	Milford, Frederick J.
1955	Clarke, Alan S. Photoproton Detection with an Ionization Chamber	Benade, A. H.
1955	Connolley, D. J. Application of an Extended Thomas-Fermi Model to the Rubidium Atom	Foldy, L. L.
1955	Daniels, William B.	
1955	Eckert, Alan C. Jr. Elastic Constants of Iron and Iron Silicon Alloys	Smith, C. S.
1955	Green, David H. The Design/Measurement of the Magnetic Fields for a New Electron Accelerator	Gregg, E. C.
1955	Klann, Paul Gerhardt Scattering of High Energy Gamma Rays	Gregg, E. C.
1955	Knorr, Thomas G. The Photodisintegration Cross Sections of Be^9, Li^7, and Li^6	Shrader, E. F.
1955	Kolb, Edwin R. Electrical Properties of Palladium Films	Hoffman, R.W.
1955	Levitt, Leo Barton The Electrical Properties of Zinc Selenide and Mercury Selenide	Gregg, E. C.
1955	Proctor, David G. Variational Methods for the Determination of Ground States of Two-Electron Systems	
1955	Romanowski, Thomas A. A General purpose Fast Coincidence and Pulse Height Analyzer	Shrader, E. F.
1955	Schoeffler, Lawrence E. Response of Organic Scintillators to Alpha Particles, Protons, and Electrons from 0-15 MEV.	Shrader, E. F.
1956	Eich, Alfred M. Jr. The Vibration Spectrum of a Square Lattice	Foldy, L. L.
1956	Glass, Solomon J. Temperature Variation of Intensity of Magnetization in Thin Nickel Films	Hoffman, R.W.
1956	Johnson, Ralph W.	
1956	Mata, S. J. The Acoustic Transit Time Error in the Ultrasonic Pulse Echo Method	Smith, C. S.

Appendix B   Graduate Degrees in Physics

1956 McKinney, Annette Irene
1956 Rose, Philip F.
Cross Section for the $Li^7(\gamma,p)He^6$ Reaction — Gregg, E. C.
1957 Cohen, D. Allen
A Calculation of the Elastic Shear Constants of the Zinc Blende Structure
Smith, C. S.
1957 Eros, Stephen
Elastic Constants of KCl and NaI — Smith, C. S.
1957 Schmunk, Richard E.
The Elastic Constants of Copper Nickel Alloys — Smith, C. S.
1957 Stooksberry, Robert William
Thermal Neutron Mean Lifetime in H_2O — Crouch, M. F.
1958 Bratton, Clyde Bruce — Weinberg, Joseph W.
1958 Finegan, Joel D.
Stress Annealing in Vacuum Deposited Iron Films — Hoffman, R.W.
1958 Greenwood, Reginald
1958 Haybron, Ronald McClure
On Statistics of Composite Quantum Mechanical Systems — Klein, M. J.
1958 Haynum, George E.
1958 Hockensmith, Duane A.
Theoretical Cross Section of the Gamma-Deuteron in Li^6 — Foldy, L. L.
1958 Petch, Hans Otto
1958 Seitz, Robert N.
1958 Walters, Virginia Fried
1959 Block, Richard B.
A Regulated Power Supply for a High Current Magnet — Gordon, W. L.
1959 Longaker, Perry R.
A Nuclear magnetic Resonance Spectrometer
1959 Molchen, Kenneth J.
Prototype Apparatus for Nuclear Excitation Studies — Benade, A. H.
1959 Schmidt, Charles T.
Photodisintegration of Li^6 at 2.62 Mev. — Benade, A. H.
1959 Weirich, Ronald G.
Design and Construction of a Neutron Spectrometer — Crouch, M. F.
1959 Wood, Van E.
Tetragonal Structure of Indium   Reitz, J. R.
1959 Yates, Edward A. C.
Photoneutron and Photoproton Production in Li^6 — Benade, A. H.
1959 Yelon, Arthur
Production of Single Crystal Copper Films at Low Temperatures — Hoffman, R.W.
1960 Blaustein, Marvin
Observation of Domain Switching in Thin Vacuum Deposited Iron Films   Hoffman, R.W.
1960 Brog, Kenneth C.
A Transistor Regulated Power Supply for a High Current Magnet — Milford, Frederick J.
1960 Finfgeld, Charles R.
The Well-Informed Heat Engine — Machlup, Stefan
1960 Garber, Donald I.
Elastic and Inelastic Scattering of Neutrons from Fluorine — Shrader, E. F.
1960 Genberg, Richard W.
Elastic Constants of Calcium Flouride
1960 Herlacher, David
The Stability of the Indium Structure — Reitz, J. R.
1960 Joseph, Alfred S.
Construction of Equipment for measuring the De Haas-Van Alphen Effect   Gordon, W. L.
1960 Kaul, Ronald Dean
Endpoint Energy of the O^{15} Beta Decay — Voelker, W. H.

1960 Kim, Hee Joong
Gamma Rays from the Proton Bombardment of Li^6 — Shrader, E. F.
1960 Meyers, Richard G.
1960 Neubert, Karl Dietrich
Experimental Investigation of the Anomalous Field Effect in Germanium — Nixon, J. D.
1960 Rosette, King H.
Temperature Variation of Spontaneous Magnetization in Thin Iron Films — Hoffman, R.W.
1960 Shoffner, Bruce M.
A Fast Neutron Detector — Shrader, E. F.
1960 Smith, Paulen Arthur
Resonance Radiation Studies of Mercury 198 — Olsen, L. O.
1960 Wolfe, Paul J.
1961 Bassichis, William H.
Kramers' Theorems and One Dimensional Wave Propagation in Periodic Potentials
Foldy, L. L.
1961 Bret, George C.
Triple Angular Correlation in Nuclear Reactions Involving Neutrons and Gamma Rays
Shrader, E. F.
1961 Chapman, Jamie C.
The Pressure Derivatives of the Elastic Stiffnesses of Silicon — Hoffman, R.W.
1961 Cook, Harlan J.
Band Structure of Black Phosphorus — Reitz, J. R.
1961 Culbert, Harvey Victor
A Proportional Counter for Low Energy Nuclear Spectra — Major, John K.
1961 Devlin, Shaun
Effect of Internal Strain on the Elastic Shear Constants of the Sphalerite Structure
1961 Freed, Norman
The Effect of Pairing Correlations on Magnetic Dipole Moments of Odd-A Nuclei
Kisslinger, L.
1961 Galloway, Louie A. III
Neutron Total Cross Sections for U^{235} and U^{238} — Shrader, E. F.
1961 Heidenreich, Rogert Darrow
1961 Helm, William J.
A Data Processing and Storage system for Dot-on-Film Pulse analysis — Gregg, E. C.
1961 Hockman, Thomas C.
1961 Jones, James Leo
A Radio-Frequency Mass Spectrometer — Voelker, W. H.
1961 McGaughy, Robert Earl
A Study of the Gamma-Ray Spectrum of Sn^{113} — Achor, W. T.
1961 Moe, Michael K.
Radiative Capture of Protons in K^{39} and K^{41} — Zimmerman, R. L.
1961 Mountain, Raymond D.
Theory of Thermal Expansion at Low Temperature — Klein, M. J.
1961 Ray, Dan Shelton
Ionization Potential of Linear Molecules Using a Delta Function Model — Machlup, Stefan
1961 Rogers, William Leslie
Elastic and Inelastic Scattering of Neutrons from Tantalum — Shrader, E. F
1961 Schiable, Paul M.
Angle of Incidence Magnetic Anisotropy in Evaporated Nickel Films — Hoffman, R.W.
1961 Smith, L. Dale
A Pulsed Vacuum Spark X-Ray Source — Benade, A. H.
1961 Stark, Royal W.
Magnetoresistance and Hall Effect of Single Crystals of Aluminum — Eck, T. G.
1961 Whitlock, Richard T.
On the Significance of Lin's Velocity Constraint in Two-Fluid Hydrodynamics
Zilsel, Paul R.

Appendix B Graduate Degrees in Physics

1962 Ali, Hyder Amir
Hartree-Fock Equations and Ionization Potential for O^{15} for a General Two Nucleon Potential
Robinson, Berol L.

1962 Babinsky, Andrew Daniel
Photo-Diffusion in Photoconductors — Machlup, Stefan

1962 Brand, Darrell A.
Deuteron Stripping reaction Evidence on Excited States of C^{12} — Shrader, E. F.

1962 Corll, James A.
Critical Evaluation of the Ultrasonic Pulse-Echo Technique in Single Crystals

1962 Harker, Yale Deon
Design, Construction, and Testing of a Large Scale Nuclear Emulsion Processing System
Frye, G. M. Jr.

1962 Lawrence, Gerald C.
Effective Range and Expansions for the Cotangent of the Scattering Phase Shift

1962 Moazed, Freydooun
Transverse Magnetoresistance in Single Crystals of Magnesium — Eck, T. G.

1962 Munsee, Jack H.
A Color Scheme for Spatial Resolution Using Liquid Scintillators — Reines, F.

1962 Nezrick, Frank A.
A Method to Detect Beta- radiation in the Presence of Gamma Rays — Reines, F.

1962 Smith, Lawrence H.
North Orienting System for High Altitude Balloon Research — Frye, G. M. Jr.

1962 Stephany, Joseph F.
A Vector Model Theory of Gravitation — Winterberg, F.

1962 Synn, Eun Hi
Angular Correlations of Gamma Rays in the Decay of Sb^{124} — Casper, Karl J.

1962 Wolfe, Elizabeth Anne

1962 Wright, Donald
Effect of Lattice Vibrations of Spin-Spin Relaxation Times in some Paramagnetic Materials
Peterson, R. L.

1963 Bambakidis, Gust

1963 Bartels, Richard A.

1963 Blood, Frank Alden Jr.
Theory of Einstein Bose Gas with an Energy Gap — Foldy, L. L.

1963 Bond, Peter D.

1963 Brown, Gerald V.

1963 Chow, Tai-Low
Inelastic Scattering of Relativistic Electrons by Nuclei — Weinberg, Joseph W.

1963 Chow, William Synjue
An Investigation of the Two-Center Integrals of Face-Centered Cubic Titanium Crystal
Reitz, J. R.

1963 Erskine, James C. Jr.
An Investigation of Surface Properties of Cadmium Sulfide — Nixon, J. D.

1963 Fair, Gale
Theory of Thin Ferrromagnetic Films by Bethe-Peierls-Weiss Method — Klein, M. J.

1963 Giltinan, David A.
Non-Local Proton-Proton Interaction — Thaler, R. M.

1963 Jackson, Jerome E.
Mean Lives of Positrons in Aqueous Solutions — McGervey, John D.

1963 Jeffers, Larry A.

1963 Kurfess, James D.

1963 Ogata, Hasashi

1963 Parker, Gerald H.
Neutron Leakage from the Ends of a Cylindrical Gaseous Fueled Nuclear Reactor

1963 Peters, Harold J.
A Preliminary Study of Thin Gadolinium Films — Hoffman, R.W.

Year	Author	Title	Advisor
1963	Plummer, Robert D.	Anharmonic Contributions to the Specific Heat of Metals	Reitz, J. R.
1963	Rote, Donald M.		
1963	Shaw, Melvin P.	Electron Spin Resonance in Lithium	Eck, T. G.
1963	Shilling, Gerd		
1963	Spencer, George L.		
1963	Sukow, Wayne W.		
1963	Tippie, John W.		
1963	Wu, Chi-Shiang	The Nuclear Matrix for the Beta-Decay of Sb^{124}	Kisslinger, L.
1964	Blaunstein, Robert P.	Comparison of Mössbauer Sources for Scattering Experiments	Major, John K.
1964	Bogan, Alexander Jr.	The Effects of Collisions upon High Frequency Plasma Oscillations	Foldy, L. L.
1964	Chen, Hsi-Tseng	Variational Calculation for Deformed Nuclei Applied to Mg-Isotopes	Tauber, G. E.
1964	Fausel, George William		
1964	Frankovsky, Frank A.	Lattice Parameters of Zinc at Low Temperatures	Smith, C. S.
1964	Gurr, Henry S.	Shortest Possible Resolution Time Achievable with a Bubble Chamber Motion Picture	Reines, F.
1964	Harpster, Joseph W. C.	Nuclear Spectroscopy with a Wide Depletion Region Silicon Radiation Detector	Casper, Karl J.
1964	Kazzaz, Abdul-Amir		
1964	Krinsky, Barnet	Temperature Variation of the Grüneisen Parameter	Foldy, L. L.
1964	Lancia, Frederick N.	Measurement of Peltier Coefficients on Bi-Te Alloys at 0 Celsius and Kelvin's Second Relation	McGervey, John D.
1964	Lawson, James R.	Magnetic Breakdown in the Long DHVA Periods in Zinc	Gordon, W. L.
1964	Lee, Wellington		
1964	Nishiyama, Tatuichiro	Analysis of an Interference Technique for Thin Film Stress Measurement	Hoffman, R.W.
1964	Riesenfeld, James	Annealing of Stress Anisotropy in Thin Iron Films	Hoffman, R.W.
1964	Ruth, Charles W.	A Self-Consistent Treatment of the Meissner Effect	Tobocman, William
1964	Sawyer, Charles M.		
1964	Stieglitz, Robert G.	An Adaptation of the Direct Reaction Calculation	Zimmerman, R. L.
1964	Walker, George E.	An Investigation of the Conditions Under Which a Kernel is a One Particle Density Matrix	Foldy, L. L.
1965	Berger, Michael A.		Shrader, E. F.
1965	Bose, Subhendu K.		Benade, A. H.
1965	Chin, Le Dinh	Exchange Effects in Scattering of Slow Electrons by Atoms	Thaler, R. M.
1965	Compton, Russell A.		Gordon, W. L.
1965	Doar, James		
1965	Fritz, Oswald G. Jr.	Decision of Ionic Changes in Stimulated Neural Elements by ATR Spectroscopy	

Appendix B   Graduate Degrees in Physics 311

	Eck, T. G.	
1965	Goff, James R.	
	Investigation of Triggering Detectors for a Balloon Borne Spark Chamber	
	Frye, G. M. Jr.	
1965	Gross, Gerard J.	
	Polarization of Protons Elastically Scattered from C^{12}	Silverstein, E. A.
1965	Hieber, Ross	
1965	Kottler, Herbert	
	D-State and Pick-Up effects in Nucleon-Deuteron Scattering	Kowalski, K. L.
1965	Kuebler, Gerard P.	
1965	Lapham, Jerome P.	
1965	Lawler, Martin T.	
	A Study of Cyclonic Two-Fluid Separation	Ostrach, Simon
1965	Le, Chinh Dinh	
1965	Lennon, Conrad	
1965	Miles, Richard G.	
	Low Temperature Thermal Expansion Coefficient of LiF	Smith, C. S.
1965	Owens, William Richard	
	Recoil Free Scattering	Major, John K.
1965	Pavelle, Richard	
1965	Pollard, Richard	
1965	Roberts, David Allen	
	DHVA Study of a Dilute Zinc in Aluminum Alloy	Gordon, W. L.
1965	Slabinski, Victor J.	
1965	Valladares C. Ariel A.	
	Low Temperature Specific Heats of Silver-Gold Alloys	Green, Ben A.
1965	Zych, Allen D.	
	Identification of Neutron Stars in Nuclear Emulsion	Frye, G. M. Jr.
1966	Armbrust, Wayne T.	
1966	Bedeson, Michael Peter	
	The Internal Compton Effect	Casper, Karl J.
1966	Braley, Richard C.	
	Study of the Gamma-Ray Correlations in the Reaction $Fe^{56}(n,n')$ $Fe^{56}*(\gamma)$ Fe^{56}	
	Nagarajan, M. A.	
1966	Caracena, Fernando	Weinberg, Joseph W.
1966	Chi, Huang Gertrude	
1966	Clune, Lavern Charles	
	Low Temperature Specific Heats of a-CuSn and a-CuZn Alloys	Green, Ben A.
1966	Geiger, Alan L. MA	
1966	Heckman, Roland V.	
1966	Kuebbing, Richard A.	
1966	Liebanauer, Paul	
	A Betatron Monitor Employing Cerenkov Radiation	Benade, A. H.
1966	Mayer, Frederick	
1966	Mayer, John T.	
1966	Scott, R. MacDonald	
	The Mott-Smith Approach Applied to a Shock Wave in a Binary Gas Mixture	
	Mawardi, O. K.	
1966	Topper, David R.	
1966	Wagner, David Loren	
1967	Chechile, Richard A.	
	Ultrasonic Equation of State of Tantalum	Smith, C. S.
1967	Dollhopf, William E.	
	A Nuclear Resonant Detector Using Mg_2Sn^{119}	Robinson, Berol L.
1967	Feldman, Lee	
	Model Calculations in Unified Nuclear Reaction Theory	Tobocman, William

1967	Ferrante, John	
1967	Fiske, John M.	
1967	Leeson, Jeffrey Sanford	
	Chemical Effects of Lead Ions on Positronium	McGervey, John D.
1967	Lin, Lu	
	On Rotating Nuclei and Coriolis Anti-Pairing Effect	Goswani, Amit
1967	Loje, Kenneth F.	
	Low Temperature Thermal Expansion Coefficient of Sodium Flouride Schuele, D. E.	
1967	Rottmeyer, Robert E.	
1967	Roy, Ranendra Kumar	
1967	Staib, Jon Albert	
	Measurement of the Cosmic Ray Gamma Ray Flux Near Sea Level Frye, G. M. Jr.	
1968	Arnold, Jeffrey Mark	
1968	Aron, Paul R.	
1968	Blackburn, David L.	Smith, C. S.
1968	Cellarosi, Mario	
1968	Chen, Ching Hsin	
1968	Chen, Ching Lu	
1968	Corlew, Gordon T.	
1968	Davis, Daniel	
1968	Gigante, Joseph R.	Huang, C. Y.
1968	Goldstein, Martin	Hoffman, R.W.
1968	Grier, Norman T.	Sparlin, D.M.
1968	Hankey, Robert E.	
1968	Haugland, Edward J.	
1968	Hoekstra, Dirk Macy	
1968	Hornbeck, Larry J.	
1968	Kisrchner, Sidney M.	
1968	Krajcik, Richard A.	
1968	Lindow, James Thomas	
	Elastic and Inelastic Scattering of 4.88 and 4.61 MeV Neutrons from Cobalt 59	Shrader, E. F
1968	Nalcioglu, Orhan	
1968	Nielsen, Philip E.	
1968	Nordgren, William E.	
1968	Roberts, Ronald	
1968	Sahr, Louis E.	
1968	Taggert, Keith A.	
1968	Tesic, Mike M.	
1968	Thurman, Ernest R.	
1968	Tobin, Thomas M.	
1968	White, David E.	
1968	Wright, David Lowell	
1968	Wu, Homei	
	Electronic Specific Heat of a-CuAl and a-CuBe Alloys at Low Temperature Green, Ben A.	
1969	Boord, Warren T.	Eck, T. G.
1969	Burke, Victor B.	
1969	Carey, Ronald F.	Dahm, A. J.
1969	Chang, Jhy Jiun	
1969	Channell, David	
1969	Deo, Priyatama	Segall, B.
1969	Doljack, Frank A.	
1969	Eshelman, James H.	
1969	Fontanella, John J.	

Appendix B Graduate Degrees in Physics   313

1969	Fukushima, Takao	
1969	Gubernatis, James E.	
1969	Gupta, Ashok Kumar	
1969	Huff, Robert James	Eck, T. G.
1969	Hunt, Ronald A.	Benade, A. H.
1969	Kovitch, James C.	
1969	Kuntz, Garland P.	
1969	Lee, Shu Tso	
1969	Logan, J. Lousi	
1969	Mathews, Alan A.	
1969	Ng, John N.	
1969	Powell, Robert E.	
1969	Reitz, Leonard M.	
1969	Smith, James B.	
1969	Varma, Matesh N.	
1969	Woodhouse, Robert L.	
1969	Worman, Walter E.	
1970	Alvarez-Elcoro, Ignacio	
1970	Anthes, William H.	
1970	Burwasser, David R.	
1970	Dinewitz, Isaac Jacob	
1970	Finkelstein, Murray M.	
1970	Harding, Thomas James	
1970	Jih, Felix R.	
1970	Kiernan, Aidin	
1970	Kim, Foon	
1970	Kogan, Alberto	
1970	Kubat, Emanuel Joseph	
1970	Reschly, Ronald R.	
1970	Wojnowski, Lawrence D.	
1971	Baltz, Anthony	
1971	Chotteau, Michael	
	The Isospectrum Clarinet System	Benade, A. H.
1971	Garfinkel, Charles L.	Gordon, W. L.
1971	Gupta, Prabhat K.	
1971	Kuo, Fu Shong	
1971	Lin, Ta Shyong	
1971	Nazaretz, Michael A.	
1971	Nerbun, Robert C.	
1971	Pak, Yue Leung	
1971	Park, Mun Whan	
1971	Sethi, Vijay K.	
1971	Sharma, Deepak Kumar	
1971	Vogel, John Stephen	
1972	Apaydin, Erdogan	
1972	Behl, Yugal	Taylor, P. L.
1972	Birn, Ivan	Taylor, P. L.
1972	Blonski, Robert P.	Gordon, W. L.
1972	Cardon de Lichtbuer, Pierre	Taylor, P. L.
1972	Cowan, John James	
	Vibrational Spectrum of the Hexagonal Close-Packed Wigner Lattice	Foldy, L. L.
1972	delaHostria, Emmanuel	
1972	Eckevit, Ahmet	
1972	Giordana, Nicholas D.	
1972	Krempasky, Jerome J.	
1972	Laplante, Paul Richard	

	Threshold-Memory and the Related Physical Changes in Amorphous $As_2Se_2Te_4$ Gordon, W. L.	
1972	Lerch, John A.	
1972	Lindberg, Vern W.	
1972	Malin, John Ralph	
1972	Nachtman, Joseph C.	
1972	Nolan, Anne M.	
1972	Ramsey, Thomas W.	
1972	Tufts, Memphis D.	
1973	Chu, Cor Man	
1973	Hunt, Willard F.	
1973	Kusner, Ronald R.	
1973	Leung, Steven	
1973	Lock, James A.	
1973	Matthews, David L.	
1973	Regone, Carl J.	
1973	Rodjak, David J.	
1973	Ryan, Robert B.	
1973	Smulowicz, Frank	
1973	Srikantia, Sriram	
1974	Boyle, Frederick P.	
1974	Goldfinger, Richard C.	
1974	Jacobs, Kenneth Roger	
1974	Puhala, Michael	
1974	Rako, John G.	
1974	Rondon Aramayo, Oscar A.	
1974	Soroka, Michael D.	
1974	Wenz, Robert P.	
1975	Acharya, Kishore	
1975	Valenzeno, Dennis P.	
1975	Wallett, Thomas M.	
1976	Jennings, Wayne D.	
1977	Chatterjee, Indira	
1977	Eckstein, David L.	
1977	Norgren, Richard Michael	
1977	Shaw, Brandon H.	
1978	Knox, Charles Anthony	
1978	Lau, Sing Choy	
1978	O'Reilly, John G.	
1978	Richards, Bruce G.	
1978	Shen, Ying	
1978	Simone, Inna	
	On the Theory of Melting in Correlated Heteropolymer Sequences	Taylor, P. L.
1978	Thompson, Stephen C.	
1979	May, Timothy E.	
1980	Bradrick, Thomas Dale	
1980	Geiger, Jack	
1980	Kordesch, Martin E.	
1980	Patrick, John	
1980	Platt, Daniel Enoch	
1980	Smith, Michael K.	
1980	Stansfield, Samuel E.	
1981	Bello, Alfredo	
1981	Lundin, Wade	
	Secondary Electron Emission from Indium Tin Oxide	Hoffman, R.W.
1982	Bilchak, Cynthia L.	

Appendix B   Graduate Degrees in Physics

1982	Fritz, William Joseph
1982	Hess, Daryll W.
1982	Holla, Jaya K.
1982	Manning, Robert M.
1982	Marino, Richard M.
1982	Middleton, Mark
1982	Stan, Mark A.
1983	Carande, Richard E.
1983	Debbe, Ramiro R.
1983	Gaitan, Frank P.
1983	Roche, James C.
1984	Abel, Phillip Benjamin
1984	Horton, Charles C.
1984	Pandya, Kaumudi Indukumar
1984	Ruan, Jian Zhong
1985	Jiang, Hongwen
1985	Kouzoupis, Spiridon N.
1986	Chang, Jing Yeu Eugene
1986	Dewhurst, Henry Samuel
1986	Fields, Jerry L.
	Interface trap properties of the metal-oxide-silicon carbide capacitor
1986	Huang, Lujia
1986	Li, Zili
1986	Rutan, Douglas M.
1986	Shih, Wen Ling
1986	Wang, Jing
1986	Wiefling, Kimberly Mae
1987	Chaffee, Kevin P.
1987	Kadi-Hanifi, Mohammed Djalil
1987	Larbi, Nacer
1987	Mathews, Paul G.
1987	Tornkvist, Ola
1987	Zypman, Fredy
1988	Jayne, Douglas T.
1988	Kell, Karsten L.
1988	Lamouri, Abbas
1988	Mearini, Gerald T.
1989	Banerjee, Shampa
1989	Chen, Liang Yu
1989	DeRose, Guy A.
1989	DiLisi, Gregory A.
1989	Krus, David
1989	Kusner, Robert E.
1989	Lee, Choon Heung
1989	Ruan, Minzi Yu
1989	Thakur, Mrinal Kanti
1989	Wang, Kui Long
1989	Wang, Yaxin
1990	Stegall, Robert N. Jr.
	Non-Linear Motion Artifacts in 2-D Fourier Transform Magnetic Resonance Imaging
	Haacke, E. Mark

Doctoral Graduates and their Dissertations
Listed by Advising Professor

The first physics doctorate granted by Case Institute of Technology was in 1949. The first at Western Reserve University was granted in 1962.

Adler, John G.
Will, Theodore A.   1968
   A Study of the Tl-Pb-Bi System Using Electron Tunneling

Akerib, Daniel S.
Perera, Thushara   2002
   The Limiting Background in a Dark Matter Search at Shallow Depth
Driscoll, Donald D.   2004
   Development and Performance of Detectors for
   the Cryogenic Dark Matter Search Experiment
Wang, Gensheng   2005
   The Cryogenic Dark Matter Search and Background Rejection
   with Event Position Information
Kamat, Sharmila   2005
   Extending the Sensitivity to the Detection of WIMP Dark Matter

Benade, Arthur H.
Chrien, Robert Edward   1958
   Photoneutron and Photoproton Production in Aluminum and Copper
Worman, Walter Elliott   1971
   Self-Sustained Nonlinear Oscillations of Medium Amplitude
   in Clarinet-Like Systems
Thompson, Stephen C.   1978
   Reed Resonance Effects on Woodwind Nonlinear Feedback Oscillations
Keefe, Douglas Howard 1981
   Woodwind Tone Hole Acoustics and the Spectrum Transformation Function
Kouzoupis, Spiridon N.   1985
   Systematic Model for the Clarinet Spectrum
Hoekje, Peter Lindsey   1986
   Nonlinear Self-Sustained Oscillations of Reed Instruments
Lindevald, Ian Morgon   1987
   The Perception of Musical Tones in the Statistical Sound Fields of Rooms

Bevington, Philip
Koral, Kenneth F.   1971
   A Search for the Bound Trineutron
Nerbun, Robert C. Jr.   1973
   A Study of the ^{24}Mg(d,n) Reaction from 2.5 to 4.0 MeV
Kogan, Alberto 1973
   A Study of the ^{28}Si(^{3}He,n)30S Reaction Between 3.7 and 4.2 MeV
Bose, Subhendu Kumar 1973
   A Study of the ^{20}Ne(^{3}He,n) Reaction from 2.5 to 4.0 MeV
Cverna, Frank H.   1978
   Pion production in proton-proton collisions at 800 MeV

Blanpied, William
Levit, Lawrence B.   1971
   A Search for a CP-nonconserving Asymmetry in the Decay $K_L \to \pi^+\pi^-\pi^\circ$

Brown, Robert W.
Martens, Michael   1991
    Gradient and rf coil issues in magnetic resonance imaging
Lai, Song   1996
    Brain Functional Magnetic Resonance Imaging Using Blood
    as an Endogenous Susceptibility Contrast Agent
Eagan, Timothy Patrick 2003
    Studies in Optimization for Magnetic Resonance Imaging
Gordon, Leon Bernard   1977
    A Study of Leptoproduction Mechanisms for Neutral Weak Bosons
Sahdev, Deshdeep   1978
    Tests for Certain Aspects of Quantum Chromodynamics
Stroughair, John D.   1982
    Tests of the Standard Model
Bilchak, Cynthia L.   1984
    A Test of Supersymmetry in the Standard Model
Zypman, Fredy R.   1988
    High Frequency Electromagnetic Fields in Magnetic Resonance Imaging
DeLaney, David 1988
    A General Low-Harmonic Cosmic String
    and Its Self-Intersection and Fragmentation
Thompson, Michael R.   1997
    Optimization Utilizing Inverse Methods for the Design
    of Magnetic Field Systems

Brown and Haacke
Boada, Fernando E.   1990
    Inverse Ultrasonic Scattering with Model Function Constraints
Petropoulos, Labros   1993
    Magnetic Field Issues in Magnetic Resonance Imaging

Brown and Krauss
Cheng, Yu-chung Norman   1998
    Physical Models from the Cosmos to the Factory: From Gravitational Lenses
    as a Probe of Cosmology to Electric Field Analysis in Industry

Brown and Shvartsman
Fujita, Hiroyuki   1998
    Theory and Design of Radio Frequency Coils in Magnetic Resonance Imaging
Willig-Onwauchi, Jacob David   2001
    Field Optimization in Magnetic Resonance Imaging:
    Striving for Perfect Shielding and Perfect Sinusoids

Brown and Yue
Liu, Jingzhi   2000
    FMRI Studies on Human Brain Function and Model of
    Muscle Activation and Fatigue

Casper, Karl
Owens, William R.   1968
    Gyromagnetic Ratio of Excited States in Iridium 191

Chandrasekhar, B. S.
Khorana, Brij M. 1968
    ac Josephson Effect in Superfluid Helium
Jackson, Jerome E.   1969
    A study of the lead-indium alloy system by electron tunneling techniques
Geiger, Alan Leslie   1969
    Inelastic Electron Tunneling in Al-Al Oxide-Metal Systems

Augis, Jacques A.   1969
     Electronic structure of bismuth alloys
Sood, Brij Raj   1971
     Phonon Spectra of Dilute Pb-In Alloys by Electron Tunneling
Gupta, Ashok Kumar   1971
     Thermal Conductivity of Superconducting Lead-Indium Alloys
Finkelstein, Murray M.   1972
     Oscillatory Magnetostriction and the De Haas-Van Alphen Effect in White Tin
Pak, Yue Leung   1974
     Superconductive tunneling in molybdenum
Haugland, Edward J.   1977
     Electrical Resistivity and Specific Heat of Nickel Near the Curie Temperature

Chottiner, Gary
Jennings, Wayne   1988
     The effects of sulfur segregation on the oxidation of iron
Eppel, Steven J.   1991
     Pt on graphite (0001): a model system for the study of physical and chemical properties on small metal islands
Wang, Kuilong   1992
     Surface Science Studies of Electrochemical Energy Storage Devices
Zhuang, Guorong   1994
     Application of Surface Analytical Techniques to the Study of the Reactivity of Lithium toward Nonacqueous Solvents

Coopersmith, Michael
Brown, Gerald V.   1967
     The Excitation Spectrum for a Boson Gas with Repulsive and Attractive Interactions
Neustadter, Harold E.   1969
     Electron Mobility in Randomly Located Hard-Core Scatterers

Crittenden, Eugene
Halteman, Eber Kingdon   1950
     Electron Diffraction Line Broadening in Nickel Films
Fawcett, Sherwood L.   1950
     The Ejection of the Electron beam from the Betatron
Hoffman, Richard Wagner   1952
     Study of Ferromagnetism by Means of Thin Films
Rasor, Ned S.   1955
     Annealing Kinetics of Lattice Defects in Vacuum Deposited Copper Films
Story, Harold S.   1956
     Stress Annealing in Vacuum Deposited Copper Films

Crouch, Marshall
Holzer, Alfred   1960
     Thermal Neutron Mean Lifetime in Water
Sobel, Henry Wayne   1969
     High Energy Gamma Ray Spectrum from Spontaneous Fission

Dahm, Arnold
Sai-Halasz, George A.   1972
     Ionic Mobilities in Solid Helium

Smith, James B. Jr. 1973
    Orthopositronium Annihilation in Liquid and Solid Helium
Carey, Ronald F. 1973
    Vortex Ring Generated Level Differences in Superfluid Helium
Lau, Shing-Choy 1978
    Ion Motion in Solid Helium
Mehrotra, Ravi 1982
    Studies on a two dimensional electron lattice
Stan, Mark Anthony 1988
    Transport measurements on a two-dimensional Wigner crystal near melting
Jiang, Hong-wen 1989
    Study of Melting and Sliding Charge-density Wave Conduction of Two-dimensional Electrons on Helium Films
Kusner, Robert Edward 1993
    Melting of Electric Dipoles in a Colloidal Monolayer
Hu, Xue Long 1994
    A study of a two-dimensional electron system in the variable-range hopping regime
Karakurt, Ismail 2000
    Dephasing Times in a Classical Two-dimensional Electron Gas
Goksu, Mehmet Ilhan 2002
    Studies on Edge Magnetoplasmon Modes

Dolan, J.F.
Slabinski, Victor J. 1970
    Induction Drag on Large Echo-Type Satellites: Observational Evidence

Eck, Thomas
Stark, Royal W. 1962
    The Galvanomagnetic Properties and Fermi Surface of Magnesium
Kaul, Ronald Dean 1963
    Level Crossings in Mercury
Brog, Kenneth C. 1963
    The Fine and Hyperfine Structure of the 2p State of Li^6 and Li^7
Wieder, Harold 1964
    Level Crossings and Anti-Crossings in the 2P state of Lithium
Shaw, Melvin P. 1965
    Cyclotron Resonance Investigation of the Fermi Surfaces for Cadmium, Zinc, and Magnesium
Sampath, Prativadi I. 1965
    Cyclotron Resonance Investigation of the Fermi Surface of Zinc
Zych, Dale A. 1967
    Cyclotron Resonance in Magnesium
Huff, Robert James 1969
    Stark-Induced Anticrossing Signals in the n=4 Term of Singly Ionized Helium
Smith, R. Lowell 1970
    Level-Crossing Signals in the 3^2D Term of Lithium
Sawyer, William Howell 1971
    A Measurement of the Hyperfine Structure of the 2P term of Lithium-6
Powell, Robert Edward 1972
    Cyclotron Resonance in Magnesium, Including Field-Normal Geometry
Kennedy, D. Munroe 1995
    Low-Energy Radio-frequency Sputtering of Cu, Anodized Al, and Kapton by Argon Plasma Ions

Farrell, David
Patrick, John Lester II 1983

A Study of Some Electrical Events in the Human Heart
Using Superconducting Magnetometry

Fickinger, William
Korpi, John L.   1972
    Some Final States with a Lambda in K⁻n Interactions at 4.9 GeV/c
Marino, Richard Matthew   1985
    Antiproton-Proton Interactions; 360 and 650 MeV/c Antiproton
    Interactions with Al, Cu, Pb

Foldy, Leslie L.
Osborn, Richard Kent   1951
    Second-Quantized Theory of Spin-1/2 Particles in Nonrelativistic Limit
Berger, Jay Menton   1953
    Deuteron Photo-disintegration at Intermediate Energies
Bing, George Franklin   1954
    The Photo-disintegration of Lithium-6: Neutron-Proton-Alpha Particle Model
Buskirk, Fred   1958
    On the Theory of Photo-disintegration of Lithium-6
Hockensmith, Duane A.   1960
    Bremsstrahlung Spectra Produced by 17.3 MeV Electrons
Haybron, Ronald McClure   1961
    Electro-disintegration of Hydrogen-3 and Helium-3
Ford, William F.   1961
    Structure of Low-Lying Levels of Lithium-6
Bassichis, William H.   1964
    An Investigation of the Boguliubov Transformation Applied to
    Bose-Einstein Particles
Blood, Frank Alden Jr.   1965
    Pair Correlation Functions in Classical Statistical Mechanics
Walker, George E.   1966
    Muon Capture and Supermultiplet Symmetry Breaking in Oxygen-16
Kottler, Herbert 1967
    Some Sequences of the Properties of the Internal
    Symmetry Crossing Matrices
Bogan, Alexander Jr.   1967
    The Systematics of Muon Capture in a Nuclear Shell Model
Braathen, Hans Jorgen 1969
    Spin Algebras and Representations of the Poincare Group
Krajcik, Richard Allen   1970
    Electromagnetic Interactions of Composite Systems
Lieh, Jong-Won   1971
    On Some Simple Lattice-Spin Systems
Prael, Richard Edward   1972
    Wigner Supermultiplet Model for Nuclear Binding Energy Formulae
Thakur, Jagannath   1973
    Multiple Scattering in Some Exactly Soluble Models
Lock, James Albert   1974
    Meson Exchange Currents in the Deuteron Electro-disintegration
Maassen, John Phillip   1978
    Dynamical Corrections to the Fixed Scatterer Approximation
Stansfield, Samuel E.   1986
    Spinless Particle in Combined Coulomb and Harmonic Oscillator Potentials
Foldy and Milford
Carome, Edward Francis   1954
    The Photo-disintegration of Lithium-6: Deuteron-Alpha Particle Model

Frye, Glenn M.
Smith, Lawrence H.   1965
    Cosmic Ray Primary Electron Flux Measurement with Rigidity Above 5 BeV
Zych, Allen D.   1968
    High Energy Albedo and Solar Neutrons Near the Top of the Atmosphere
Oran, William A.   1968
    Flux and Energy Spectrum of Cosmic Ray Electrons at a
    Geomagnetic Rigidity of 4.5 BeV
Staib, Jon Albert   1969
    Measurement of the Cosmic Ray Gamma-Ray Energy Spectrum
    at Balloon Altitudes
Kuo, Fu-Shong   1973
    Diffuse Cosmic Radiation Between 5 MeV and 20 MeV
Nazaretz, Michael A.   1976
    Atmospheric Neutrons at Geomagnetic Rigidities of 0.4, 4.5, and 11.5 GV
Vogel, John Stephen   1977
    Measurement of 10 MeV Atmospheric Gamma Rays at a High Magnetic Latitude
Pendleton, Geoffery   1988
    Predicted Performance of a Prototype Solar Neutron Detector

Gordon, William L.
Joseph, Alfred S.   1962
    The Low Field De Haas-Van Aphen Effect in Zinc
Genberg, Richard W.   1964
    A Determination of the Fermi Surface of Beryllium by the
    deHaas-vanAlphen Technique
Plummer, Robert D.   1965
    Magnetic Interaction in the DHVA Effect in Beryllium
Larson, Curtiss O.   1965
    A Low Field DHVA Study of the Fermi Surface of Aluminum
Beasley, J. Donald   1967
    Pulsed Field De Haas-Van Alphen Effect in Zinc
Everett, Paul Marvin   1968
    The Fermi Surfaces of Beryllium and of Dilute Alloys of Copper in Beryllium
Lawson, James R.   1969
    Effect of increased c/a ratio through alloying on the magnetic
    susceptibility and Fermi surface of zinc
Wagner, David Loren   1970
    Pulsed Field De Haas-Van Alphen Effect in $MgCu_2$
Fiske, John Mayo   1971
    The Fermi Surfaces and Electron Scattering of Dilute Beryllium:
    Transition Metal Alloys
Park, Jea Kook   1972
    Fermi Surfaces, Electron Scattering; Magnetic Breakdown; Annealed and
    Aged Dilute Zn-Cd Alloys
Hornbeck, Larry Joseph 1974
    The Effect of Alloying on the Fermi Surface of Magnesium
Fung, Wai Keung   1976
    Fermi Surface Changes in Dilute Mg Alloys:
    A Pseudopotential Band Structure Model
Bachmann, Michael A.   1981
    Characterization and Phase Transition Studies of Poly(Vinylidene Flouride)
Thakur, Mrinal Kanti   1983
    Structure Determination of Novel Polydiacetylene Materials
Ruan, Jian Zhong   1986

Soluble Ladder Aromatic Polymers: Synthesis and Characterization; Electronic Properties
Miranda, Felix Antonio 1991
   Microwave Response of High Transition Temperature Superconducting Thin Films
Zhong, Zheng-Zhong 1993
   Dielectric Relaxations in Side-Chain Liquid Crystalline Polymers

Green, Ben A.
Culbert, Harvey Victor 1964
   Low Temperature Specific Heats of Silver-Tin Alloys
Clune, Lavern Charles 1968
   Low Temperature Specific Heats of Dilute Lead-thallium and Lead-bismuth Alloys

Gregg, Earl C.
Tucker, Benson Leland 1953
   Cross Section Resonances in the $Li^7(\gamma,p)He^6$; the High Energy Bremsstrahlung Curve
Robson, John W. 1955
   Energy Spectrum of the Case Betatron Measured with a Compton Spectrometer
Stearns, Robert L. 1956
   Energy Spectrum of Electrons Produced by Bremsstrahlung in Aluminum
French, Park 1958
   Bremsstrahlung Spectra Produced by 17.3 MeV Electrons

Heuer, A.H.
Park, Mun-Whan 1975
   Spinodal Decomposition in the TiO_2-SnO_2 System

Hoffman, Richard W.
Knorr, Thomas G. 1959
   Geometric Dependence of Magnetic Anisotropy in Thin Iron Films
Yelon, Arthur 1961
   Fiber Texture and Magnetic Anisotropy in Evaporated Iron Films
Rosette, King H. 1961
   Temperature Dependence of Magnetization in Thin Nickel Films
Finnegan, Joel Dean 1961
   Stress and Stress Anisotropy in Iron Films
Myers, Richard G. 1964
   Electrical Resistivity of Thin Nickel Films Deposited in Ultra-High Vacuum
Riesenfeld, James 1965
   Temperature Dependence of Stress and Stress Anisotropy in Iron Films
Baron, Reinhart 1969
   Saturation Magnetization and Perpendicular Anisotropy of Nickel Films
Zuppero, Anthony C. 1970
   Mössbauer effect of ultra thin iron films
Rottmayer, Robert E. 1970
   Structure and Intrinsic Stress of Platinum Thin Films
Varma, Matesh Narayan 1971
   UHV Mossbauer Emission Spectra in Thin Iron Films
Doljack, Frank Anthony 1971
   The Origins of Stress in Thin Nickel Films
Springer, Robert W. 1973
   Growth Effects in Thin Nickel Films
Kwan, Mackenzie M. L. 1974

Ferromagnetism in Thin Amorphous Cobalt Films
Alexander, Pierre Martin 1975
    The Effect of Impurities on Intrinsic Stress in Thin Nickel Films
Wenz, Robert Paul   1977
    The Intrinsic Stress and Microstructure of ZnS and TiO_x Thin Films
Jansen, Frank   1977
    Investigation of the Surface Roughness of Evaporated Copper Films by Optical Scattering
Anderson, Charles R., Jr.   1978
    Mössbauer Effect Emission and AES Study of the Ni (111) Surface
Chase, Richard E.   1979
    Secondary Electron Emission from Aluminum Alloy Surfaces
Hagerling, Carl William   1980
    The Tensile Properties of Polyethylene Single Crystals
Krainsky, Isay   1981
    Plasmon Excitation in Alkaline Earth Metals
Kordesch, Martin E.   1984
    Conversion Electron Mössbauer Spectroscopy; X-Ray Absorption Fine Structure Spectroscopy
Eldridge, Jeffrey I.   1984
    A Mössbauer Spectroscopy Investigation of Passivated Iron Films
De Melo, Alfeu V.   1984
    High Resolution Electron Energy Loss Spectroscopy Studies of Surface Reactions on Pt(111)
Natarajan, Chitra   1985
    Extended Energy Loss Fine Structure Investigations
Abel, Phillip Benjamin   1985
    A Study of Forces Developed During Oxidation of Iron Thin Films
Yang, Kai-Yueh   1986
    Electron Yields and Escape Depths from Spacecraft Materials
Pandya, Kaumudi I.   1987
    EXAFS Investigations of Nickel Hydroxides and Nickel Oxide Electrodes
Horton, Charles Clarke   1988
    Kinetics of Interaction from Low Energy Ion Bombardment of Surfaces
Sanders, John N.   1989
    Dynamic Properties of Langmuir Films by Laser Light Scattering
Wang, Ya-Xin   1990
    Electron Energy Loss Analysis for Diamond and Diamond-like Carbon Materials
Chen, Hsiung   1990
    Preparation, properties, and structure of hydrogenated amorphous carbon films
Lamouri, Abbas   1991
    Low-Energy Sputtering of Teflon by Oxygen Ion Bombardment
Krus, David Jr.   1991
    Finite Element Analysis of Thin Film Mechanical Properties
Chaffee, Kevin P.   1991
    Ion beam analysis of diffusion in diamond-like carbon films
Mearini, Gerald T.   1992
    Mechanical Properties of Thin Aluminum/Alumina Multilayer Films
DeRose, Guy Arthur   1992
    X-Ray Absorption Fine Structure Strain Determination in Thin Films
Shiao, Jeansong   1993
    Properties of Diamond-like Carbon and Nitrogen Containing Diamond-like Carbon Films
Heidger, Susan   1993
    The Mechanical Properties of Diamond-like Carbon Films

Zorman, Christian A. 1994
    Annealing of Diamond and Diamond-like Carbon Films:
    An Ion Beam Analysis Study
Chen, Liang-Yu 1994
    Secondary Ions Sputtered by Low Energy Ion Bombardment of
    Cu and Al Surfaces

Huang, Chao-Yuan
Sugawara, Kazushi 1974
    Electron Paramagnetic Resonance of Rare-Earth Ions
Rachford, Frederic J. 1975
    Microwave studies of superconducting magnetometers

Jenkins, Thomas
East, Larry V. 1965
    Search for the Double Beta Decay of Nd^{150}
Munsee, Jack H. 1968
    Search for Disintegration of Deuterons by Electron Antineutrinos
Dix, Fred W. 1970
    Search for proton decay as a test of baryon conservation
Smith, Gary R. 1971
    Search for the Double Beta Decay of Nd^{150} Using a
    Magnetic Spectrometer Detector
Kirby, Allan Robert 1971
    High Energy Eta Meson Production at Large Momentum Transfers
Brockett, William S. 1971
    Pi⁻ P Charge Exchange Differential Cross Section at
    Intermediate Momentum Transfers
Delsignore, Kenneth W. 1996
    June 1991 Solar Flares by the Oriented Scintillation
    Spectrometer Experiment (OSSE) Above 6 MeV

Jha, Shacheenatha
Boolchand, Punit 1969
    Electric quadrupole interactions in single crystals of
    Te-METAL, Hf-METAL and $CaWO_4$ by Mössbauer effect
Bond, Peter Danford 1969
    Nuclear Structure and Hyperfine Field Studies with Molybdenum-95
Arnold, Jeffrey Mark 1970
    Study of the (He^4, He^6) Reaction on s-d Shell Nuclei

Johnson, W. B.
Spencer, George L. 1966
    An Experimental Investigation of Shock Waves in
    Highly Ionized Helium

Kantor, Paul B.
Taggart, Keith A. 1970
    Two Problems of the 3π decays of the eta meson

Kash, Kathy
Dyck, Jeffrey S. 2000
    Indium Nitride and Gallium Nitride Grown from the Melt
    at Subatmospheric Pressures

Kisslinger, Leonard

Ogata, Hisashi   1963
    Non-Harmonicity of Vibrations of Spherical Nuclei
Freed, Norman   1964
    A Self-Consistant Field Treatment of Spherical Nuclei
Wu, Chi-Shiang   1966
    Study of Collective States of Nuclei by Inelastic Electron Scattering
Rote, Donald Milton   1967
    Quasiparticle Random-Phase Approximation in Odd-Odd Nuclei
    and the Neutron-Proton Interaction
Thompson, Richard H.   1968
    A Relativistic Two Body Theory
Huang, Wei-chung   1971
    A Study of the Equations of Motion Method for Nuclear Structure Physics
Fukushima, Takao   1971
    Proton Deuteron Elastic Scattering at High Energy
Smith, Robert Stephen   1951
    Some Studies In Collective Electron Ferromagnetism

Klein, Martin J.
Rosenbaum, Burt M.   1957
    Statistical Foundations of Irreversible Thermodynamics
Glass, Soloman J.   1959
    Investigations on the Third Law of Dynamics
Mountain, Raymond D.   1963
    Statistical Mechanics of Adiabatic Processes
Wright, Donald   1965
    Properties of Bose Gases Using Reduced Density Matrices

Kowalski, Kenneth L.
Krauss, Jeffrey   1970
    Spin-Dependent Three Particle Scattering
Sharma, Deepak Kumar   1973
    Completely Unitary Approximations in Three-Particle Scattering
Cognion, Rita   2001
    Electromagnetic Wave Propagation through Dense Media
    Exhibiting Intrinsic Optical Bistability

Krauss, Lawrence M.
Liu, Hong   1997
    Quantum Hair, Magnetic Monopoles and Topology in Quantum Field Theory

Lambrecht, Walter
Kim, Kwiseon   1998
    First Principles Studies of Wide Band Gap Semiconductors and Their Alloys
Limpijumnong, Sukit   2000
    Theoretical Study of Some Aspects of Polytypism in Silicon Carbide
Jiang, Xiaoshu   2005
    First Principles Study of Electronic Structures of Defects in $ZnGeP_2$ and Defect Chalcopyrites

Leff, Harvey
Millard, Kenneth Y.   1971
    An Exact Solution of the Anisotropic Heisenberg Model
    With Long-Range KAC Interactions
Czika, Joseph Jr.   1971
    The Statistical Mechanical Basis of Local Pressure

Mathur, Harsh
Herman, Damir   2003
> Interplay of Disorder and Interaction in Electron Gas on Liquid Helium and in Quantum Dots

Mawardi, O. K.
Frank, Eugene N.   1965
> Transport properties of turbulent Lorentz gas

Scott, R. MacDonald   1967
> Downstream Plasma State in Hydrogen Produced by an EM Coaxial Shock Tube

Mayer, Frederick J.   1968
> Two Aspects of Transverse Charged-particle Diffusion in the Positive Column

McGervey, John D.
Walters, Virginia Fried   1965
> Correlation of Positron Lifetime with the Angle between the Annihilation Gamma Rays

Erksine, James C. Jr.   1966
> Electron Momentum Distribution in Silicon and Germanium by Positron Annihilation

Murray, Brian W.   1970
> Positron annihilation in copper and copper alloys

Lindberg, Vern W.   1976
> A positron annihilation study of neutron-irradiated aluminum

Panigrahi, Nokuleswar   1987
> Positron annihilation studies of cyclic fatigue damage in metals and aging in polymers

Meyer, Edwin F.   1988
> Monitoring the free volume changes in polyvinylacetate using positron annihilation, fluorescence, and electron spin resonance

Ruan, Minzi Yu   1992
> Positron Annihilation Spectroscopy; Microscopic Structure and Physical Aging

Yu, Zhibin   1994
> Positron and Positronium Annihilation Lifetime, and Free Volume in Polymers

Olson, Brian G.   2004
> Positron Annihilation Lifetime Studies of Polymers

Nagarajan, M. A.
Stieglitz, Robert G.   1968
> The Effects of a Spin Orbit Potential on the Reaction $C^{12}(d,p)C^{13}$

Hanna, John S.   1968
> A Shell-Model Calculation for the Reaction $N^{15}(p,n)O^{15}$

Olson, Leonard O.
Cunningham, Donald E.   1959
> Collision Processes in Mixtures of Mercury Vapor and Foreign Gases

Huang, Cheng-chung   1973
> Comparison of GaAs and CdTe Crystals for High Frequency Intracavity Coupling

Pao Y. H.
Boord, Warren Timothy   1973
> Far Infared Isolator

Appendix B Graduate Degrees in Physics

Petschek, Rolfe G.
Hinshaw, George A. Jr. 1988
    Phase Transitions; Modulated States in Ferroelectrics and
    Chiral Smectic C Liquid Crystals
Sones, Richard 1995
    Microphase Segregation and Twisting Transitions in Liquid Crystal Polymers
Stott, Jonathan 1998
    Theoretical Study of Dendrimeric and Heliclinic Liquid Crystals

Reichert, Jonathan F.
Usmani, Zahhiruddin   1969
    Electric Field Effects in ENDOR Spectroscopy of F-Centers

Reines, Frederic
Giamati, Charles C.   1962
    An Experimental Test of the Law of Conservation of Nucleons
Kropp, William R. Jr.   1964
    Neutrino Flux Limits and an Experimental Test of Baryon Conservation
Nezrick, Frank A.   1965
    Positron Spectrum from Fission Antineutrinos on Protons
Moe, Michael K.   1965
    The Electron Lifetime, an Experimental Test of Charge Conservation

Reitz, John R.
Tomasch, Walter J.   1958
    The Thermoelectric Power of Dilute Indium-Lead and Indium-Thallium Alloys
Unruh, Henry   1960
    Spin Wave Calculations for Ferrimagnetics, Antiferromagnetics and Ferromagnetics
Cohen, Donald A.   1960
    Calculations on the Group Theory and Band Structure of Cadmium Sulfide
Wood, Van E.   1961
    Electronic Band Structure of Cesium Gold
Michael, James Allen   1963
    Ionization Cross Section of Cesium in the Near Threshold Region
Block, Richard B.   1963
    The Onset of an Electromagnetically Driven Shock Wave
Devlin, Shaun S.   1964
    The Galvano-Magnetic Transport Coefficients of Semiconductors
    with Optical Mode Scattering
Allen, Gabriel   1964
    One-Dimensional Model for Magnetic Breakdown
Chow, William Synjue   1964
    Statistical Mechanics for a Charged Boson Gas
Cook, Harlan J.   1969
    Helium metastable cross-section measurements in a He and He-Hg plasma
    with a two wavelength laser heterodyne interferometer

Rix, John R.
Chow, Gee-Yin   1969
    Elastic Diffraction Scattering of Hadrons at High Energies

Robinson, Berol
Shera, E. Brooks   1962
    Search for a Spinless Excited State in Strontium-88
McGaughy, Robert Earl 1966

    Photochemical Studies of the Frog Retina
Schmidt, Stanley A.   1969
    Mössbauer effect studies of some intermetallic compounds of tin
Kuebbing, Richard A.   1970
    Measurements of Transition Probabilities in Some Middle Weight Nuclei

Robinson, D. Keith
Burdick, Bernard J.   1970
    Single Particle Production in K⁻n Interactions at 5 BeV/c with a Visible Λ or K
Gourevitch, S. A.   1971
    K p Final States in K⁻d Interactions at 4.9 GeV/c
Matthews, David L.   1976
    A Study of Strange Particles at 6 GeV/c in π^+n Interactions
Rondon-Aramayo, Oscar   1978
    Neutron Polarization in K^+N Interactions at 800 MeV/c Incident K^+
Debbe, Ramiro Rolando 1985
    Study of Pion Inclusive Momentum Distributions from Antiproton Interactions in Deuterium

Rosenblatt, Charles S.
Li, Zili   1990
    A Magneto- and Electro-optic Study of Ferroelectric Liquid Crystals
DiLisi, Gregory Anthony 1992
    Oligomeric Liquid Crystals: Viscoelastic Properties and Surface Interactions
Lu, Minhua   1993
    Elasticity and Polarizations in Ferroelectric Liquid Crystals
Tripathi, Sanjay   1996
    Surfaces and Symmetries in Liquid Crystals
Crandall, Karl A.   1996
    Chirality and Interfaces in Liquid Crystals
Li, Jian-Feng   1997
    Optical Studies of Molecular Architecture and Symmetry in Liquid Crystals
Kang, Daeseung   1999
    Chirality and Confinement in Liquid Crystal
Harrison, Daniel E.   1999
    A Light Scattering Study of Involucrin
Mahajan, Milind P.   2000
    Interfacial and Gravitationally Related Properties of Liquid Crystals and Other Fluids
Zhong, Shiyoung   2001
    Electro-optical Switching of Antiferroelectric Liquid Crystals
Wen, Bing   2003
    Surface Control for Liquid Crystal Alignment
Daj, Mohammad Reza   2003
    Electro-optical Investigations of Liquid Crystalline Mixtures
Patel, Neha Mehul   2004
    Electro-optic Studies of Liquid Crystalline Phase and Magnetically Levitated Liquid Bridges

Scharenberg, R. P.
Tippie, John W.   1965
    The Gyromagnetic Ratio of the First Excited Rotational States in Ytterbium
Wolfe, Paul J.   1966
    The Nuclear g-Factors of the First Excited States

in Sm 152 and 154, and Gd 156, 158, and 160
Schilling, Gerd   1967
    Static Electric Quadrupole Moment of the 0.558 MeV state in $^{114}Cd_{48}$

Schuele, Donald E.
Wong, Chuen   1968
    The Pressure and Temperature Derivatives of the
    Elastic Constants of BaF_2 and CaF_2
Roberts, Ronald   1969
    Ultrasonic Parameters in the Equation of State of the
    Sodium and Potassium Halides
Loje, Kenneth F.   1969
    The pressure and temperature derivatives of the
    elastic constants of AgBr and AgCl
Hankey, Robert Ernest 1969
    Third Order Elastic Constants of Aluminum Oxide
Fontanella, John J.   1969
    The dielectric properties of some alkali halides
Swyt, Dennis A.   1971
    The Grüneisen parameter of some semiconductors from elasticity data
Debesis, John R.   1971
    The Elastic Properties of Copper and Copper-Nickel Alloys
Andeen, Carl G.   1971
    Accurate determination of dielectric properties
Bello, Alfredo   1984
    The Dielectric Properties of Poly(Vinylidene Flouride)
Akins, Robert   1991
    Dielectric investigation of double glass transitions in polymers

Segall, Benjamin
Deo, Priyatama   1972
    Electronic Structure of thallium using the
    Relativistic Green's Function Method
Szmulowicz, Frank   1977
    Calculation of Optical and X-Ray K-Absorption Spectra of Aluminum
Lee, Choon-Heung   1993
    Theoretical Study of Diamond-Like Carbons and Nucleation of Diamond

Shankland, Robert S.
Gregg, Earle Covington Jr.   1949
    A Flux-Forced Field-Biased Electron Induction Accelerator
Voelker, William Henry 1952
    Angular Distribution of Compton Scattered Gamma Rays

Shrader, Erwin
Fleisher, Harold   1950
    The Angular Distribution of Protons from the
    Photodisintegration of Deuterium
Rosenblum, Earl Sobel 1951
    Gamma Ray Absorption Coefficients in Lead and Uranium at 5 and 10 MeV
Warner, Raymond M. Jr.   1952
    Angular Distribution of Bremsstrahlung Measured
    With an Energy-Selective Detector
Krohn, Victor E. Jr.   1952
    The $Cu^{63}(\gamma,n)Cu^{62}$ Cross Section
Romanowski, Thomas A.   1957

Ahmed, Nighat M. 1960
    The Photoneutron Cross Sections of Li^6 and Li^7
Green, David H. 1961
    Particle Recognizing Apparatus and Photoproton Processes in Li^6
Eckert, Alan C. Jr. 1961
    Gamma Ray Attenuation Coefficients for Helium from 5 MeV to 19 MeV
Kim, Hee J. 1962
    Nuclear Resonant Absorption of Gamma Radiation by Calcium-40
Garber, Donald I. 1964
    A Study of Neutron Producing Reactions by Deuteron Bombardment of $Boron^{11}$
Galloway, Louie A. III 1966
    Polarization of Neutrons from (d,n) Reactions on Carbon
Rogers, William Leslie 1967
    Neutron Total Cross Section Measurements Using a White Neutron Source
Kurfess, James. D. 1967
    Scattering of Polarized 2.33 MeV Neutrons from
    C^{12}, $Fe^{54,56}$, $Ni^{58,60}$ and Cu^{65}
Lindstrom, Walter W. 1970
    The Nuclear g-Factors of the 2^+ Rotational States
    in Nd^{150}, Dysprosium and Erbium
Lindow, James Thomas 1970
    Polarization of Neutrons from the C^{13}(d,n) Reaction
    Elastic and Inelastic Neutron Scattering Cross Sections
    of ^{54,56}Fe, ^{58,60}Ni, and Natural Carbon

Silverstein, Edward A.
Liebenauer, Paul H. 1970
    Deuteron induced reactions on C^{13}

Singer, Kenneth D.
Andrews, James Herbert 1995
    Third-Order Optical Nonlinearities in Organic Chromophores
Kowalczyk, Anthony 1996
    Second Order Optical Nonlinearities in Polymeric Waveguides
Hubbard, Steven F. 1997
    Measurements of Second Harmonic Susceptibility in
    Organic Nonlinear Optical Systems
Dureiko, Richard D. 1998
    Relaxation Processes in Poled Electro-optic Polymer Films
Ostroverkhova, Oksana 2001
    Nonlinear Optical Probes and Processes in Polymers and Liquid Crystals
Ostroverkhov, Victor 2001
    Chiral Second-order Nonlinear Optics
Kurti, R. S. 2004
    Pulse Compression in a Mid-infrared Synchronously Pumped Optical Parametric Oscillation

Singer and Brown
Dai, Tehui 2000
    Studies in Nonlinear Optics and Functional
    Magnetic Resonance Imaging

Smith, Charles S.
Van Horn, David Downing 1949
    An Experimental and Theoretical Study of the
    Solid Solubility of Tungsten in Iron
Neighbours, John R. 1953

Elastic Constants of Copper Alloys
Long, Thomas R.   1956
    The Elastic Constants of Magnesium Alloys
Bacon, Roger   1956
    The Elastic Constants of Silver Alloys
Winder, Dale Richard   1957
    The Elastic Constants of Indium
Nash, Harry C.   1958
    Single Crystal Elastic Constants of Lithium
Daniels, William B.   1958
    The Pressure Derivatives of the Elastic Constants of
    Cu, Ag, Au to 10,000 Bars
Scmunk, Richard E.   1959
    The Pressure Derivatives of the Elastic Constants
    of Al and Mg to 6,500 bars
Eros, Stephen   1960
    Low-temperature Elastic Constants of Magnesium Alloys
Trivisonno, Joseph   1961
    The Elastic Constants of Lithium-Magnesium Alloys
Corll, James A.   1962
    Pressure Derivatives of the Elastic Constants of Cadmium
Schuele, Donald E.   1963
    The Thermal Expansion at Low Temperatures of Rubidium Iodide
Miller, Roger Allen   1963
    Pressure Derivatives of the Elastic Constants of LiF and NaF
Smith, Paulen Arthur   1964
    Pressure Derivatives of the Single Crystal Elastic Constants of Potassium
Bartels, Richard A.   1964
    The Pressure Derivatives of the Elastic Constants
    of NaCl and KCl at 295K and 195K
Rotter, Carl A.   1966
    Ultrasonic Equation of State of Iron
Pauer, Lyle Anthony   1968
    Pressure Derivatives of the Elastic Constants of Rubidium
Hill, Edwin R.   1968
    Pressure Variation of the Elastic Constants
    of f.c.c. Indium-Thallium Alloys

Starkman, Glenn D.
Stojkovic, Dejan   2001
    Neutrino Mass and its Implications for the Zero Mode
    and Vacuum Structures of the Standard Model and its Extensions

Taylor, Cyrus C.
Gates, Evalyn I.   1990
    Irreducible Representations from the Quantization
    of Constrained Dynamical Systems
Convery, Mary E.   1997
    A Disoriented Chiral Condensate Search at the Fermilab Tevatron

Taylor, Philip L.
Bambakidis, Gust   1966
    A Study of the Electronic Structure of Liquid Metals
Hosack, Harold   1969
    Energy Levels of Bloch Electrons in Magnetic Fields
Fair, Gale   1969

>Theory of Galvanomagnetic Effects in Metals
Nielsen, Philip Edward   1970
>Theory of Thermoelectric Power in Metals and Alloys
Gubernatis, James E.   1972
>Properties of Disordered Systems
Gupta, Prabhat K.   1972
>An investigation of phase separation theories in glass forming systems
Cardon de Lichtbuer, Pierre   1974
>The magnetoresistance of cubic metals
Boyle, Frederick Phillip   1977
>A Continuum Model for the Prediction of Thermodynamic Properties in Simple Polymers
Ferrante, John   1978
>Adhesion at a Bimetallic Interface
Purvis, Carolyn K.   1981
>The Influence of Dipole Fields on Orthorhombic Ferroelectric Crystals: Application to PVF2
Good, Brian Scott   1983
>Theory of Structure, Morphology and Conductivity in Polymers
Banerjea, Amitava   1983
>Devil's Staircase In a One-Dimensional Model
Heinonen, Olle Gunnar   1985
>Equilibrium and Non-Equilibrium in the Quantum Hall Effect
Worrell, Gregory A.   1987
>Electronic Properties of Conducting Polymers
Zhang, Renshi   1991
>Theory of PVF2 and it's Ferroelectric Random Copolymers
Wang, Xin-yi   1996
>Theory of Phase Stability in Polymers and Liquid Crystals
Krupenkin, Tom Nikita   1997
>Microscopic Mechanisms of Polymer Fracture
Doerr, Timothy P.   1998
>Mechanics of Complex Molecules
Tsige, Mesfin   2001
>Theory and Simulation of Soft Condensed Matter

Thaler, Raphael M.
Schneider, Ronald E.   1964
>Are Low Energy Scattering Data Consistent with Charge Independence?
Holdeman, Jones T. Jr.   1966
>Electromagnetic Scattering of Charged Particles with Spin
Weiss, Douglas Louis   1976
>Elastic Scattering of One Fermion from a Finite Collection of Fermions in a 2nd Quantized Form
Friedenberg, Robert A.   1976
>A Field Theoretic Treatment of Pion-Nuclear Matter Elastic Scattering

Tobocman, William
Ruth, Charles W.   1967
>The Theory of the Meissner Effect
Purcell, James E.   1967
>Applications of Generalized R-matrix Theory to Simple Scattering Models
Giltinan, David A.   1968
>Distorted Wave Theory of the One Meson Exchange Reaction
Baltz, Anthony John   1971
>A Shell Model Calculation of Ti^{49}

Gintner, Henry   1972
    A Study of the Continuum States of the Beryllium-7 System Below 10 MeV
Rodjak, David Joseph   1975
    Test of the Shell Model Approach to Nuclear Reaction Theory
Goldfinger, Richard Carl 1976
    Analysis of Exchange Effects in the Reactions
    $^{16}O(^6Li, ^4He)^{18}F$ and $^{16}O(^3He,p)^{18}F$
Lewanski, Andrew J.   1978
    Test of Many Body Scattering Formalisms by
    Application to an Exactly Soluble Model
Chulick, Gary S.   1988
    One Boson Exchange Model in the Tobocman-Chulick Formalism
Driscoll, Diana   2001
    Free of Speckle Ultrasound Images of Small Tissue Structures

Vachaspati, Tanmay
Pogosian, Levon   2001
    Formation and Interactions of Topological Defects
    and Their Role in Cosmology

Voelker, W. H.
Proctor, David G.   1958
    Photodisintegration of Li^6

Wagner, Glenn
Longaker, Perry R.   1962
    Electron Spin Resonance in Donor Doped Graphite

Weinberg, Joseph W.
Bratton, Clyde Bruce   1964
    Nuclear Magnetic Resonance Studies of Living Muscle
Pereira, Carlos Martin 1967
    Relationship Between Space-Time Distance and
    the Conservation of Energy-Momentum
Pentz, Norman E.   1968
    Correlations in boson production of high multiplicity
Caracena, Fernando Jr. 1968
    The Electrodynamics of Vector Bosons

Willard, Harvey B.
Kosiara, Andrezej   1972
    A Search for Parity and Isobaric Spin Violating Decay
    of the 3.56 MeV Level in ^6Li
Anderson, Byron Don   1972
    A Study of the $^{16}O(d,n)$ Reaction from 3 to 4 MeV

Winterberg, Friedwart
Seitz, Robert N.   1964
    A Method of Obtaining Exact Solutions of the
    Boltzmann-Vlasov Equations

Woods, Robert
Reilly, Terence Douglas 1970
    Muon Decay Deep Underground

Zilsel, Paul A.

Whitlock, Richard T.   1963
    On the Pseudospin Model for Hard-Core Bosons with Attractive Interaction
Jun, Jau S.   1969
    Phase diagram of a quantum lattice gas with repulsive interaction
Roy, Ranendra Kumar   1970
    Quantum Lattice-Gas Model of Liquid Helium-4

Appendix C Alphabetical List of Faculty Appearing in Book

These are the faculty members mentioned in the book. They include, hopefully, all who were assistant professor and above, and who remained three years or more.

		born	died	degree	from	to	at	interest	chap
Adler	John G.	1935		Alberta	1965	1968	WRU	lo temp expt	15
Akerib	Daniel	1962		Princeton	1991	1996	CWRU	expt astro	18
Albats	Paul	1941		Cornell	1973	1978	CWRU	expt astro	8
Albright	John G.			Chicago	1921	1943	CSAS	lightning	4
Baer	Helmut	1939	1991	Michigan	1974	1978	CWRU	expt nuc part	16
Barrows	Allen C.			WRC	1866	1869	WRU		5
Benade	Arthur H.	1925	1987	WashSt.L.	1955	1987	CSAS	acoustics	12
Beth	Richard	1908		Frankfurt	1946	1957	WRU	positrons	10
Bevington	Philip R.	1933	1980	Duke	1970	1980	CWRU	expt nucl	16
Blanpied	William	1933		Princeton	1966	1972	CIT	expt part	16
Brown	Robert W.	1941		MIT	1970		CWRU	part theory	13
Casper	Karl Jos.	1932		Ohio SU	1960	1967	WRU	expt nucl	14
Chandrasekhar	Bellur S.	1928		Oxford	1963	1987	WRU	expt cm	15
Chew	Herman W.	1935		Chicago	1964	1967	WRU	wk em int'ns	11
Chottiner	Gary S.	1952		Maryland	1980		CWRU	expt cm	12
Coopersmith	Michael H.	1936		Cornell	1964	1969	CIT	stat mech	13
Covault	Corbin	1962		Harvard	2001		CWRU	expt astro	18
Crittenden	Eugene C.			Cornell	1938	1956	CSAS	expt nucl	7
Crouch	Marshall	1920		Wash St. L.	1952	1987	CSAS	expt cosm	8
Curtis	Cassius W.			Princeton	1937	1941	WRU		5
Dahm	Arnold J.	1932		Minnesota	1968	2001	CWRU	expt cm	15
Eck	Thomas G.	1929		Columbia	1957	2002	CIT	atomic expt	12
Eisner	Robert	1942		Purdue	1974	1977	CWRU	expt part	16
Emerson	Alfred			Yale	1853	1856	WRC		1
Farrell	David E.	1939		London	1966		CWRU	expt cm	15
Fickinger	William	1934		Yale	1967	1999	CWRU	expt part	16
Foldy	Leslie L.	1919	2001	U. Cal. Berk	1948	1990	CSAS	theory	9
Freeman	Spencer H.	1855	1886	Hopkins	1882	1886	WRU		5
Frisken	William R.	1933		Birm'ham	1966	1971	CIT	expt part	16
Frye	Glenn M.	1926		Michigan	1960	1987	CIT	cosmic expt	8
Gordon	William L.	1927		Ohio State	1955	1994	CSAS	expt cm	12
Goswami	Amit	1936		Calcutta	1963	1969	WRU	nucl theory	11
Green	Ben Arthur	1930		Hopkins	1961	1967	WRU	expt cm	15
Gregg	Earle C.			CSAS	1945	1961	CSAS	expt nucl	7
Hodgman	Charles D.	1881		CSAS	1906	1952	CSAS	photography	4
Hoffman	Richard W.	1927	2002	CIT	1952	1992	CSAS	expt cm	12
Huang	Chao-Yuan	1935		Harvard	1966	1974	WRU	expt cm	15
Jenkins	Thomas L.	1927		Cornell	1960	1997	CIT	expt cosmic	8
Jha	Shacheen.	1918		Edinburgh	1966	1969	WRU	expt nucl	14
Kantor	Paul	1938		Princeton	1967	1974	CWRU	part theory	13
Kash	Kathleen	1953		MIT	1994		CWRU	cm expt	18
Kikuchi	Tadashi	1934		Duke	1967	1970	CWRU	part expt	16
Kisslinger	Leonard	1930		Indiana	1956	1969	WRU	part theory	11
Klein	Martin J.	1924		MIT	1949	1967	CSAS	stat mech,hist	9
Koga	Rokutaro	1942		U. Cal. Riv'sd	1974	1980	CWRU	expt cosmic	8
Kowalski	Kenneth L.	1932		Brown	1963		CIT	part theory	13
Krauss	Lawrence	1954		MIT	1993		CWRU	part theory	18
Lambrecht	Walter R.	1955		Ghent	1996		CWRU	cm theory	18
Leff	Harvey	1937		Iowa	1964	1971	CIT	cm theory	17
Loomis	Elias	1811	1889	Yale	1834	1844	WRC	expt astron	1
Machlup	Stefan	1927		Yale	1956	2000	WRU	cm theory	11
Major	John Keene	1924	2003	Paris	1955	1966	WRU	cm & nucl	10

Surname	Given	Born	Died	School	Start	End	Affiliation	Field	Ref
Mathur	Harsh	1965		Yale	1995		CWRU	theory	18
McCarthy	John T.	1912		Yale	1937	1956	WRU	expt nucl	10
McGervey	John D.	1931	2001	Carnegie	1960	1999	WRU	expt cm	14
Meeks	Wilkison	1915		Northwestern	1948	1955	WRU	speech syn	10
Michelson	Albert A.	1852	1931	USNA	1882	1889	CSAS	optics expt	3
Milford	Frederick	1926		MIT	1952	1959	CSAS	nucl theory	9
Miller	Dayton C.	1866	1941	Princeton	1890	1941	CSAS	ether, acous	4
Mountcastle	Harry Wm.	1875	1955	Hopkins	1907	1945	WRU	expt atomic	5
Nagarajan	Mangalam	1933		Calcutta	1964	1969	CIT	nucl theory	13
Nooney	James			Yale	1844	1848	WRC		1
Nusbaum	Christian	1882		Harvard	1922	1953	WRU	expt cm	4
Olsen	Leonard O.	1910	1981	Iowa	1937	1960	CSAS	expr atomic	7
Pearle	Philip M.	1936		MIT	1966	1969	CIT	theory	13
Petschek	Rolfe G.	1954		Harvard	1983		CWRU	cm theory	17
Reichert	Jonathan	1931		Wash. St. L.	1966	1970	WRU	expt cm	14
Reid	Harry F.	1859	1944	Hopkins	1893		CSAS	seismology	3
Reines	Frederick	1918	1998	NYU	1959	1966	CIT	expt part	8
Reitz	John R.	1923		Chicago	1954	1965	CSAS	cm theory	17
Rix	John R.	1938		Harvard	1967	1970	CWRU	part theory	13
Robinson	Berol Lee	1924		Hopkins	1960	1970	WRU	nucl expt	14
Robinson	D. Keith	1932		Oxford	1966	2002	WRU	expt part	16
Rosenblatt	Charles	1952		Harvard	1987		CWRU	expt cm	18
Ruhl	John			Princeton	2002		CWRU	expt astro	18
Scharenberg	Rolf Paul	1927		Michigan	1961	1965	CIT	expt nucl	7
Schuele	Donald E.	1934		Case	1963		CIT	expt cm	12
Segall	Benjamin	1925		Illinois	1968	2000	CWRU	cm theory	17
Shakin	Carl M.	1934		Harvard	1970	1973	CWRU	nucl theory	13
Shan	Jie			Columbia	2002		CWRU	expt cm	18
Shankland	Robert S.	1908	1982	Chicago	1930	1976	CSAS	nucl expt	6
Shrader	Erwin F.	1916	1989	Yale	1940	1969	CSAS	nucl expt	7
Shutt	Thomas			U. Cal. Berk	2005		CWRU	expt astro	18
Silverstein	Edward A.	1930		Wisconsin	1964	1969	WRU	nucl medicine	7
Silvert	William L.	1937		Brown	1966	1969	WRU	cm theory	13
Singer	Kenneth D.	1952		Penn	1990		CWRU	expt cm	19
Smith	Charles J.			WRC	1870	1881	WRU		5
Smith	Charles S.	1916		MIT	1942	1968	CSAS	expt cm	7
Sparlin	Don Merle	1937		Northwestern	1964	1968	WRU	cond mat	15
Starkman	Glenn D.	1962		Stanford	1995		CWRU	astr theory	18
Sullivan	Charles A.	1934	1973	Vanderbilt	1969	1973	CWRU	expt part	16
Tauber	Gerald E.	1922		Minnesota	1954	1968	WRU	theory	11
Taylor	Cyrus C.	1958		MIT	1988		CWRU	part expt	18
Taylor	Philip L.	1937		Cambridge	1964		CIT	cm theory	17
Thaler	Raphael M.	1925		Brown	1960	1981	CIT	part theory	9
Tobocman	William	1926		MIT	1960	2001	CIT	part theory	9
Vachaspati	Tanmay	1959		Tufts	1996		CWRU	astr theory	18
Wallace	Clarence W.			CSAS	1926	1960	CSAS	editor	4
Wang	Chia Ping			Singapore	1966	1970	CIT	cosmic expt	8
Weinberg	Joseph W.	1917		U. Cal. Berk	1959	1969	WRU	gravity	11
Whitman	Frank P.	1853		Hopkins	1886	1919	WRU	photometry	5
Willard	Harvey	1925		MIT	1967	1981	CWRU	expt part	16
Winter	Rolf G.	1928		Carnegie	1951	1954	WRU	nucl theory	10
Winterberg	Friedwart	1929		Göttingen	1959	1963	CIT	theory	9
Woods	Robert			Michigan	1964	1970	CIT	cosmic expt	8
Wright	Elizur Jr.	1804	1885	Yale	1829	1833	WRC		1
Young	Charles A.	1834	1908	Dartmouth	1856	1866	WRC	solar phys	1
Zilsel	Paul R.			Yale	1960	1970	WRU	theory	11

The papers presented at this sectional meeting of the APS provide an interesting snapshot of the research being done in 1949 in both the Case and Western Reserve physics departments.

Meeting of the Ohio Section of the American Physical Society
5 November 1949

1. Apparatus for Acoustical Measurements with Pulse-modulated Ultra-sonic Waves
    Yeager, Chessin, Bugosh, Hovorka; CIT
2. A High-frequency Barium Titanate Hydrophone
    Bugosh, Yeager, Hovorka - CIT
3. A High-frequency Electro-acoustic Effect and Its Utilization in the Construction of Hydrophones
    Bugosh, Yeager, Hovorka - CIT
4. The Condensation of Evaporated Metals on Surfaces
    Olsen, Crittenden, Hoffman - CIT
5. Use of WWV Signals to time Pendulums
    McCarthy - WRU
6. A Novel Magnetostriction Effect
    Janofsky, Melamed, Beth - WRU
7. Report of Results of the science Abstracting Survey of APS-AIP
    Hutchisson – CIT
8. An Adjustable Mounting for Large Mirrors
    Haynie, OSU
9. Monochromatic Electron Groups in Long-life Manganese
    Emmerick, Ballweg, Kurbatov – OSU
10. Radiations Emitted y Long-life Species of Nickel
    Thomas, Kurbatov – OSU

Invited Paper: The Case Betatron and Plans for Its Use
    Gregg – CIT

Modern Physics in America
A Michelson-Morley Centennial Symposium
30-31 October 1987

In celebration of the centennial of the Michelson-Morley ether drift experiment, the university, along with twelve other Cleveland educational and cultural institutions, organized "Light, Space, and Time – A Cleveland Festival 1987". The event was chaired by Dorothy Humel Hovorka, assisted by physics Professor Philip Taylor The final event of this seven-month long celebration was a two-day symposium entitled "Modern Physics in America", organized by the Physics Department under the direction of co-chairmen, Professors William Fickinger and Kenneth Kowalski.

*The unique audience of the symposium included about six hundred physics graduate students who were bussed in from major universities within a 200-mile radius of Cleveland, from Toronto to East Lansing, from Pittsburgh to Cincinnati. Among the speakers, chairpersons, and guests-of-honor were nine Nobel Laureates (a few of whom would be so honored after 1987). All the expenses for the students and other guests were provided for by generous gifts from the 1525 Foundation (a special friend of the University), by the General Electric Foundation, and by NASA. The proceedings, including text of all the talks, have appeared in the AIP Conference Proceedings **169**, editors Fickinger and Kowalski.*

Here is a list of the lectures:

1. High Energy Colliders and Exploration of Small Distances: What are the Limits?
    Wolfgang K. H. Panofsky
2. Reminiscences of My Father
    Dorothy Michelson Livingston
3. Atoms, Molecules and Light
    Arthur L. Schawlow
4. The Life of the Stars
    Hans A. Bethe
5. Neutrinos from the Atmosphere and Beyond
    Frederick Reines
6. The Search for gravitational Waves: Probing the Dynamics of Space-time
    Peter F. Michelson
7. Chaos and Turbulence: An Experimental View
    Albert J. Libchaber
8. A Physicist's View of Biology
    Ivar Giaever
9. Strange Insulators, Strange Semiconductors, Strange Metals:
        High T_c as a Case History in Condensed Matter Physics
    Philip W. Anderson
10. Grand Challenges to Computational Science
    Kenneth G. Wilson
11. The Supercollider: Assault on the Summit
    Leon Lederman
12. Superstrings
    Murray Gell-Mann
13. SN1987a: The Supernova of a Lifetime
    Robert Kirshner
14. The Discovery and Physics of Superconductivity above 100K
    Ching Wu Paul Chu

The Michelson Lectures and Awards

*The Michelson Lecture series, in its various incarnations, has brought eminent scientists and engineers to the CWRU campus for the past four decades. The series began in 1963 when the Trustees of the Case Institute of Technology established the **Michelson Award**.*

1963	J. H. van Vleck	"Father of Modern Magnetism"
1964	H. K. Hartline	"Pioneering Biophysicist"
1965	Luis W. Alvarez	"Leading Nuclear Physicist"
1966	Edwin H. Land	"Lightness, Brightness and Reality"
1967	Martin Schwarzschild	"Structure and Evolution of the Stars"

*The series was renamed the **Michelson-Morley Award** after the Federation of Case and Western Reserve, and continued almost every year until the 1990's.*

1968	John Bardeen
1970	Charles H. Townes
1976	John D. Roberts
1977	Gene M. Amdahl
1978	Harry G. Drickamer
1979	Hans Wolfgang Liepmann
1980	Frank Albert Cotton
1981	Francis H. C. Crick
1982	Michael Ellis Fisher
1983	Subrahmanyan Chandrasekhar
1984	Paul C. Lauterbur
1985	Paul A. Fleury
1986	Richard N. Zare
1987	George A. Olah
1988	John J. Hopfield
1989	Herman F. Mark
1990	Frederick Reines
1991	John Cahn
1992	Watt W. Webb

*The **Michelson Lectures** were established by the Physics Department in 1995.*

1995	Joseph H. Taylor	"Binary Pulsars and Relativistic Gravity"
1995	Frank Wilczek	"Black Holes and Quantum Mechanics: Trouble on the Horizon?"
1996	Sheldon L. Glashow	"The Universe and the Particle: All Features Great and Small"
1997	Robert C. Richardson	"The Superfluidity of Helium-3"
1999	Michael E. Fisher	"Phase Transitions and Our Understanding of the Physical World"
2000	Gerhardus 't Hooft	"A Confrontation with Infinity"
2003	Stephen Chu	"What can we learn from looking at biological processes, one molecule at a time?"

*In 2002, the **Michelson-Morley Award** was re-established. The award was made possible through an endowment from the 1987 Michelson Morley Centennial Celebration Committee, Dorothy Humel Hovorka, chairperson.*

2002	Frank Wilczek	"The World's Numerical Recipe"
2003	Stephen Hawking	"Brane New World"

The Michelson Post-doctoral Prize Lectureship

This prize, established in 1997 by Profs. Lawrence Krauss and Glenn Starkman, is awarded annually to a junior scholar active in any field of physics. Each Fall, nominations are solicited from advisors and mentors at institutions throughout the U.S. Each Spring, a young scholar is chosen by the MPPL Committee based upon merit and recommendation, and is invited to spend one week in residence at CWRU, presenting one Colloquium and three Seminars.

The participants have been the following:

April 1998   Thomas Walther   Texas A&M University
Applications of Laser Spectroscopy

February 1999   Christopher Fuchs   California Institute of Technology
Quantum Information Theory

April 1999   Joe Mohr   University of Chicago
Cosmic X-rays, Galaxy Clusters, and Cosmology

May 2000   Keith Schwab   NSA, University of Maryland
Macro- and Mesoscopic Quantum Effects

April 2001   Jonathan Feng   Massachusetts Institute of Technology
Supersymmetry, Dark Matter, and the Cosmological Constant

April 2002   Re'em Sari   California Institute of Technology
Gamma Ray Bursts, Extrasolar Planets

April 2003   Brian DeMarco   National Institute of Standards and Technology
Quantum Behavior of an Atomic Fermi Gas

April 2004   Karsten Heeger   Lawrence Berkeley Laboratory
Recent Discoveries in Neutrino Physics

May 2005   Yaroslav Tserkovnyak   Harvard University
Collective Spin Dynamics in Magnetic Nanostructures

The Physical Review 100th Anniversary Volume
Compilation of CWRU Related Papers
Leslie L. Foldy   Fall 1995

The year 1993 was the one hundredth anniversary of the founding of the Physical Review. To celebrate this occasion, the American Physical Society and the American Institute of Physics decided to join in producing a centennial collection of noteworthy articles from The Physical Review and Physical Review Letters to be published as a book and a CD-ROM under the title: "The First Hundred Years, A Selection of Seminal Papers and Commentaries." Approximately 200 of the articles are reproduced in the book along with commentaries, while these and the remainder of the thousand-odd papers are contained on the CD-ROM. The joint collection was published in 1995.

While the number of papers in the collection is described at "more than a thousand," my own count of them is somewhat less, namely 979. The number of authors (including co-authors) is 1725. The selection of papers from those nominated was overseen by an expert, with aid and assistance from others in a dozen different fields of physics. The experts each also introduced the material in a chapter with an appropriate commentary. The Editor of the book was H. Henry Stroke of New York University with assistance from various individuals and particularly, Maria Taylor, Publisher of the AIP Press at the American Institute of Physics. The book itself contains 1266 printed pages; the CD-ROM more that 7500 pages. While most the papers are printed in full, some of the older ones had to be abbreviated because of their length.

Of the 979 papers, I found that at least 21 of them had an author or authors who had a connection with CWRU in that they had been students, professors, or research associates at CWRU or one of its predecessor institutions. The total number of these authors was 16. Prof. A. A. Michelson was not among these because his early papers were published in the American Journal of Science.

ABBREVIATIONS :
- o   CIT: CASE INSTITUTE OF TECHNOLOGY *(previously Case School of Applied Science)*
- o   WRU: *Western Reserve University*
- o   CWRU: *Case Western Reserve University (formed from merger of CIT and WRU)*

PAPERS: *All papers were originally published in The Physical Review or Physical Review Letters between the years 1893 and 1985, with the exception of the last paper in the list (Bratton, et al) on the observation of neutrinos from the Supernova 1987. References are listed according to standard Physical Review convention (Journal, Volume, Pages, Year (in parentheses)). The papers are arranged chronologically.*

AUTHORS: *Authors and co-authors are listed in the same order as in the original publication. Authors who have a "CWRU connection" have their names in capitals and are followed by square brackets containing information about their CWRU connection.*

CWRU CONNECTION: *To qualify as having a CWRU connection it is required that an author, at some time, received a Baccalaureate or Graduate Degree from CWRU or one of its predecessor institutions (not including honorary degrees), and/or served or is serving as a regular Faculty member, a Visiting Faculty member, an Emeritus Professor, or a Research Associate in the Department of Physics. (Adjunct faculty are not included though we are not aware that any such person was included as an author in PR100). Attribution of an institution to each author who was a faculty member or a research associate is given as CWRU if he is still associated with CWRU or if his association terminated after merger (1967). Otherwise it is given as CIT or WRU, whichever was appropriate at the time of termination. For authors who were students at CWRU, CIT or WRU, the brackets referred to above contain the degree or degrees earned, the institution which granted the degree, and the year it was granted.*

CWRU RELATED PAPERS CONTAINED IN THE PUBLICATION: THE PHYSICAL REVIEW, THE FIRST HUNDRED YEARS.

Piezoelectric and allied phenomena in Rochelle salt, Phys. Rev. **15** 537-538(A) (1920). J. VALASEK, [BS, CIT, 1917].

Diatomic molecules according to the wave mechanics. II. Vibrational levels, Phys. Rev. **34**, 57-64 (1929). P. M. MORSE [BS, CIT, 1926].

Direct Detection of the Angular Momentum of Light, Phys. Rev. **48** 47 (1935) R. A. BETH, [PROFESSOR AND CHAIRMAN, DEPARTMENT OF PHYSICS, WRU, 46-55].

The quantum-mechanical basis of statistical mechanics, Phys. Rev. **37**, 1146-1164 (1939). E. C. KEMBLE [BS, CIT, 1911].

On the neutron-proton interaction, Phys. Rev. **59**, 436-452 (1941). W. RARITA [VISITING PROFESSOR, CIT, 1955-56] and J. Schwinger.

On the theory of particles with half-integral spin, Phys. Rev. **60**, 61 (L) (1941). W. RARITA [VISITING PROFESSOR, CIT, 1955-56] and J. Schwinger.

Precision measurement of the ratio of the atomic 'g values' in the $2P_{3/2}$ and $2P_{1/2}$ states of gallium, Phys. Rev. **72**, 1256-1257 (L) (1947). P. KUSCH [BS, CIT, 1932] and H. M. Foley.

On the Dirac theory of spin 1/2 particles and its non-relativistic limit. Phys. Rev. **78**, 29-36 (1950). L. L. FOLDY [BS, CIT. 1941; FACULTY, CWRU, 1948-PRESENT] and S. A. Wouthuysen.

Some effects of ionizing radiation on the formation of bubbles in liquids. Phys. Rev. **87**, 665 (L) (1952). D. A. GLASER [B., CIT, 1946].

Bubble Chamber Tracks of Penetrating Cosmic Ray Particles. Phys. Rev. **91** 62-763 (L) (1953). D. A. GLASER [BS, CIT, 1946].

Fluctuations and irreversible processes, Phys. Rev. **91**, 1505-1512, (1953). L. Onsager and S. MACHLUP [FACULTY, CWRU, 1956-PRESENT].

Detection of the free neutrino. Phys. Rev. **92** 830-831 (L) (1954). F. REINES, [PROFESSOR AND CHAIRMAN, DEPARTMENT OF PHYSICS, CWRU, 1959-1966], C. L. Cowan, Jr.

Conservation of the number of nucleons. Phys. Rev. **96** 1157 (L) (1954). F. REINES, [PROFESSOR AND CHAIRMAN, DEPARTMENT OF PHYSICS, CWRU, 1959-1966], C. L. Cowan, Jr. and M. Goldhaber.

Application of formal scattering theory to many-body problems. Phys. Rev. **105** 1099-1100 (1957). L. L. FOLDY [BS, CIT, 1941; FACULTY, CWRU, 1948-PRESENT] and W. TOBOCMAN [RESEARCH ASSOCIATE, CIT, 1956-1957; FACULTY, CWRU, 1960-PRESENT].

Demonstration of parity non-conservation in hyperon decay. Phys. Rev. **108** 1353-1355 (L) (1957). D. A. GLASER [BS, CIT, 1946] and 19 coauthors.

Observations of the Failure of Conservation of Parity and Charge Conjugation in Meson Decays: the Magnetic Moment of the Free Muon. Phys. Rev. Lett. **105** 1415-1417 (1957). R. L. GARWIN [BS, CIT, 1947], L. M. Lederman, and M. Weinreich.

Free antineutrino absorption cross-section. I. Measurement of the free antineutrino absorption cross-section by protons. Phys. Rev. **113** 273-279 (1959). F. REINES, [PROFESSOR AND CHAIRMAN, DEPARTMENT OF PHYSICS, CWRU, 1959-1966], C. L. Cowan, Jr.

Phase-parameter representation of proton-proton scattering from 9.7 to 345 Mev. Phys. Rev. **120** 2227-2249 (1960). G. Breit, M. H. Hull, Jr., K. E. LASSILA [RESEARCH ASSOCIATE, CIT, (WITH L. L. FOLDY), 1961-63], and K. D. Pyatt (1960).

Cosmological density fluctuations produced by vacuum strings. Phys. Rev. Lett. **46** 1169-1172 (1981). A. VILENKIN [RESEARCH ASSOCIATE, CWRU (WITH P. L. TAYLOR), 1977-78].

Detection of the free antineutrino. Phys. Rev. **117** 159-173 (1960). F. REINES, [PROFESSOR AND CHAIRMAN, DEPARTMENT OF PHYSICS, CWRU, 1959-1966], C. L. Cowan, Jr., F. B. Harrison, A. D. Mcguire, and H. W. Kruse.

Observation of a Neutrino Burst in Coincidence with Supernova 1987 in the Large Magellanic Cloud, Phys. Rev. **58**, 1490-1493 (1987). C. B. BRATTON [BA, WRU, 1956, MA, WRU, 1958, PhD, WRU, 1964], M. CROUCH, [FACULTY, CWRU, 1950-PRESENT], D. Casper, W. R. KROPP [PhD, CIT, 1964], F. REINES, [PROFESSOR AND CHAIRMAN, DEPARTMENT OF PHYSICS, CWRU, 1959-1966], H. W. SOBEL [PhD, CIT, 1969] and 31 other co-authors.

Appendix F. Departmental Chairs and Endowed Chairs

Chairmen

Western Reserve College / Western Reserve University

Charles A. Young 1856-1866
Spencer H. Freeman 1866-1871
Charles J. Smith 1871-1881
Spencer H. Freeman 1882-1886
Frank Perkins Whitman 1886-1918
Harry William Springsteen Mountcastle   1918-1945
Richard Beth 1946-1955
John K. Major 1955-1964
Gerald E. Tauber 1964-1965
Bellur S. Chandrasekhar 1965-1967

Case School of Applied Science / Case Institute of Technology

Albert A. Michelson 1882-1889
Harry F. Reid 1889-1893
Dayton C. Miller 1895-1941
Frank Hovorka, *acting chair during the war*
Robert S. Shankland 1944-1958
Frederick Reines 1959-1966
Martin Klein *interim chair* 1966-1967

Case Western Reserve University

Harvey Willard 1967-1971
Kenneth L. Kowalski   1971-1976
Donald E. Schuele 1976-1981
William L. Gordon 1981-1993
Lawrence M. Krauss 1993-2005
Cyrus C. Taylor 2005-

Endowed Chairs

The Simon Perkins Professorship in Physics

This endowed chair was established at Western Reserve University in 1865 with a gift from Joseph Perkins, honoring his father, Simon. The elder Perkins, originally from Connecticut, arrived in the Western Reserve in 1798. He served in the War of 1812, rising to the rank of general. He was president of the Western Reserve Bank from 1813 until 1836. The professorship was originally called the Perkins Professorship in Mathematics and Natural Philosophy. In 1882 it was renamed the Perkins Professorship of Physics and Astronomy.

Charles A. Young 1865
Allen C. Barrows 1866
Charles J. Smith 1870
Spencer H. Freeman 1882
Frank Perkins Whitman 1886

Harry William Springsteen Mountcastle 1914
Richard Beth 1946
John K. Major 1957
Bellur S. Chandrasekhar 1966
Philip L. Taylor 1988

The Ambrose Swasey Chair of Physics

This chair was established in 1926 by the Board of Trustees of the Case School of Applied Science with a gift from Ambrose Swasey. Originally from New Hampshire, Swasey and his partner, Worcester R. Warner, opened a machine tool shop in Cleveland in 1881. Swasey was a member of the Corporation of the CSAS from 1922 until 1937 and a generous supporter of the school. His intention in establishing this chair was to allow Miller (see Chapter 4) to continue his search for ether drift.

Dayton Clarence Miller 1926
Robert Sherwood Shankland 1944
Richard W. Hoffman 1981
Lawrence Maxwell Krauss 1993

The Institute Professorship

This professorship was established in 1966 by the Board of Trustees of the Case Institute of Technology.

Leslie L. Foldy 1966
Robert W. Brown 1991
Arnold Dahm 2000

The Albert A. Michelson Professorship

This chair was funded in 1987 by a gift from Roger and Anne Clapp in celebration of the 100th anniversary of the 1887 Michelson-Morley ether-drift experiment.

Donald E. Schuele 1987-2005

Origins of the FW Transformation: A Memoir

This five-page typewritten document was found among the papers of Professor Foldy by Robert W. Brown who organized them for the Foldy family. It had several small penciled corrections on it. It has probably not been published elsewhere. I thank Professor Foldy's widow, Roma Foldy, for permission to include it here.

Leslie L. Foldy
Department of Physics, Case Western Reserve University
Cleveland, Ohio 44106

I have chosen to discuss in this memoir some of the circumstances surrounding the discovery of the Foldy-Wouthuysen (FW) transformation, the work for which I am probably best known. This work is concerned with the relationship between the Dirac equation which describes an electron relativistically and the Schrödinger equation which gives the more familiar non-relativistic description of its behavior. This problem had been explored by both Pauli and Schrödinger, and by Dirac himself but the resolution gave rise to various mysterious concepts like an imaginary electric dipole moment for the electron. In addition, the equation itself in a conventional interpretation predicts that any component of the electron's velocity is equal to the speed of light, yet has the beneficial predictions that the electron has spin ½ as well as a magnetic moment, but with an "anomalous" gyromagnetic ratio, and that the electron has an anti-particle: the positron. The FW transformation clears up most of these problems at a single stroke. But our discovery of it did not arise out of an attempt to exorcise these "ghosts'; it arose as a by-product of an attempt to solve a very practical problem.

Siegfried (Sieg) Wouthuysen (pronounced Vout'-high-sen), a Dutch student of H. A. Kramers (the "father of renormalization" in the context of quantum field theory) before World War II, was in hiding in Belgium during the Nazi occupation, and then came to the University of California/Berkeley at the time that I was a graduate student there. He, I and another graduate student, Harold Lewis, were Oppenheimer's last three doctoral students and we accompanied him to the Institute for Advanced Study in Princeton in 1947 when he became its Director. While Sieg and I were good friends we did not work together there. I became involved in a problem on renormalization in classical electrodynamics with Bram (Abraham) Pais. We came close but did not succeed in solving this problem (this was done substantially later in a very elegant fashion by Fritz Rohrlich) but in the process I learned a great deal about canonical transformation theory from the study of Schwinger's papers of that period. The three of us received our PhD.'s at the end of that academic year. While I had been offered (and had accepted) a position at Case Institute of Technology (now part of Case Western Reserve University) for the following academic year, I had no position for the intervening summer. Bob Marshak, then head of the physics department at the University of Rochester had visited the Institute and offered me a summer position at Rochester to work with him on a problem—namely meson production in nucleon-nucleon collisions. Sieg, who was married that summer, had accepted a position at Rochester for the coming academic year. I worked that summer on the meson-production problem, which Marshak and I had decided to do in the approximation that the nucleons are treated non-relativistically and succeeded in getting some interesting results. Actually some sign errors sneaked through (which were not decisive) and it was only some time afterwards that it was realized that important isotopic spin selection rules played a role in getting results that some of the meson-production cross-sections vanished. In any case I spent the next nine months at Case working on some other problems.

During the summer of 1949, Richard Feynman was scheduled to give lectures on his novel approach o quantum electrodynamics at the annual Theoretical Physics Summer School at the University of Michigan so I arranged to attend these and then to spend some time again at Rochester. While in Ann Arbor I met another young theoretical physicist, Cecile Morette (later Cecile Morette DeWitt), and learned that she also had worked on the meson production problem getting results different from ours which she attributed to our having dropped some terms which she claimed contributed substantially. I contended that these were relativistic terms but I did not know how to establish this. In this case it was nucleons rather than electrons which were under consideration. (Of course it was not then known that the proton was made up of constituent particles.) I was still preoccupied with this problem of separating relativistic from non-relativistic terms when I arrived later in Rochester. Arthur Wightman, who had just received his Ph.D. at Princeton that summer, was there and I talked about the problem with him. He reminded me of Pauli's method of handling the problem but when I went off to think about it I realized that it simply would not work well in the context of the problem with which I was concerned. The reason is that essentially the Dirac equation involves a four-component wave function and the Schrödinger equation involves a two-component wave function. For a non-relativistic particle the two "upper" components of the wave function are large and the two "lower" components are small (of order v/c compared to the upper components). Pauli used a method of eliminating the two lower components in terms of the upper components to get a Schrödinger-like equation correct to second order in v/c. Unfortunately the resultant "effective Hamiltonian" of this equation contains the "imaginary" terms referred to earlier, and is therefore non-Hermitian with the unfortunate consequence that a "Schrödinger interpretation" of the resultant wave function leads to non-conservation of probability.

It was at this point that the light flashed on in my mind: the solution was to eliminate coupling terms between the upper and lower components by canonical transformations patterned after the transformations used by Schwinger to eliminate certain coupling terms between matter and the electromagnetic field – namely those corresponding to so-called "virtual processes". In fact, the elimination of those terms in the Dirac Hamiltonian which couple upper and lower components in the Dirac four-component wave function also corresponded to elimination of a type of virtual transition: in this case virtual transitions of the Dirac electron from positive to negative energy states, or in Dirac's theory of the filled sea of negative energy states where an empty negative energy state corresponds to a positron, virtual transitions of an electron from a negative energy state to a positive energy state or what we would now call an electron-positron pair state. The problem with Pauli's method (which can be repaired, but with considerable effort) is that one essentially throws away the lower components and with it conservation of probability. This is why it is expedient to perform a canonical transformation which eliminates the terms coupling upper and lower components and not just to discard the lower components of the wave function. From what I had learned about the classical case and Schwinger's methods it was an easy matter to construct the required canonical transformation to eliminate such terms to any order of v/c or equivalently in inverse powers of c. For the case of a free particle I obtained for the Hamiltonian just the expansion of the usual classical relativistic energy in powers of the momentum with the first term just the rest energy, the second the usual non-relativistic kinetic energy and then terms of fourth and higher order in the momentum p. Working out the same result for the case of an electron in an electromagnetic field I again obtained the proper Pauli Hamiltonian with the well-known second order relativistic corrections: the magnetic moment-magnetic field interaction the spin-orbit coupling and the so-called Darwin term to second order, and could get the higher order terms straightforwardly. (This Darwin is Charles G. Darwin, a physicist, and grandson of the more famous Charles G. Darwin of the Origin of Species). When I started to obtain these last results, I told my friend Sieg about what I had found in the hope of getting his help in carrying out and checking these and planned future calculations. A day or two later he came beck to me with the result that he could obtain the canonical transformation for a free particle in closed form (rather than as an infinite series) which would completely decouple the upper and lower components of the Dirac wave-function. It contained a square-root of an operator involving the momentum! I cannot recall whether Sieg appreciated its importance but I soon realized (though I do not recall whether this was minutes or days later) that here was the key to most of the puzzles about the Dirac equation. I will explain this in a moment but first I want to say something else.

Ever since our collaboration Sieg has always deprecated his contribution to this work, regarding it as a rather trivial extension of my own, and I have tried to convince him of the important role this contribution played. I do not know whether I have really succeeded. The last time we had occasion to see each other (about eight years ago) he was somewhat upset that Max Dresden in his biography of Kramers had at one point in the text credited our work to him without mentioning me (though Dresden's citation to the work contains both our names). He had told Max that this was a distorted view of what occurred and Max had promised to correct this in a second edition, should one ever be published. During that meeting Sieg and I put forward our recollections of our collaboration. There were various discrepancies in what I think were details but no real disagreements of substance, even after forty years. Actually, I discovered later that the transformation for a free particle had been discovered previously by Smio Tani during the war and published in the Japanese journal "Progress of Theoretical Physics" and even earlier by the German theoretical physicist R. Becker (but in this case only to second order in v/c). *(ed. note: Tani, who became Foldy's research associate at Case in 1957, has recently pointed out that his discovery was actually made after the war.* Prog. Theor. Phys. (Japan) **6** *267 1951)* Neither dealt with the problem of interaction with an external field. Now to return to my principle subject, the importance of the form for the canonical transformation obtained by Sieg was that it is essentially a non-local transformation on the wave function with the non-locality of the order of the Compton wavelength of the electron. What this means is that the transformed wave function at any point depends on the values of the original wave function on a set of points lying within a distance of the order of the Compton wave-length of the electron from the original point. Another way to express this fact is that the operator which represents the position of an electron in the usual form of the Dirac equation represents a different observable after the canonical transformation, an observable which we called the mean position. For a free particle the mean position moves with constant velocity equal to the momentum of the particle divided by its energy, as we would expect from classical relativity. On the other hand the original (Dirac) position moves with a velocity of the order of the velocity of light over a region of radius of the order of the Compton wave-length (the so-called zitterbewegung). Since the different components of the operator representing the Dirac velocity do not commute with one another in general it is not possible to measure more than one component at a time and such measurement will always yield a value equal to the velocity of light. With this picture we can understand qualitatively the features of the interaction of a particle with an external field exhibited by the Dirac equation, and in particular, the origin of the magnetic moment, the resultant spin-orbit coupling, and the Darwin term which has just the form of an interaction of an extended spherically-symmetric charge distribution of Compton wave-length radius with an electrostatic field. These ideas are described in substantial detail in our paper. With this methodology available I could immediately see that I had not dropped any non-relativistic terms in our meson-production calculation, regardless of what other errors or oversights that paper may contain. In fact it soon became clear that the perturbation method on which virtually all such calculations, including ours, depended at that time was quite unreliable for hadronic quantities since the meson-nucleon coupling was so strong.

This is not really quite the end of the story. I presented a paper on this work at a meeting of the American Physical Society which was held at Columbia University in February of 1950 (the society was very much smaller then than it is now). As I was waiting to be called I noticed with some trepidation that Pauli was in the room. Pauli was as expert as anyone on relativistic quantum mechanics and the Dirac equation and he was also the *enfant terrible* of physics. I recognized Pauli since I had heard him deliver some lectures at Princeton during the war and I knew his reputation so I was, to say the least, quite distressed to see him there. Shortly after I started delivering my paper, Pauli's head began to nod back and forth and I felt relieved that he seemed to be accepting what I had to say. He made no comment on the paper but I learned later that this nodding was a well-known tic of Pauli's when he was in an audience, but of this I was not aware. I have often wondered how I would have felt and responded if his tic consisted of shaking his head from side-to-side. In 1963-64 I spent a year at CERN in Geneva and discovered that Pauli had left his collection of journals to CERN. I looked up the FW paper to see whether had had made any notes beside it. I found some pencil marks, which at lest suggested that he had read it, but there were no remarks of any substance, as I recall. The paper seems to have been fairly well received. My first graduate student, Richard Osborne, wrote a thesis on extending the transformation to a system of two particles and some formal difficulties arose which we did not resolve but were soon after clarified by Z. V. Chraplevy and a student W. A. Barker, at St. Louis University. Someone, though I cannot recall who, once told me that Dirac had known of, but did not publish, the FW transformation, presumably again for a free particle. I would find it hard to believe that had he been aware of the transformation for a particle interacting with a field that he would not have presented it in his book in place of the rather unaesthetic treatment he does give which involves the imaginary electric dipole moment term in the effective Hamiltonian.

The work contained in the FW paper had some far-reaching consequences for me, personally. A year after its publication I discovered an application of the transformation to understanding the problem of the so-called "electron-neutron interaction". The result of that work was "instant fame" of a sort for reasons I cannot go into here but hope to discuss in another memoir. A year after that, while returning from Copenhagen by ship, I realized that the FW transformation decomposed the Dirac equation into irreducible representations of the Poincaré group, which I soon after discovered were explored by Wigner and his student, T. D. Newton. This ultimately led me to a line of investigation which I continued to follow well into the mid-1970's and was linked to another influential paper of Dirac's published not long after the end of World War II. This also I hope to recount in another memoir.

Appendix H   Table of Figures

*The principal sources of the figures in this book are the departmental archives, the **Case Western Reserve University Archives**, and the various technical journals in which the faculty published their research. I thank the CWRU Archives for their generous help in finding interesting documents.*

*All the figures from journals are protected by copyright. The figures from the **Physical Review, Physical Review Letters,** and the **Reviews of Modern Physics** are copyrighted by the **American Physical Society**. Those from the **Review of Scientific Instruments** and **Journal of Applied Physics** are copyrighted by the **American Institute of Physics**.*

The list shows the chapter and figure number in this book, and its source.

1	1	Western Reserve Historical Society, Cleveland Ohio		
1	2	CWRU Archives		
1	3	Yale University website		
1	4	CWRU Archives		
2	1	CWRU Archives		
2	2,3,4	Smithsonian vol 980 1895		
3	1	Phys. Archives		
3	2	CWRU Archives		
3	3	W. Fickinger		
3	4,5,6	U. S. Naval Observatory Nautical Almanac		
3	7,8,9,10	Amer. Jour. Sci. **34** 273		
3	11,12	Phys. Archives		
3	13	CWRU Archives		
4	1	CWRU Archives		
4	2 to 6	Phys. Archives		
4	7,8,9	Rev. Mod. Phys. **5** 203 1933	Figs. 7,21,22	
4	10,12,13,17	Phys. Archives		
4	11,14,15,16	CWRU Archives		
4	18,19	Miller's Soundwaves	© Macmillan 1937	
4	20,21,22	Phys. Archives		
4	23	W. Fickinger		
4	24,26,27	Jour. Franklin Institute 1916	Lippincott 1916	
4	25	R. Hanson, U. Northern Iowa		
4	28,30,31	Phys. Archives		
4	29	W. Fickinger		
5	1,2,3	CWRU Archives		
5	4	Phys. Rev. **3** 241 1896	Fig. 1	
5	5,6	Phys. Rev. Ser I **21** 41 1905	Figs. 2,3	
6	1,2	Phys. Rev. **49** 8 1936	Figs. 2,1	
6	3,4	Phys. Rev. **52** 414 1937	Figs. 1,2	
6	5	Phys. Archives		
6	6	J. Acous. Soc. Am. **43** 426 1968	Fig. 6	© Acoustical Society of America
6	7	Phys. Archives		
6	8,9	Phys. Rev. **72** 1131 1947	Figs. 1,3	
6	10	Rev. Mod. Phys. **5**	cover page	
6	11,12,13,14	Rev. Mod. Phys. **27** 167 1955	Figs. 1,2,3,4	
6	15	Phys. Archives		
6	16	Eleanor Shankland		
7	1	Phys. Archives		
7	2	Rev. Sci. Instr. **22** 176 1951	Fig. 6	
7	3,4,5,6	Phys. Archives		
7	7	Amer. J. Roentgenology **76** 979 1956		© American Roentgen Ray Society

Appendix H List of Figures

7	8	Phys. Rev. **87** 685 1952	Fig. 4	
7	9	Phys. Rev. **88** 612 1952	Fig. 1	
7	10	Phys. Rev. **102** 1 1956	Fig. 4	
7	11	J. Applied Physics **27** 697 1956		
7	12,13	Phys. Archives		
7	14	Rev. Mod. Phys. **25** 310 1953	Fig. 5	
7	15,16,17	Phys. Archives		
7	18	Phys. Rev. **129** 1275 1963	Fig. 2	
7	19	Nuc. Inst. & Meth. **85** 151 1970		© Elsevier
7	20,21,22	Phys. Archives		
8	1	Phys. Archives		
8	2	Phys. Rev. **76** 1134 1949	Fig. 1	
8	3	Nuc. Sci. & Engr. **2** 631 1957	Fig. 1	© American Nuclear Society
8	4,5	Phys. Archives		
8	6	Phys. Rev. B **137** 740 1965	Fig. 1	
8	7	Rev. Sci. Instr. **35** 370 1964	Fig. 4	
8	8	Phys. Archives		
8	9	Phys. Rev. C **4** 1344 1971	Fig. 1	
8	10	Rev. Sci. Instr. **35** 370 1964	Fig. 1	
8	11,12	Sci. Am. **214** 2 40 1966		© Scientific American
8	13	Phys. Rev. Lett. **17** 733 1966	Fig. 2	
8	14	Nature **223** 1320 1969	Fig. 2	© Nature Publishing Group
8	15	Nature **233** 456 1971	Fig. 1	© Nature Publishing Group
8	16	Astrophys. J. **158** 925 1969		© American Astronomical Society
8	17	Phys. Archives		
8	18,19	Nature **240** 221 1972		© Nature Publishing Group
8	20, 21	J. Geophys. Res. **79** 929 1974	Figs. 3, 5	
8	22	Nuc. Inst. & Meth. **144** 183 1977	Fig. 2	© Elsevier
8	23	Phys. Archives		
8	24	Phys. Rev. **D18** 2239 1978	Fig. 9	
8	25	Phys. Rev. **D5** 2667 1972	Fig. 1	
8	26	Phys. Rev. **D5** 2667 1972	Fig. 9	
9	1,2,4,5	Phys. Archives		
9	3	Phys. Rev. **D17** 3065 1978	Fig. 1	
9	6	Phys. Rev. **136** 1682B 1964	Fig. 7	
9	7	Phys. Archives		
9	8	Phys. Rev. Lett. **52** 978 1984	Fig. 1.	
10	1	CWRU Archives		
10	2,3	Phys. Rev. **53** 30 1938	Figs. 1,2	
10	4	Phys. Rev. **95** 447 1954	Table I	
10	5,7	Phys. Archives		
10	6	Phys. Rev. **50** 115 1956	Fig. 1	
10	8	Rev. Sci. Instr. **25** 603 1954	Fig. 3	
10	9,10	Phys. Archives		
11	1,2,4,5,6	Phys. Archives		
11	3	Rev. Mod. Phys.**35** 853 1963	Fig. 2	
12	1	J. Acous. Soc. Amer.**31** 137 1959		© Acoustical Society of America
12	2	J. Acous. Soc. Amer. **31** 866 1959		© Acoustical Society of America
12	3	J. Acous. Soc. Amer.**38** 780 1965	Fig. 7	© Acoustical Society of America
12	4	J. Acous. Soc. Amer. **37** 679 1965	Fig. 1	© Acoustical Society of America
12	5	Scientific American July 1973		© Scientific American
12	6,7,9	Phys. Archives		
12	8	Proceedings of the 1962 8th Vacuum Symposium		© American Vacuum Society
12	10	Phys. Rev. **126** 489 1962	Fig. 13	
12	11,12	Phys. Rev. **133** A443 1964	Figs. 2,5	
12	13,14	Phys. Rev. **142** 399 1966	Figs. 2,3	
12	15	Phys. Rev. **153** 91 1967	Fig. 1	
12	16	Phys. Rev. Lett. **22** 319 1969	Fig. 2	
12	17,24	Phys. Archives		

12	18,19,20,21	J.Phys.ChemSolids **25** 801 1964	Figs. 4,1,5,8	© Elsevier
12	22	J.Phys.ChemSolids **31** 647 1970		© Elsevier
12	23	Phys. Rev. **B6** 582 1972	Fig. 4	
12	25	J. Applied Physics **41**2 1623 1970	Fig. 3	
12	26	J. Vacuum Technology **16-2** 466 1979	Fig. 1	© American Vacuum Society
12	27,28,29	Phys. Archives		
13	1 to 7	Phys. Archives		
13	8,9	Rev. Sci. Instr. **62** 2639 1991	Figs. 3,11	
13	10,11	Phys. Rev. Lett. **83** 1946 1999.	Figs. 1,2	
14	1,6,7	Phys. Archives		
14	2	Phys. Rev. **180** 1158 1969	Fig. 6	
14	3,4	Nucl. Phys. **72** 106 1965	Figs. 1,5	© Elsevier
14	5	Phys. Rev. **138** B1378 1965	Fig. 1	
14	8,9,10	Phys. Rev. **B2** 2421	Figs. 1,4,5	
14	11	Phil. Mag. **36** 117 1977	Fig. 2	© Taylor and Francis Group
14	12,13,14	Phys. Archives		
15	1	CWRU Archives		
15	2	Phys. Rev. **150** 519 1966	Fig. 2	
15	3,7	Phys. Archives		
15	4	Phys. Rev. Lett. **11** 331 1963	Fig. 1	
15	5	Phys. Rev. Lett. **16** 53 1966	Fig. 1	
15	6	Phys. Rev. **188** 1130 1969	Fig. 4	
15	8	Phys. Rev. **176** 562 1968	Fig. 3	
15	9	Phys. Rev. **B5** 3523 1972	Figs. 4 & 5	
15	10	Phys. Rev. **B7** 3037 1973	Fig. 1	
15	11	Phys. Stat. Solidi **20** 419 1973	Fig. 4	© Wiley-VCH Verlag
15	12	Phys. Rev. Lett. **38** 788 1977	Fig. 1	
15	13	David Farrell		
15	14	Phys. Archives		
15	15	Phys. Rev. Lett. **28** 1244 1972	Fig. 2	
15	16	Phys. Rev. Lett. **31** 873 1973	Fig. 2	
15	17	Phys. Rev. B15 **2630** 1977	Fig. 11	
15	18,19	J. Low Temp. Phys. **23** 477 1975		© Springer Science
15	20	Phys. Rev. **B15** 1378 1977	Fig. 1	
15	21	Phys. Rev. **B40** 8995 1989	Fig. 1	
16	1,2,3,4,7,9	Phys. Archives		
16	5,6	Phys. Rev. **C6** 1513 1972	Figs. 1,3	
16	8	Nucl. Phys. **A175** 156 1971		
16	10	Phys. Rev. **C14** 1545 1976	Fig. 4	
16	11	Phys. Rev. Lett. **41** 384 1978	Fig. 1	
16	12	Phys. Rev. **C21** 2535 1980	Fig. 5	
16	13 to 18	Phys. Archives		
16	19	Nucl. Phys. **B41** 45 19	Fig. 11	© Elsevier
16	20	Phys. Rev. Lett. **31** 562 1973	Fig. 1	
16	21	Phys. Archives		
16	22	Phys. Rev. **C30** 1080 1984	Fig. 1	
16	23	Phys. Rev. Lett. **54** 518 1985	Fig. 3	
16	24	Phys. Rev. **D34** 3332 1986	Figs. 12,13	
16	25	Phys. Rev. C3**0** 1080 1984	Fig. 3	
16	26,27	Phys. Rev. Lett. **56** 211 1986	Figs. 1,5	
16	28,29	Phys. Rev. Lett. **63** 1352 1989	Figs. 1,2	
16	30	Z. für Phys. **C42** 173	Fig.2	© Springer Science
16	31	Phys. Rev. **C44** 956 1991	Fig. 1	
17	1,2	Physics Archives		
17	3	Phys. Rev. Lett. **43** 456 1979	Fig. 1	
17	4 to 10	Physics Archives		
18	1 to 12	Physics Archives		

Index

Abrikosov, Alexei A., 216
ACBAR experiment, 289
acoustics, architectural, 147
Adhikari, S.K., 179
Adler, John G., 213
AIP Shankland interview, 57, 59, 61
Akerib, Daniel, 287
Akins, Robert B., 165
Albats, Paul, 105, 109
Albright, John G., 35
ALCOM, 158, 273
Allen, Chris, 221
Alvarez, Luis, 53, 101
Andeen, Carl, 163
Anderson, B.D., 240
Andrews, James, 178, 189
angular momentum of light, 133
Angus, John, 169, 173, 275
architectural acoustics, 58
Argonne National Lab ZGS, 234
Argonne National Laboratory, 249,
Athenaeum, 5, 47
atomic masses, Morley, 7
Auger Electron Spectroscopy, 168
Auger Observatory, 288
Bachman, Michael, 157
Baer, Helmut, 240
ballistics photography, Miller, 43
balloon-borne detectors, 101
band structure, 154
Barrows, Allen C., 46
BCS theory, 213
Bell, Alexander Graham, 14
Bello, Alfredo, 164
Benade, Arthur H., 76, 89, 146
beta decay, double, 132, 134
beta decay, 198
betatron, 70
Beth, Richard A., 132
Bevatron, 101
Bevington, Philip R., 238
Bilchak, Cynthia L., 186
biological systems, 141
Bio-physics, 168
Bjorken, J. D., 283
Blanpied, William A., 235
Bohm, D., 113
Bohr, Niels, 120
Bond, Peter, 200

Boomerang experiment, 289
Bragg scattering, 161
brasses, acoustics of, 150
Bratton, Clyde, 142
Breit, Gregory, 142
Brillouin, Leon, 112
Brittenham, Gary M., 220
Brookhaven National Laboratory, 250
Brown, Robert W., 178, 183
bubble chamber experiments, 245
Burdick, Bernard, 248
carbon films, 169
Carnahan, Walter, 247
Case, Eckstein, 37
Casper, Karl J., 197
CDMS, 287
CERCA, 284
Chandrasekhar, B. S., 100, 210, 232
Chemical Rubber Handbook, 35
Chen, An-Ban, 269
Cheng, Y.C., 193
Chew, Herman W., 144
Chottiner, Gary S., 169
Chow, Gee-Yin, 181
Chulick, Gary, 129
CMB, 289
Cold Dark Matter Search, 288
Compton, A. H., 53
Compton Papers, Shankland, 67
Compton scattering, 54, 76
Convery, M.E., 188, 189
Coopersmith, Michael H., 266
Copi, Craig, 284
cosmic microwave background, 286
cosmic strings, 185
cosmological defects, 285
cosmological symmetry breaking, 187
Covault, Corbin, 287, 288
Cowan, Clyde, 91
Cramer, C. H., 1, 7, 13
Crittenden, Eugene C., 70
cross sections, 75
Crouch, Marshall F., 88
crystallography, 79
Curtis, Cassius W., 51
Cverna, Frank, 240
Dahm, Arnold, 204, 206, 223
Davis, Ray, 92
de Haas–van Alphen, 154

Debbe, Ramiro, 253
defects in superconductors, 219
DelSignore, Kenneth, 107
dephasing, quantum, 277
deuteron, photodisintegration, 72
diamonds, materials theory, 275
Diaz Bejarano, José, 249
DiBianca, Frank, 249
dielectric properties of crystals, 163
dielectric relaxation spectroscopy, 165
disordered chiral condensate, 283
disordered systems, 261
dispersion, 18
Dix, Fred, 94
Doan Brook Project, 69
double beta decay, 94
East, Larry V., 94
Eck, Thomas G., 146, 154, 157
Eddy, Henry T., 26
Ehrenfest, Paul, 122
Einstein, Albert, 31
Einstein, Shankland visits, 64, 66
Eisenmann, John, 15
Eisner, Robert L., 249
electron spin resonance, 226
Emerson, Alfred, 5
emulsions, nuclear, 73
Entrepreneurship Program, 283
Eppell, Steven J., 172, 173
Ernst, David J., 128
Eros, Stephen, 258
ether, 14
ether-drift, Miller, 27
Everett, Paul M., 155
Farrell, David E., 216
Fawcett, Sherwood L., 71
Fermi surface, 153, 158
Fickinger, William, 179, 243
fine structure crossing, 159
Finkelstein, Murray, 216
fission, nuclear, 77
Fleisher, Harold, 75
flicker photometry, 49
flute acoustics, 149
flute collection, Miller, 45
Focke, Theodore Moses, 50
Foldy, Leslie L., 60, 70, 90
Foldy-Wouthuysen transformation, 115
Fontanella, John, 163
Freeman, Spencer H., 46
Frisken, William R., 233

Frye, Glenn M., 93
funding for research, 68, 121, 179
Fung, Wai K., 155
gamma-ray sources, cosmic, 103
Garwin, Richard L., 73, 85
Gates, Evalyn, 282
Geiger, Alan, 215
General Electric, 69
geomagnetism, 5
Giamati, Charles G., 92
Giltinan, David A., 125
Giltinen, Frank, 127
Glaser, Donald A., 80, 246
Glass. Solomon J., 121
Glennan, T. Keith, 111
Goldfinger, Richard, 125
Goldflam, Rudy, 177
Gordon, William L., 146, 153
Gordon, Leon, 186
Goswami, Amit, 144
Gourevitch, Sergei Al, 248
"Green, Jr.", Ben A., 209
Greenslade, Thomas B., 37
Gregg, Earle C., 71, 74, 76, 111
Grüneisen parameter, 161
Gubernatis, James, 261
Guenin, B. M., 230
Gupta, Ashok, 215
Gurr, Henry, 97
Haacke, E. Mark, 192
Hall effect, 263
Halteman, E. I., 80
Hamerla, Ralph R., 8
Hanson, Roger, 40
Harris, John W., 220
Haugland, Edward, 220
heavy ions, 127
Heinonen, Olle, 263, 265
helium, electrons in liquid, 226
helium, neutral excitations in, 228
helium, solid, 224, 229
helium, superfluid, 225
helium surface, 229
Helmholtz harmonic synthesizer, 42
Henrici harmonic analyzer, 30, 39, 112
Herman, Damir, 231
Hibbin, Samuel G., 29
Hinckley, Larry, 83, 237, 251
Hinshaw, George, 274
history of physics, 122
Hitchcock, Henry, 5

Hodgman, Charles D., 35
Hoffman, Richard W., 80, 146, 151
Hopfinger, Anton, 262
Hornbeck, Larry J., 155
Hotes, S.A., 189
Hovey, Ralph F., 38
Hovorka, Frank, 131, 132
Howe, Charles S., 25, 37
HREELS, elect'n en'gy loss spec'scopy, 171
Hrushka, August Gus, 72, 93, 155
Hu, Xue Long, 231
Huang, Chao-Yuan, 206
Huang, Whittack, 227
Huterer, Dragan, 286
hyperfine coupling in solids, 206
hyperfine structure, 205
hypernuclei, 139
impulse approximation, 176
interferometer, 13
IRAS, IR reflec-absorption spectroscopy, 171
iron in humans, measurement of, 221
Jackson, J.D., 189
Jackson, Jerome, 215
Jansson, Erik V., 150
Jenkins, Thomas L., 93, 233
Jennings, Wayne, 171
Jha, S., 196, 197, 199
Jiang, Hong Wen, 231
Joseph, Alfred S., 154
Kalogeropoulos, Ted, 250, 253
Kash, Kathleen, 281
Kelvin, Lord, 27
Kelvin harmonic synthesizer, 41
Kemble, Edwin C., 44
Kernan, Peter, 284
Khorana, Brij M., 212
Kikuchi, Tad, 247
Kisslinger, Leonard, 138, 200
Klein, Martin J., 100, 122
Knepley, M.G., 189
Koenig, Rudolph, 16, 36
Koenig, Jack, 156
Koga, Rokataro, 107
Kogan, Alberto, 240
Kohl, Max, 36
Koral, Kenneth, 239
Korpi, John, 248
Kowalski, Kenneth L., 128, 175, 283
Krainsky, Isay, 169

Krajcik, Richard, 119
Krauss, Lawrence M., 284
Kropp, William, 97
Kuerti, Gustav, 61
Kurie plot, 198
Kurtay, M., 193
Kusch, Polycarp, 71, 78, 157
Kusner, Robert, 231
Lambrecht, Walter, 275
LAMPF, 128, 241
Landau domains, 218
Lando, Jerome, 156
Larson, Curtiss O., 155
Lawrence, Ernest O., 113
LEED, low energy electron diffraction, 170
Leff, Harvey, 266
Leone, F. C., 61
Leskovec, Robert A., 82, 237, 241
Levit, Larry, 94, 235
Lewanski, Andrew, 125
light, velocity of, 14
lightning bolts, 35
liquid crystal polymers, 165
liquid crystals, 264, 272, 279
Liu, Hong, 284
localization, 277
Lock, James, 206
London, Fritz, 143
Loomis, Elias, 2
Loomis Observatory, 3
Lorentz, H. A., 19, 31
Machlup, Stefan, 139
magnetic declination, 3
magnetic domains, 217
magnetic resonance imaging, 142, 189
magnetization of thin films, 166
magnetometry, torque, 222
magnetoresistance, 156
magnetostriction, oscillatory, 211
Major, John K., 135
Major, John, 194, 211
Malko, John, 249
Mann, J. A., 231
Marshak, Robert, 114
Martens, Mike, 189
Martin, John Richard, 35
Mather, Samuel, 48
Mathur, Harsh, 231, 277
Matthews, David, 249

Maynard, William J., 24
McCarthy, John, 130
McCuskey, Sidney, 61
McGervey, John D., 196, 201, 220
Mearini, G. T., 169
Meeks, Wilkison W., 133
Mehrotra, Ravi, 230
meteor trails, 2
meteorology, 3
Michelson, Albert A., 11, 13, 32
Michelson Livingston, Dorothy, 22
Michelson-Morley experiment, 19
Mikaelian, Karnig O., 185
Milford, Frederick W., 119, 122, 258
Miller, Dayton C., 11, 22, 51
Miller ether-drift, Shankland, 61
Millikan, Robert A., 44
Millis, John S., 211
MiniMax, 180, 283
mobilities, ions in helium, 224
Moe, Michael, 96
Morich, M.A., 189
Morley, Edward W., 7, 18, 25
Morse, Philip M., 44
Morton Salt Mine, 92
MOSFETs, 277
Mössbauer effect, 195
Mössbauer spectroscopy, 166
Mountcastle, Harry W. (Springsteen), 50
Mountcastle, Harry, 130
Muons, cosmic, 89
Nagarajan, M. A., 179
nanotechnology, 281
nanotensilometers, 168
Nassau, Jason, 44
Nath, Kashi, 264
negative temperatures, 140
Neighbours, J. R., 78
neutral currents, 185
Neutrinos, atmospheric, 109
neutrinos from a reactor, 91
neutron chopper, 60
neutrons in water, 90
Newcomb, Simon, 13
Nezrick, Frank, 99
non-linear optics, 178, 280
Nooney, James, 5
NSF Science Development Program, 233
nuclear lifetimes, 195
nuclear magnetic resonance, 142
Nusbaum, Christian, 35

Offner, Abe, 51
off-shell scattering, 176
Olsen, Leonard O., 69
Onsager, Lars, 140
Oppenheimer, J. Robert, 113
optical model, 176
optics, non-linear, 280
organ pipes, absorption by, 150
Osborn, Richard, 119
Owens, Jeff F., 249
Pantalony, David, 42
particle accelerators, 113
Patrick, John, 189
Pauli, Myron, 125
Pearle, Phillip, 180
Peierls, R. F., 117
Peltier effect, 258
Perkins, Simon, 5
Petropoulos, L.S., 189
Petschek, Rolfe, 271
Pfeuty, Pierre, 272
phonodeik, 37
photon-photon interactions, 185
photons, 53
physics building, WRU, 48
Picklesimer, Alan, 127, 177
Pines, Vloadimir, 264
Pogosian, Levon, 286
polarized protons, 249
polymers, 262, 272
positron annihilation in helium, 228
positronium, 201
positrons, 201
positrons in liquid helium, 228
Primakoff, Henry, 112
Proctor, David G., 76
proton decay, 92
proton scattering, 59
proton-proton elastic scattering, 241
Prout, William, 8
pseudospin model, 143
quantum computer, 232
quantum dots, 278
quark search, cosmic, 110
qubits, 232
Rachford, Frederic, 207
radiative processes, 255
Rayleigh, Lord, 14, 21
Razor, Ned S., 81
reciprocity theorem, 113
Reichert, Jonathan F., 205

Reichert, Jonathan, 226
Reid, Harry F., 22
Reines, Frederick, 22, 91
Reitz, John R., 79, 122, 154
relativity, general, 137
reverberation, 58
Rix, John, 127
Roberts, Lee, 250, 254
Robinson, Berol L., 136, 194, 200
Robinson, D. Keith, 243
Rockefeller Building, 33, 83
Romanowski, Thomas, 76
Rondon Aramayo, Oscar, 251
Rosenblatt, Charles, 274, 279
Rosenblum, Earl S., 75
Roy, Ranendra, 143
Ruhl, John, 287, 289
Sabine, Wallace, 44
Sahdev, Deshdeep, 187
Sakitt, Mark, 250
Salant, Edward O., 243
Sampath, Prativadi, 158
Sard, Robert D., 88
Savannah River, 99
scattering, multiple, 112
scattering theory, 117
Scharenberg, Rolf P., 86
Scherson, Daniel, 172
Schick, Michael, 143
Schuele, Donald E., 146, 157, 160
Segall, Benjamin, 268
separated beam, 244
Severance Hall, 43, 147
Shakin, Carl M., 127, 182
Shan, Jie, 281
Shankland, Robert S., 22, 53, 77
Shankland, Eleanor, 65, 67
Shaw, Melvin P., 158
shell model, 179
Shepherd, John P.G., 157
Shera, E. B., 136
Shrader, Erwin F., 71, 86
Shutt, Thomas, 287
Shvartsman, Shmaryu, 188, 191, 282
Siciliano, E. R., 128, 177
Silverstein, Edward A., 86
Silvert, William L., 267
SIMS, secondary ion spectroscopy, 170
Singer, Kenneth L., 178, 274, 280

Smith, Charles Josiah, 46
Smith, Charles S., 77, 160
Smith, Gary R., 94
Smith, Lawrence H., 101
Smith, Robert S., 121
solar neutrinos, 95
solar physics, , 6
Sones, Richard, 275
Sood, Brig Raj, 215
spark chamber, 234
Sparlin, Don M., 212
specific heat of alloys, 210
spectrometer, pair, 75
spectroscopy, time domain, 281
Spremulli, Paul, 52
Springsteen, Harry W. (Mountcastle), 49
SQUID, 221
Sreedhar, V. A., 251
STACEE, 288
Staib, Jon, 103
Staley, Cady, 25
Stan, Mark A., 230
Stansfield, Sam, 120
Stapleton, Darwin, 69
Stark, Royal W., 155
Starkman, Glenn, 285
Stecker, Floyd W., 186
Stockwell, John, 13
Stojkovic, Dejan, 286
Stone, Amasa, 48
Stooksberry, Robert, 90
Story, Harold S., 81
Stott, Jonathan, 275
Strelzoff, Alan, 234
stripping reactions, 124
Strough, Robert, 73
Stroughair, John D., 186
Sugawara, Kazushi, 207
Sullivan, Charles, 234, 249
superconductivity, 213
superconductivity, high-Tc, 222
superconductors, type II, 216
superfluidity theory, 143
supershielding, 191
susceptometry, biological, 221
SUSHI, 192
Swenson, Loyd S., 57
tachyon search, 110
Taggert, Keith, 182

Tandy, Peter, 127
Tani, Smio, 349
Tauber, Gerald E., 137
Taylor, Cyrus C., 178, 256, 282
Taylor, Philip L., 259
Teller, Edward, 101
Terentjev, Eugene, 274
text books, 47
Thaler, Roy, 127
thermal field theories, 178
thin films, 80, 151, 165
Thompson, Michael R., 192
Thompson, Richard, 198
three body scattering, 177
Thwing, Charles, 11
Tilger, Clarence, 247
Tobocman, William, 74, 123, 179
Tomasch, W. J., 258
torque, magnetic measurement, 134
TPD, temp. programmed desorption, 171
tri-neutron, 239
Tripp, John, 155
Trodden, Mark, 284
tunneling, 214
ultrasonic imaging, 126
Ultrasonics, 74
Ultrasound measurements, 162
undergraduate research, 188
Usmani, Zahiruddin, 206
Vachaspati, Tanmay, 285
van de Graaff, 82, 237
van Keuls, F. W., 231
Venkatesan, R., 192
Voelker, William.H., 76
voids, detection of, 203
vortices, quantized, 225
W boson, 184
Wagner, David, 156
Waite, Frederick C., 1, 46
Wallace, Clarence W., 36
Walters, Virginia, 202
Wang, C.P., 104

Wang, Chia Ping, 102
Wang, Kui Long, 173
Wang, Kuilong, 173
Wang, Yaxin, 169
Warner, Raymond M., 75
Warner and Swasey Co., 48
weather map, 4
Weber, Joseph, 108
Weinberg, Joseph, 137, 141
Wheeler, John C., 272
White, Herbert E., 51
Whitlock, Richard, 143
Whitman, Frank Perkins, 46
"Wick, Jr.", Dudley B., 26
Wickenden, William E., 111
Wiefling, Kimberly, 272
Wilczek, Frank, 284
Willard, Harvey, 236
Wilson, Robert R., 59
Winter, Rolf G., 134
Winterberg, Friedwart, 123
Witwatersrand, 97
Wood, R. W., 50
Woods, Robert M., 96
woodwinds, acoustics of, 149
Wouthuysen, Siegfried. A., 115
Wright Jr., Elizur, 1
Wright Sr., Elizur, 2
WWV timing signal, 131
XENON experiment, 290
XPS, X-ray photoemission spectroscopy, 170
x-rays, Miller, 25
X-rays, Shankland, 53
Yaeger, Ernest, 131
YBCO, 222
Young, Charles A., 5, 24
Zhong, Zhengzhong, 165
Zilsel, Paul R., 142
Zorman, Chris, 169
Zych, Alan, 103
Zych, Dale A., 158